AMPLIFYING INFORMAL SCIENCE LEARNING

This collection explores the broad landscape of current and future out-of-school science learning environments. Written by leading experts and innovators in informal science learning, these thoughtful and critical essays examine the changing nature of informal institutions such as science museums, zoos, nature centers, planetariums, aquaria, and botanical gardens and their impact on science education. The book examines the learning opportunities and challenges created by community-based experiences including citizen science, makerspaces, science media, escape rooms, hobby groups, and gaming.

Based on current practices, case studies, and research, the book focuses on four cross-cutting themes – inclusivity, digital engagement, community partnerships, and bridging formal and informal learning – to examine how people learn science informally.

The book will be of interest to STEM (science, technology, engineering and math) educators – both in and out of school – designers of science and experiential education programs, and those interested in building STEM learning ecosystems in their communities.

Judy Diamond PhD is Professor of Libraries and Curator at the University of Nebraska State Museum and a leader in the field of informal science education. She is lead author of the books *Practical Evaluation Guide: Tools for Museums and Other Informal Educational Institutions* (2016) and *World of Viruses* (2012) and co-author of several popular science books, including *Parasites: The Inside Story* (2022), *Thinking Like a Parrot: Perspectives from the Wild* (2019), *Concealing Coloration in Animals* (2013), and *Kea: Bird of Paradox* (1999).

Sherman Rosenfeld PhD has worked as a researcher and science educator at the Weizmann Institute of Science, in Rehovot, Israel, since 1982. His research interests include bridging the gap between formal and informal science learning, teacher professional development, and Project-Based Learning (PBL). He directed a science museum in California, led informal science education programs for youth in Israel's institutions of higher learning, spearheaded the introduction of PBL in Israel, and participated in multi-national research projects funded by the European Union.

AMPLIFYING INFORMAL SCIENCE LEARNING

Rethinking Research, Design, and Engagement

Edited by Judy Diamond and Sherman Rosenfeld

Routledge
Taylor & Francis Group

NEW YORK AND LONDON

Designed cover image: © Getty Images

First published 2023
by Routledge
605 Third Avenue, New York, NY 10158

and by Routledge
4 Park Square, Milton Park, Abingdon, Oxon, OX14 4RN

Routledge is an imprint of the Taylor & Francis Group, an informa business

ISBN: 978-0-367-70276-2 (hbk)
ISBN: 978-0-367-70275-5 (pbk)
ISBN: 978-1-003-14538-7 (ebk)

DOI: 10.4324/9781003145387

Typeset in Bembo
by KnowledgeWorks Global Ltd.

This book is dedicated to Watson "Mac" Laetsch (1933–2020), a pioneer in the field of informal science learning.

CONTENTS

ILLUSTRATIONS

Figures

Tables

FOREWORD

One evening in 1982 I attended a "Speaking of Music" event at the Exploratorium in San Francisco. A composer friend was the guest speaker on this series of programs where contemporary musicians were interviewed about their creative processes and influences, and as a musician and educator myself I was interested in such things. At the time I didn't know what an "Exploratorium" was, but the handful of interactive exhibits that surrounded the "museum of science, art, and human perception" theater that night piqued my curiosity enough for me to come back for more. This was before there was a *STEAM* (science, technology, engineering, arts, and math) acronym, let alone *STEM* (science, technology, engineering, and math), and today we might say that my experience at the Exploratorium that night ultimately proved an "entry" point into another a set of other interests I didn't even know I had.

Five years later, following many visits to the Exploratorium and a stint as a field trip explainer, I was deep into a new career trajectory, co-leading the museum's High School Explainer program. That's when I first became aware of the work of one of this volume's co-editors and contributing authors, Judy Diamond, through a study that she and Mark St. John conducted on the positive and surprising long-term impacts of the explainer program on its participants. That article was my introduction to the ways in which research could both learn from and inform practice, an idea that was just gaining ground in the informal science education (ISE) community and one that further solidified my decision to pursue a career in the field.

Fast forward to 2010, when after working in a variety of national and international ISE organizations, I had the opportunity to become Project Director of the Center for Advancement of Informal Science Education (CAISE), the US National Science Foundation (NSF)-funded resource center for awardees of the agency's ISE program. By then I had been experiencing the ever-expanding range of ISE settings, approaches, and activities being designed, implemented, and studied across the US and globally. The NSF *Framework for Evaluating Impacts of Informal Science Education Projects* had recently been released, providing ISE professionals with informed guidelines for measuring their progress, as had the National Research Council's *Learning Science in Informal Environments: People, Places and Pursuits* consensus report, which had surveyed the research literature to identify specific learning strands that participants in informal contexts can experience. Knowledge-generation from scholarship and wise practices, including from many authors in this volume, has continued to grow and build on these documents and findings since then.

This timely book arrives at a moment when people around the world are getting used to interacting and reconnecting with each other in person, traditionally a primary mechanism for facilitating what we now call informal STEM learning. The global pandemic and racial reckoning crises that we have all been experiencing, albeit unevenly, have had profound effects on the professional field, particularly the place-based organizations that rely on visitors and program participants to keep the doors open and their staff employed. But as you'll see through these essays thoughtfully curated by Judy and Sherman Rosenfeld the field has also continued to innovate, adapt, and leverage new technologies and ways of thinking to create more equitable and inclusive opportunities for people of all ages and backgrounds to engage with and understand natural, physical, and technological phenomena as they choose to, with their whole selves, at their own pace, in their own time.

I invite you to delve into this rich, diverse collection of perspectives on how the informal science, or now almost ubiquitously called the STEM education field of practice, research, and evaluation has been evolving, where it came from, and where it might, and hopefully will, be going. In it you will learn how constructs such as speculative and embodied design, futures literacy, deep canvassing, placemaking, holistically based knowing, relational scaffolding, navigational knowledge, and empathy, to name just a few, are enriching how we think about, design, facilitate, and study STEM learning in informal settings.

And while we both reflect and look ahead, let's think about how we can continue to evolve and diversify this professional community so that the next book that looks back contains more new and different voices from cultures and countries whose ISE work is either not yet recognized, or still emerging, on their unique experiences of working in this field. If we keep our eyes and ears open to the lived experiences, ways of knowing, and practices of those who will make our work more accessible, inclusive, equitable, and as a result – stronger, the better chance we have of science, technology, engineering, and math enjoying their rightful place in our human culture.

Jamie Bell
Center for Advancement of Informal Science Education (CAISE),
Association of Science and Technology Centers (ASTC)

PREFACE

Many people form their attitudes toward and understanding of science, not from school, but from their engagement with the activities and programs of informal science education. Described as free-choice and experiential, the field of informal science attests to helping people develop sustained interest in science. This collection of essays reflects on the current state of informal science education. Some contributors, through their many years of involvement, helped shape the practices that are now common in our museums and science centers. Others in this volume help us reexamine our practices through the lenses of indigenous cultures. Still others bring new challenges and perspectives from the arts and humanities and from science and social science research. Together, these dedicated innovators are challenging established notions of how to teach and learn science in out-of-school settings, fostering new and diverse communities of science learners.

The essays in this book are divided into the following four sections: Exhibits and informal science, engaged with Earth stewardship, places and spaces for informal science learning, and rethinking informal science learning. These sections celebrate the diversity of activities, programs, and perspectives in the field of informal science. Based on current research and practice, the book explores the broad landscape of out-of-school science learning environments. These thoughtful and critical essays examine the state of informal science institutions, such as science museums, zoos, nature centers, planetariums, aquariums, and botanical gardens. They also examine the opportunities and challenges provided by community-based experiences, such as citizen science, makerspaces, science media, libraries, escape rooms, and hobby groups. These writers focus on diversity, inclusivity, and the importance of recognizing and accommodating the differences between learners in educational settings. They examine how informal science learning environments incorporate educational technologies, via games, simulations, curated digital resources, and immersive technologies. They recognize how informal science learning environments can develop reciprocal and mutually beneficial partnerships with community groups. And they underscore the important collaborations between schools and informal science learning environments.

We started working in the field of informal science education almost 50 years ago, when the field was emerging and exhilarating. The Exploratorium in San Francisco, founded by Frank Oppenheimer, was becoming recognized as a world leader, melding together art and science to help the public understand natural phenomena. Across the Bay, science professors at the University of California at Berkeley started a graduate group in Science and Mathematics Education (known as SESAME) that would train a new kind of researcher who would integrate a knowledge of their scientific discipline, psychology, and education to study how people learn science. SESAME is still an inter-disciplinary graduate program leading to a doctoral degree in science, mathematics, technology, and engineering education, and back then it launched the careers of several of the authors contributing to this book, including its two editors. During the first decade of the program, graduate students in the program clustered around emphases that ranged from cognitive science, physics problem-solving, and informal science education. The informal science education graduate students worked under the direction of Watson 'Mac' Laetsch, UC Berkeley professor of Biology and Director of the Lawrence Hall of Science, in whose memory this book is dedicated. These graduate student researchers – including the two of us – studied science centers, museums, zoos, aquaria, and nature centers, examining how these institutions influenced science learning on the part of their visitors. At the time, we felt we were breaking new ground by investigating how people learned in informal science settings. Today, researchers from a wide range of disciplines within the social sciences and education make informal science learning their chosen research focus, and the field has grown immensely.

The essays in this book attempt to articulate the state of informal science education today and the prospects for its future. The field is no less exhilarating, but it is no longer emerging. Fifty years of growth and experimentation in how science is presented in informal settings have solidified the field, establishing guidelines for our programs and institutions. We hope now that the newly entrenched nature of informal science institutions will consider opportunities for reconsidering, rethinking, and reckoning.

Judy Diamond and Sherman Rosenfeld, editors

PART I

Exhibits and Informal Science Learning

1

TRANSFORMING LEARNING LANDSCAPES

A Radical Model for Informal Science Learning

Joshua P. Gutwill, Shawn Lani, and Hsin-Yi Chien

What is the purpose of a museum? According to the American Alliance of Museums, the overriding goals are public service and education.[1] As such, museums have a responsibility to offer educational opportunities to diverse audiences and provide shared spaces in which all communities feel welcomed. Despite the increasing awareness of museums' duty to dismantle educational and social barriers to participation, most science museums today – including ours – remain elitist, exclusive institutions that attract a predominantly wealthy, culturally dominant, highly educated audience. The people who are most harmed by educational and societal inequities may never visit science museums. Statistics comparing census information[2] with data collected across dozens of science museums[3] bear out this view: While only 46% of people in the United States earn over $75,000 per year, 60% of science museum visitors reported earning more than that amount; and although only 32% of American adults acquire a bachelor's degree or more education, 73% of adult visitors to science museums have done so (for a more extensive review of international data on museum visitorship, see Dawson[4]). The gap between science museums' status quo and the responsibilities we owe to the public suggests the need for a new, inclusive approach unencumbered by museum visitation. Our proposal calls into question the very idea of visiting a museum. In this chapter, we present an evolving, tripartite model with which we hope to help transform the future landscape of informal science learning.

Working outside our museum, a brief history

The Exploratorium's Studio for Public Spaces, led by second author Lani, has been developing informal science learning experiences in public outdoor spaces for over a decade, meeting people where they are rather than asking

DOI: 10.4324/9781003145387-2

them to come to us. Creating learning experiences in outdoor community spaces both reduces barriers such as entry and transportation fees and motivates authentic partnership opportunities between museums and communities. The Exploratorium's pedagogy emphasizes learning about real phenomena, which, in outdoor work, leads to a focus on phenomena indigenous to a place: Highlighting the social and geographical elements that make a given place unique and helping people discover their relationship to each other and that place.[5] In planning for place-focused learning, the site itself is both substrate – the environment in which learning happens – and subject – a potential topic of noticing, investigation, and self-reflection. Accordingly, work in any new location needs to start with an evaluation of what's available. What is the context of the place? Who is in the space? What is the potential raw material for inquiry and investigation? The practice has evolved over time, broadening from a focus on physical phenomena to one that includes using social interactions as the subject matter.[6]

In 2017, the Exploratorium received funding from the National Science Foundation to create a social science exhibition and install it in San Francisco's Civic Center. That neighborhood is one of the city's most ethnically diverse and among its most vulnerable. More than a quarter of its 33,000 residents live below the federal poverty line, and illness and chronic disease rates in the area rank among the city's highest. To help alleviate challenging social issues in the area such as intensifying racial and economic division, we created Middle Ground, an exhibition that explores human social interactions and fosters empathy for others (see Figure 1.1). The project crystallized for us a new model for working in outdoor spaces that we call Transforming Informal Learning Landscapes (TILL).

After creating, implementing, and studying the Middle Ground exhibition (see Box 1.1: The Middle Ground Experience), we have come to realize that its impacts on visitors stem from the weaving together of three core project components. We have reflected on how these components fit together and developed the TILL model that we believe may be applied to virtually any science topic area to turn outdoor urban spaces into informal science learning landscapes.

Transforming learning landscapes

The TILL model endeavors to change underutilized public spaces into pleasurable community areas to spend time learning. Rather than requiring members of the public to enter the museum building, museum staff work in partnership with community groups to revamp and outfit public spaces with freely offered exhibitions. The museum also partners with community groups that facilitate learning at the exhibitions. The exhibition sites and facilitators are both chosen to embody the content offered. In this model, funding may come from multiple sources, including foundations, donors, and city budgets.

FIGURE 1.1 (a) Urban Alchemy facilitator Robert Dixon dances with a family at hands-on music. (b) The Middle Ground exhibition was located on the front plaza of the San Francisco Main Library.

The integration of three components – science content, "placemaking," and facilitation – creates new public spaces where community members may learn germane science skills and concepts by interacting with exhibits and human facilitators. These three components of the model are shown in the inner circle of Figure 1.2, while the outer ring represents attributes of the experience – learning-centered, relevant, and humanizing – that emerge from the interplay of the components. The following sections describe each of the components of the TILL model, followed by a discussion of how the

BOX 1.1

THE MIDDLE GROUND EXPERIENCE

Imagine walking through San Francisco's Civic Center, a vast open space, surrounded by imposing gray buildings. People experiencing homelessness or addiction sit or lie on benches and sidewalks while others hurry past on their way to and from work. On the plaza in front of the main library, you notice an array of 14 bright yellow columns, each outfitted with some kind of interactive element – a screen, push buttons, stickers, or even a set of ropes for pulling. A uniformed person greets you with a smile and asks if you'd like to hear a joke. After a good laugh at the *Share a Joke* column, you talk together about the importance of humor in social situations, but also how joking can inadvertently harm others. Then you and the facilitator move on to *Making It in America*, where you try to guess the percentage of people who make it out of poverty every year. You're asked, what about when they work 60 hours per week? What about when they get a college degree? You learn that not only do you overestimate how easy it is to move up economically, but that wealthier people are worse at estimating than poorer people. To take a break, you buy a cup of coffee at the *Pay It Forward Café*, where you generously decide to purchase a half-price token so someone else can enjoy a free cup later on. Chatting with the personable facilitator, you learn that he was incarcerated for the past 36 years and was released only recently. You realize that your surprise reflects your own stereotypes about people who have ended up in prison. You and the facilitator move to *Unseen Stories* to reflect further on how we pigeonhole others, where you read and share personal stories about stereotyping. Then you notice the *My-Side Bias* experience, where you see how your own confirmation bias works as you try to take in new information. Finally, you hold hands with the facilitator and another stranger at *Hands-On Music* and dance to the ensuing music before heading on your way. The whole encounter leaves you feeling energized yet introspective. As you walk down the street, you feel a bit more compassionate toward the people around you, especially those living on the margins. Over the next few days, you reflect on how your mind works, how we all make snap judgments but can slow down, think twice, and question our own beliefs. (For digital versions of many of the experiences, go to www.exploratorium.edu/middleground.)

FIGURE 1.2 Transforming Informal Learning Landscapes (TILL) model.

components work together to realize the attributes that we believe are essential to transforming informal learning.

Placemaking

Meeting people where they are by bringing exhibitions into public spaces is challenging, because public areas are not typically designed to support learning. Outdoor urban places often feel like thoroughfares, where people rush on their way to indoor locations for work, home, or play. To transform them into informal learning environments where people slow down, engage, and interact, we often must create a comforting redesign of the space, for example by providing tables and chairs, sufficient lighting, and protection from harsh weather. Without such efforts, outdoor museum exhibitions may be ignored or misperceived as unwanted material. For this reason, the TILL model begins with placemaking.

> Outdoor urban places often feel like thoroughfares. To transform them into informal learning environments, we must create a comforting redesign of the space.

Placemaking is a broadly defined movement most often led by city planners, designers, and community activists to increase social interaction and build community in urban areas.[7] Why is placemaking important to cities? Cities require positive social dynamics in order to provide safe and welcoming neighborhoods to their increasingly diverse communities. To this end, metropolitan areas have reinvested in urban design over the last decades to foster pleasant social spaces, such as farmer's markets, pedestrian-friendly downtowns, and community parks. Henry Shaftoe coined the term "convivial spaces" to describe "open, public locations (usually squares or piazzas) where citizens can gather, linger or wander through."[8]

We support the idea that everyone has a right to learn about and participate in the improvement of public spaces they use every day. The Exploratorium's learning-centered placemaking practice utilizes informal learning as a tool for supporting the community-centered, codesign methods pioneered by Project for Public Spaces in New York City. According to Project for Public Spaces,[9] a leader in the placemaking movement, convivial outdoor spaces require:

Uses and activities to attract people and engage them. In the Exploratorium's learning-centered placemaking work, exhibitions have served this purpose, focusing on phenomena that are readily available at the site of the exhibition.[10] For Middle Ground, this meant offering activities about the science of stereotyping, confirmation bias, compliance with authority, bystander effect, and other concepts relevant to the Civic Center neighborhood.

Community guidance. Such guidance, which inspires people to collectively reimagine and reinvent public spaces at the heart of every community, strengthens the connection between people and the places they share. In Middle Ground, we partnered with the Public Library to gain perspective on the redesign of its front plaza, the City Planning department to better understand the Civic Center Commons area, and Urban Alchemy to learn more about the social needs of the community.

Comfort and positive image projected by the amenities offered. Our outdoor exhibitions over the years have utilized bright colors, artworks, and attractive exhibits to capture attention and promote welcome and positivity.[11] Visitors to the brightly colored Middle Ground exhibition often cited the look and feel of it when explaining their reason for stopping to use it.[12]

Sociability by gathering people for pleasant social interactions and encouraging return visits. Outdoor exhibitions often have included socially oriented design elements such as curved benches or "whisper dishes" that allow people to hear one another across distances of 10 meters.[13] Middle Ground featured exhibits like *Hands-on Music* where even strangers held hands and the *Pay It Forward Café* where people could sit, relax, and socialize. The café created a lively environment that afforded low-risk "intergroup contact" among people of different walks of life[14] as strangers talked to one another over coffee.[15] In addition, the facilitators of the Middle Ground greeted visitors and passersby with warmth in order to convey the message that this was a space where all were welcome to linger and explore.

Access and linkages that make the space easy to find and use. All our installations have been placed in high traffic zones. Middle Ground was located on the front plaza of the Main Library at San Francisco's Civic Center, a transit hub for the entire Bay Area.

Utilizing these principles, we have sought to undertake learning-centered placemaking to achieve radical accessibility, conviviality, and science learning in outdoor installations.

Learning-centered placemaking differs from typical placemaking designs by focusing on science inquiry as the main activity, both through interactive exhibits and human facilitation. Offered for free and tailored to fit the social, cultural, and geographical needs of each site, the installations produced so far have provided access and relevance unmatched by most indoor museum exhibitions. Audiences have included people from all walks of life, especially those traditionally underserved in science. Each installation site has become a place where members of the public dawdle and socialize through pleasurable learning activities. The placemaking aspects of Middle Ground – bright colors, tables and chairs, a café, and an art installation – as well as the facilitation of the exhibits by trained stewards of public space helped transform the front plaza of the library from a hallway to a living room and provided the foundation for science learning. Moreover, the content was particularly germane to the behaviors of people in that location, strengthening learners' engagement and interest.[16]

Science content

Inside science museums, visitors are often confronted with exhibits and programs that bear little resemblance to the issues in their daily lives.[17] There are many reasons for this, starting with museum design. The museum itself is a decontextualized container, built to house ever-changing exhibitions. In addition, science exhibits may be stripped of context in order to focus on a particular concept, phenomenon, or message. But museum design is only part of the problem. The topics and their forms (environments, exhibits, and imagery) used in museums are often irrelevant to the millions of people who do not ordinarily visit.[18]

> To engage the public in learning the social science underlying contentious societal issues, we leveraged the natural social interactions endemic in the unsettled urban context of San Francisco's Civic Center.

For science museum visitorship to become more inclusive, broadening beyond wealthier, highly educated people, we contend that museums must become more relevant to their communities. Relevance boosts meaning, value, and importance and ultimately increases interest.[19] The second part of the TILL model involves linking the science content with the phenomena in the space itself to create contextually relevant learning experiences. We believe that this model could be used with nearly any science content, as long as the topics are relevant to the installation site. So far, the Exploratorium has created exhibitions about physics, environmental science, and social science. Whatever the content domain, we have endeavored to create experiences that draw upon both the science and the placemaking to enhance relevance of the science learning.

In the Middle Ground project, our goal was to engage members of the public in learning the social science underlying contentious societal issues. We leveraged the natural social interactions endemic in the unsettled urban context of San Francisco's Civic Center. For example, at *Unseen Stories*, visitors learned the science of stereotyping. They wrote on stickers, responding to prompts directing them to share stereotypes that others hold about them (Because I…People think…But actually…) and those they hold for other people (Because some people…I think they're…But maybe…). They then placed their stickers on the exhibit for future visitors to read. In this way, people interacted asynchronously, challenging themselves and each other to engage in metacognitive reflection on stereotypes they may hold.

Here are some examples of stickers written at the exhibit:

Because I…	People think…	But actually…
have tattoos	I play in the NBA	I'm a librarian
look OK	I'm well	I'm fighting unwanted cancer
am a black woman	I am from a broken home, grew up in the ghetto, and have kids outside of wedlock	I am from a two parent military family, am well-traveled, cultured extensively, and have not had any children

Beyond combating stereotypes, the exhibit illustrated the concept that people tend to assume dispositional rather than contextual motivations for others' behavior.[20] For example, when a stranger cuts in line, we often think they are inherently selfish or dishonest, rather than considering contextual factors that may be affecting their behavior, such as failing to notice the queue or being terribly late. *Unseen Stories* encouraged people to become aware of their own attributional tendencies by asking them to share their explanations for other people's actions. Below are examples of this second type of sticker:

Because some people...	I think they're...	But maybe...
approach me as a "stranger"	creepy	they just need someone to respond to them today/ need a smile
have no awareness of those around them	privileged & self-righteous	they just don't understand/ have yet to learn
don't use their turn signals	MONSTERS!	they are in a rush to exit and forget

Another exhibit, *My-Side Bias*, made use of a repurposed library catalog to invite inquiry into how confirmation bias can shape how we seek information about important issues. The exhibit asks questions (e.g., "Should the United States have universal health care?") and offers two catalog drawers (YES and NO) for each question, full of arguments pro and con. Which drawer do people reach for first? Often, they pull out the one with arguments supporting their beliefs and that's an example of confirmation bias.

From a place-based learning perspective, the location of the experience can be leveraged to emphasize the relevance of the content or even provide fuel for the interaction. When placed in free public spaces, where strangers can observe and socialize with each other, experiences like *Unseen Stories* and *My-Side Bias* may become more powerful than in a museum context. Reading stereotypes of the people in the Civic Center and thinking about confirmation bias seemed to help people see their fellow city-dwellers in a new light. Summative evaluation found that these and other exhibits in the installation promoted metacognitive self-reflection. Not only did visitors think about "cognitive shortcuts" like stereotyping and confirmation bias, but they reflected on how their own thinking often leads to inaccurate conclusions.[21] In addition, the experiences themselves, and the kinds of self-reflection they promoted, appeared to help make the area a more comfortable place to mix and spend time. One of the strengths of the TILL model to create outdoor learning experiences lies in the heightened relevance of the content to members of the community.

Contextually relevant science content and learning-focused placemaking may be further enhanced by the people who set the tone in the space, namely facilitators.

Facilitation

The final component of the model is facilitation that deeply humanizes the content of the learning experiences. Authentic sharing of lived experience through human facilitation can develop common ground and meaningful emotional connections to the exhibition content.[22] We believe that learning experiences will be more inclusive – meaningful to people from a much broader range of backgrounds – when facilitated in this way. To that end, the facilitators of the Middle Ground hailed from a partner institution, Urban Alchemy (UA). UA employs former long-term offenders who "have dedicated their lives to healing and redemption through service" as stewards of public spaces and strives to "transform people and urban spaces with respect and compassion in order to heal communities at the intersection of extreme poverty, addiction, mental illness and homelessness."[23]

> Authentic sharing of lived experience can develop common ground and meaningful emotional connections to the exhibition content.

The Middle Ground facilitators learned the science content of the exhibits, offering their own stories of prison and redemption as real-life examples of the impact of cognitive biases. This level of personal, authentic self-disclosure humanized the social science of bias. Armed with strong emotional intelligence to set boundaries and de-escalate conflicts, the facilitators also ensured conviviality, safety, and cleanliness in the area, and as importantly, communicated that everyone is welcomed and valued. For them, it was important to behave as courteously with people experiencing homelessness as with people working at City Hall. Their facilitation strategies focused on inviting members of the public to engage with the exhibition, collaborating with visitors to use exhibits, and encouraging people to try as many exhibits as they wished. For example, a woman who interacted with an Urban Alchemy facilitator reported:

> He knew a lot about the exhibits. He knew what they all did and what the purpose was for them and how it might help you in your life. That was pretty cool. He was easy to talk to and engaged me in telling my own story as well as sometimes, his own story. Right off the bat. And he had really good proximity. His body language was excellent. It wasn't flirtatious or anything. And it was all professional, yet still warm. And I thought he was very sincere.

In the model, facilitation that humanizes the content works together with place-making to transform urban areas into informal science learning landscapes. We have found that the three components mutually reinforce one another.

Whole greater than the parts

According to a comparative research study and a summative evaluation, Middle Ground fostered conceptual learning about social science, metacognitive self-reflection about how one thinks about others, and feelings of connection for people who are different from oneself.[24] Visitors learned social science concepts of fast- and slow-thinking, cognitive biases, and stereotypes and saw the relevance to making snap judgments of others. They had feelings of welcome and intellectual engagement, along with positive interactions with facilitators. Moreover, the exhibition won multiple design awards as well as praise from colleagues in the museum field.[25] What led to this apparent success? We believe the answer lies in the mutual support that all three components of the model – placemaking, science content and facilitation – can provide each other in creating equitable, inclusive learning opportunities.

> *Placemaking + Facilitation → Learning-centered.* The key to creating a more inclusive museum may be to bring exhibitions to public spaces and meet people where they are. Public spaces, however, are not ideal learning environments as they do not offer enough physical and social comfort for people to engage. Facilitation and placemaking thus are critical to slowing people down and establishing the norms that knowledge exploration and exchange of ideas are expected in this space. Placemaking and facilitation are synergistic in Middle Ground, as placemaking supports facilitation by encouraging lengthy conversations through amenities such as coffee, tables, and chairs. Facilitation in turn reinforces placemaking by providing a warm welcome, compassionate listening, and a meticulously maintained space. This combination of the two components creates a *learning-centered* space.
>
> *Science Content + Placemaking → Sense of relevance.* When science content and placemaking designs fit together well, the *relevance* of the content is made plain and binds the two components together. Science ideas and inquiry activities drawn from the site may enhance placemaking, as people begin to see the space through the lens of scientific understanding. In the other direction, open, inclusive designs support people in trying more than one activity, deepening connections between the content and the space.
>
> *Facilitation + Science Content → Humanizing experience.* We have found that facilitators who bring relevant lived experience *humanize* the content in a powerful way. These two components – facilitation and content – can play off one another. For example, facilitators with relevant backgrounds may help learners link science concepts to the real world. As one participant said of a facilitator, "It was a real interaction. I mean he even told us pieces of his life and himself, you

know, that were personal and intimate to him." In Middle Ground, the facilitators served as living examples of many of the science ideas in the exhibition, like stereotyping. In the reverse direction, the science exhibits may serve as rich props for facilitation, offering fun, hands-on activities to spark interest as well as raise ideas for sustained conversation. One facilitator, referring to the *Share a Joke* exhibit, said, "My favorite part of [facilitation] is [working with] the homeless, you know because they're dehumanized. A lot of people don't listen to them. And when they come up, and they want to know what's going on with the exhibits, if I can just put a smile on one person's face and make them laugh, then I'm doing something good." The synergy of facilitation and content may have been most apparent at *Unseen Stories*, which the facilitators used as a platform for discussions of cognitive biases that often involved swapping personal stories. Working with Urban Alchemy taught us that when facilitators have a personal connection to the content, it becomes easier for them to use the exhibits to create engaging learning experiences and for visitors to connect that learning with human issues.

Discussion and future directions

The apparent success of Middle Ground indicated that this model, applied in the domain of social science inquiry, shows promise, yet its validity and applicability need to be tested in other projects, in other sites, and with other topics. As a next step to expand and refine the TILL model, we plan to create free, outdoor science learning experiences around local *environmental justice* issues such as sea-level rise and air quality. In our next projects, we hope to go beyond the limitations in Middle Ground that circumscribed the community's involvement in the project. Having learned more about co-creative, community-based design, we wish we had more fully codeveloped the exhibition with community partners and members. We involved community partners in conversations that inspired Exploratorium exhibit developers, and we invited community members to participate in formative evaluation of exhibit prototypes, but these methods did not go far enough. We produced community-informed exhibits, but not ones that were collaboratively created. To improve our approach, we have begun working with environmental justice partners in a process of codeveloping all three aspects of the model – science content, placemaking, and facilitation – to ensure that the informal learning experiences offered will be more culturally relevant to the communities living near our installations. We are excited to explore how our model will evolve with this environmental project.

We are developing and testing the TILL model with the goal of increasing the inclusivity and relevance of informal science learning. Involving community members in urban design practices, meeting the public where they are rather than expecting them to come to us, and offering a scientific lens for important, relevant topics seems a credible approach for reconstructing the relationship between museums and their surrounding communities. We hope that by expanding learning opportunities that address local challenges, the TILL model will ultimately enhance civic agency and participation. Ideally, outdoor informal science learning environments will support members of San Francisco communities in becoming more involved in neighborhood and citywide affairs that impact their lives and future. The model has already transformed our thinking; now we're hoping to use it to transform both the physical and the learning landscapes of many of the communities that museums have failed to serve.

Acknowledgments

This article is based on work supported by the National Science Foundation (#1713638) and Science Sandbox, an initiative of the Simons Foundation. Any opinions, findings, and conclusions or recommendations expressed here are those of the authors and do not necessarily reflect the views of the National Science Foundation or Science Sandbox.

We would like to thank the members of the Middle Ground project team: Barakah Aly, Josh Bacigalupi, Emma Bailey, Julie Berger, Eileen Campbell, Randy Carter, David Chang, Amy Cohen, Toni Dancstep, Robert Dixon, Gabriel Ehrlich, Adam Esposito, Julie Flynn, Thomas Fortin, Cecilia Garibay, Steve Gennrich, Allen Gerrue, Dana Goldberg, Louie Hammonds, Neil Hrushowy, Meghan Kroning, Manny Lee, Cynthia Lee, Kevin Lee, Lena Miller, Greg Nottage, Romie Nottage, Sue Pomon, Tom Rockwell, Phoebe Schenker, Timothy Smith, Doug Thistlewolf, David Torgersen, Jenny Villagran, Heike Winterheld, and Allison Wyckoff. We are deeply grateful to our advisors Athena Aktipis, Larry Bell, Chris Cardiel, Elena Madison, Hugh McDonald, Jeff Risom, and Nina Simon.

Notes

1 American Association of Museums (Ed.). (1992). *Excellence and Equity: Education and the public dimension of museums: A report from the American Association of Museums. American Association of Museums.*

2 US Census. (2019). *Income and Poverty in the United States: 2019.* The United States Census Bureau. https://www.census.gov/data/tables/2020/demo/income-poverty/p60-270.html.

3 COVES. (2019). *Understanding our Visitors: Multi-Institutional Museum Study (July 2018–June 2019).* Collaboration for Ongoing Visitor Experience Studies. http://www.understandingvisitors.org/wp-content/uploads/2019/09/COVES_2019_AggregateReport.pdf.

4 Dawson, E. (2014). Equity in informal science education: Developing an access and equity framework for science museums and science centres. *Studies in Science Education, 50*(2), 209–247. https://doi.org/10.1080/03057267.2014.957558.

5 Semken, S. (2012). Place-based teaching and learning. In *Encyclopedia of the Sciences of Learning* (pp. 2641–2642). Springer.

6 Campbell, E., & Lani, S. (2021). Public spaces and potential places. *Exhibition, Fall*, 30–39.

7 Gehl, J., & Matan, A. (2009). *Two Perspectives on Public Spaces*. Taylor & Francis; Madden, K. (2018). *How to Turn a Place Around: A Placemaking Guidebook* (2nd ed.). Project for Public Spaces; Shaftoe, H. (2012). *Convivial Urban Spaces: Creating Effective Public Places*. Earthscan; Whyte, W. H. (2001). *The Social Life of Small Urban Spaces* (8th ed.). Project for Public Spaces.

8 Shaftoe, H. (2012), p. 14.

9 Project for Public Spaces. (2019). *What Makes a Successful Place?* https://www.pps.org/article/grplacefeat.

10 Semken, S. (2012). Place-based teaching and learning. In Seel, N. M., ed., *Encyclopedia of the Sciences of Learning*. New York: Springer, pp. 2641–2642; Semken, S., & Freeman, C. B. (2008). Sense of place in the practice and assessment of place-based science teaching. *Science Education*, *92*(6), 1042–1057.

11 Cardiel, C. L. B., Pattison, S. A., Benne, M., & Johnson, M. (2016). Science on the move: A design-based research study of informal STEM learning in public spaces. *Visitor Studies*, *19*(1), 39–59. https://doi.org/10.1080/10645578.2016.1144027.

12 Garibay, C. (2021). *Middle Ground Summative Evaluation* [Summative Evaluation]. Garibay Group. https://www.informalscience.org/street-smarts-experiments-urban-social-science.

13 Campbell, E., & Lani, S. (2021). Public spaces and potential places. *Exhibition, 32* (Fall), 30–39.

14 Stephan, W. G., & Stephan, C. W. (2017). Intergroup threat theory. In Y.Y. Kim (Ed.), *The International Encyclopedia of Intercultural Communication* (pp. 1–12). Wiley Online Library. https://onlinelibrary.wiley.com/doi/10.1002/9781118783665.ieicc0162.

15 Wise, K. (2019). Middle Ground: The Exploratorium's exhibition at Civic Center Plaza, San Francisco. *Informal Learning Review*, *150* (November/December), 13–16.

16 Garibay, C. (2021). *Middle Ground Summative Evaluation* [Summative Evaluation]. Garibay Group. https://www.informalscience.org/street-smarts-experiments-urban-social-science.

17 Dawson, E. (2014). "Not Designed for Us": How science museums and science centers socially exclude low-income, minority ethnic groups. *Science Education*, *98*(6), 981–1008. https://doi.org/10.1002/sce.21133.

18 Dawson, E. (2014). "Not Designed for Us": How science museums and science centers socially exclude low-income, minority ethnic groups. *Science Education*, *98*(6), 981–1008. https://doi.org/10.1002/sce.21133; Garibay, C. (2009). Latinos, leisure values, and decisions: Implications for informal science learning and engagement. *Informal Learning Review, Jan–Feb*, 10–13.

19 Anderson, A., Kollmann, E. K., Beyer, M., Weitzman, O., Bequette, M., Haupt, G., & Velázquez, H. (2021). Design strategies for hands-on activities to increase interest, relevance, and self-efficacy in chemistry. *Journal of Chemical Education*, *98*(6), 1841–1851. https://doi.org/10.1021/acs.jchemed.1c00193; Keller, J. M. (1987). Development and use of the ARCS model of instructional design. *Journal of Instructional Development*, *10*(3), 2. https://doi.org/10.1007/BF02905780.

20 Kahneman, D. (2011). *Thinking, Fast and Slow*. Macmillan; Ross, L. (1977). The intuitive psychologist and his shortcomings: Distortions in the attribution process. In *Advances in Experimental Social Psychology* (Vol. 10, pp. 173–220). Elsevier.

21 Garibay, C. (2021). *Middle Ground Summative Evaluation* [Summative Evaluation]. Garibay Group. https://www.informalscience.org/street-smarts-experiments-urban-social-science.

22 Cohen, O., & Heinecke, A. (2018). Dialogue exhibitions: Putting transformative learning theory into practice. *Curator: The Museum Journal, 61*(2), 269–283. https://doi.org/doi:10.1111/cura.12254.

23 Urban Alchemy. (2020). *Urban Alchemy Responsibilities Training Document*. Page 1.

24 Chien, H.-Y., Dixon, R., Gutwill, J. P., & Hammonds, L. (2020, October 21). *Creating Middle Ground: Transforming Urban Outdoor Spaces with Social Science Exhibits and Facilitation about Biases*. Annual meeting of the Association of Science-Technology Centers. Virtual meetinghttps://www.exploratorium.edu/sites/default/files/pdfs/Middle%20Ground%20ASTC%20Presentation.pdf; Chien, H.-Y., & Gutwill, J. P. (in review). Measuring the impact of facilitation at an outdoor social science exhibition. *Visitor Studies*; Garibay, C. (2021). *Middle Ground Summative Evaluation* [Summative Evaluation]. Garibay Group.

25 Design Week. (2020). 2020 Recipients. *San Francisco Design Week*. https://sfdesignweek.org/award-winners-2020/; Mora, J. G., & Russick, J. (2021). *AAM Excellence in Exhibition Label Writing Competition 2021*; Schwarz, T. (2020). Shedding light on shared humanity. *Exhibits Newsline, Spring*, 10–11; Wise, K. (2019). Middle Ground: The Exploratorium's exhibition at Civic Center Plaza, San Francisco. *Informal Learning Review, 150* (November/December), 13–16.

2

WHAT ABOUT AIDS? THE CASE OF THE NATIONAL AIDS EXHIBIT CONSORTIUM AND ITS TRAVELING EXHIBITION

Sherman Rosenfeld, Martin Weiss, Roberta Cooks, Larry Bell, and Wendy Pollock

When a new human disease was first identified in 1981 by the Centers for Disease Control and Prevention (CDC), its long-term consequences were unknown. But soon it became clear that the disease, acquired immunodeficiency syndrome (AIDS), was both contagious and deadly and that it was caused by a virus called human immunodeficiency virus (HIV). The virus spread rapidly, leading to the deaths of over 100,000 people in the USA by 1990.

In the early 1990s a group of eight American science museums and two public health institutions formed the National AIDS Exhibit Consortium (NAEC), with the main goal of assisting and supporting public science museums "to inform and influence the American public" about the science and health decisions regarding HIV/AIDS. By the end of 1993 the NAEC had designed, fabricated, and launched three copies of a 2,500 sq. ft. national traveling exhibition, "What About AIDS?" that reached 26 cities within the following three years. In addition, individual science museums in the Consortium developed many other educational products, so that by the end of that decade, millions of Americans of all ages learned about the science and public health of HIV/AIDS through the Consortium's work.

How was the Consortium established? How was its traveling exhibition developed? What was its final form? How were controversial issues navigated? What was the exhibition's impact on the host science museums and their visitors? And what lessons might be useful for the Informal Science Learning (ISL) community to prepare the public to learn about and respond to other emerging pandemics, such as COVID-19, as well as other public health issues?

DOI: 10.4324/9781003145387-3

A Consortium is formed

In the early 1990s science museum staffs around the USA were grappling with how to approach the AIDS epidemic. Because of the fear and misinformation about the science relating to the disease and its mode of transmission, the media were filled with articles about viruses, immunology, and especially AIDS activists[1] who demonstrated against the inactivity of the government to help them fight the disease. It made sense that science museums, trusted institutions that serve young people and families, could have an important role in translating the science of HIV and AIDS to these audiences. At the same time, the CDC was looking for new ways to educate the public about this new public health crisis and had funds to offer institutions that could design and deliver effective educational materials.

The CDC and eight science museums joined forces in 1990 to create the National AIDS Exhibit Consortium (NAEC).[2] Its founding members were the Museum of Science and Industry in Chicago, the Museum of Science in Boston, the California Museum of Science and Industry in Los Angeles, the Exploratorium in San Francisco, the Franklin Institute in Philadelphia, the Maryland Science Center in Baltimore, the National Museum of Health and Medicine in Washington, DC, the New York Hall of Science in New York City, and the American Medical Association. In September 1991, the Museum of Science and Industry in Chicago submitted a proposal on behalf of the Consortium and was awarded a total of $2 million from the CDC's National AIDS Minority Education and Information Program. Members of the Consortium met regularly from 1990 to 1993, to work on developing a traveling exhibition and to share the educational work of the different institutions on HIV-related topics. Toward the end of 1993, the Association of Science-Technology Centers (ASTC) facilitated a three-year tour of three copies of the traveling exhibition throughout North America.

The development of the "What About AIDS?" traveling exhibition

The members of the Consortium formed bylaws that created a framework for managing the organization; they met for about a year to discuss the conceptual design of the exhibition. Afterwards they decided that the exhibition would be developed and fabricated by one of the museums – The Franklin Institute of Science in Philadelphia – based on other successful models of creating traveling exhibitions, such as the Science Museum Exhibit Collaborative. Some of the smaller HIV-related projects developed by other museums were eventually folded into this larger, traveling exhibition.

A non-profit research and development organization, Science Learning, inc (SLi),[3] was hired to conduct front-end evaluation of the prospective audiences (to inform exhibit design), formative evaluations of prototypes (to improve the exhibition), remedial evaluation of the full exhibition (to make final adjustments before traveling), and a summative evaluation (to measure visitor outcomes). The front-end evaluation found that while the main target group (young adults aged 10–26) generally knew about AIDS health messages, they did not know about the science of viruses nor how the science connected to the health messages. The Consortium saw this gap as a unique educational opportunity and formulated three interrelated types of educational goals: cognitive, affective, and behavioral.

1. Cognitive: to provide information about the science of the HIV virus, how it is transmitted and how it attacks the human body to cause AIDS
2. Affective: to convey personal stories of people whose lives were affected by the disease to encourage compassion toward AIDS sufferers and receptivity toward AIDS information
3. Behavioral: to present public health messages about effective protection measures against the disease

The physical exhibition was designed in three concentric circles, each based on one of these goals. The outer circle presented the science of HIV/AIDS, the middle circle dealt with stories about people affected by the disease, and the inner circle focused on protection measures, particularly on the rationale of using condoms during sex. A "PG-10" sign was displayed before the entrance to the inner circle, so parents would be able to decide whether or not to allow their children to enter the area that had explicit information on safe sex and intravenous drug use.

To make needed changes in the exhibition, five formative research studies were conducted on prototype versions of the exhibition, or parts of it, before its final version was released in 1993. A comprehensive "remedial evaluation" concluded that visitor responses to the exhibition "were overwelmingly positive" and that it "accommodated diverse knowledge levels and interests," with close to 80% of the visitors reporting they learned something new or gained new insights. In addition, the above three educational goals "often created a synergistic effect" since the dramatic human stories motivated people to learn about the science of HIV and about how to prevent the spread of AIDS.[4]

The formative studies made several recommendations. Problematic aspects of the exhibition's organization were modified in its final version. A children's area adjacent to the main exhibition was added that included puzzles and carefully selected books about viruses (see Figure 2.1). Based on the finding that visitors were emotionally engaged in the exhibition and were interested in expressing their reactions, they were given an opportunity to write about their reactions to the exhibition, creating a dialogue with other visitors and with the museum staff.

FIGURE 2.1 The children's area adjacent to the "What About AIDS?" traveling exhibition.

Another study conducted at the New York Hall of Science investigated the hypothesis that while early adolescents (aged 10–16) knew that condoms can be useful to prevent the spread of AIDS, "there was not a widespread comprehension of how condoms function in this regard." The study compared the learning outcomes of this target group for two versions of an exhibit on the sexual transmission of HIV: an original version that provided text with abstract graphics and a revised version with the same text but with more explicit graphics. Based on the results of the study, the version with the more explicit graphics was used in the final exhibition.[5]

At the Franklin Institute, where the prototype exhibition was developed, many different questions, fears, emotions, and controversies arose among the museum staff and members of the board of directors, as well as the local religious and activist communities. For example, a *New York Times* article quoted the coordinator for the Philadelphia Archdiocese of the Catholic Church who "praised the exhibit in general," but "wished the discussion of condom use could have been 'a little less blatant.' He also said that he thought that the exhibition 'encouraged, rather than discouraged' teen-agers from engaging in sexual activities."[6] Due to comments like these, an additional panel was added – "A Hundred Ways of Making Love without Doing It" – that highlighted abstinence as the safest protective measure against AIDS and also a viable way to be affectionate with someone without having sexual intercourse.

The components of the final exhibition were varied. To explore the science of HIV and AIDS, the exhibit included a highly accurate model of HIV that people could hold in their hands, giant electron micrographs of the virus attacking white blood cells and an illustrated timeline of the developing science of this new and deadly disease. To present personal stories, poignant photo portraits were combined with audio recordings of adults and children who spoke about how living with HIV and AIDS impacted their lives. Another panel presented the voices of the exhibition designers and the reasons they thought the exhibition was important.

To present a protection measure against the virus, the exhibit included a huge wall of red, yellow, and blue condoms that looked like birthday balloons. To illustrate the connection between the science of HIV/AIDS, personal behavior, and the risk of being infected, visitors interacted with a giant game where three different dice had the image of a skull and crossbones on none, one, or more sides. Depending on the risky behavior a person chose to engage in, the visitors picked which of the dice to roll and took their chances of being infected with HIV, indicated by the skull and crossbones graphic. People of all ages spent a great deal of time rolling the dice and learning about sex, needles, and the science of HIV risk (see Figure 2.2).

The exhibit also included a computer kiosk in English and Spanish where the actor Edward James Olmos and a group of teenagers discussed the use and proper placement of condoms, as well as spermicides, safe sex, and abstinence.

At the end of the exhibit a notebook was available for visitors to write down the name of someone with AIDS whom they wanted to remember; note cards were available for visitors to write their thoughts, feelings, and questions after

FIGURE 2.2 The interactive What's My Risk? game in the exhibition.

experiencing the exhibit. The note cards were placed in a box; volunteers read these note cards and selected some of them to post on an AIDS bulletin board. In this way, visitors could express their own positive and negative emotions with each other and with the museum, while ensuring individual privacy.

Dealing with controversial issues by developing community involvement

Due to the sensitive nature of the exhibition, the design team at the Franklin Institute engaged in a rigorous process of presenting the exhibition's content and design to its board of directors, as well as to other community entities such as The Archdiocese of Philadelphia and AIDS activist organizations. A committee was formed at to discuss and develop the best way to ensure the exhibit would be successful when it opened on the exhibit floor. The committee met biweekly and included a representative from every department in the museum, including exhibitions, education, security, maintenance, membership, public relations, volunteers, and fundraising. Based on the committee's recommendation, special HIV Red Cross training was required for all museum personnel. The exhibit was prototyped over a three-month period. During this time, staff, selected visitors, and representatives from the Archdiocese and the AIDS community were invited in to give their comments and critiques. This challenging process created camaraderie and ownership among museum staff when the exhibit finally opened to the public.[7]

Based on this experience, the following guidelines were developed to deal with sensitive and controversial issues:

1. Believe in what you're doing.
2. Prepare your museum (get all departments involved).
3. Reach out to your community. Identify community partners.
4. Host a preview and invite your potential enemies. Be ready to make changes.
5. Learn from the stories of people who have already hosted the exhibit.[8]

These guidelines and the same processes of institutional development and community involvement developed by the Franklin Institute were shared with other museums that hosted the traveling exhibition, via two-day preparatory workshops, funded by the Consortium and organized by the National Museum of Health and Medicine in association with ASTC. The training addressed issues of staffing, instruction, community participating and response, as well as working with museum and local school boards.[8]

By all accounts, these preparatory workshops led to positive outcomes in the host museums. Participants reported that being exposed to potential challenges – such as dealing with bad press, dealing with prejudice, and forming community alliances – along with being exposed to real-life solutions, was invaluable. For example, the director of education at the Cranbrook Institute of Science said that hosting the exhibition was "a transforming experience" that gave her museum the confidence to tackle other potentially difficult subjects. "Because of this exhibition, we're much more open and accepting of a much broader-based population than ever before." In Charlotte, North Carolina, school board officials previewed the exhibition at the local science museum. Afterwards an official, up for reelection, objected to a video display about condoms and decided that students in the city should be prevented from seeing the exhibition. The museum director called the media, the story was presented in the local press, and the resulting community protest sparked a decision to require all secondary school teachers to visit the exhibition with their students and to enroll them in a minicourse about how the human immune system works.[9]

Exhibition outcomes

The traveling exhibition catalyzed the 26 host science museums to integrate across departmental divisions and develop new models of community involvement. Positive visitor outcomes were also demonstrated. A comprehensive summative evaluation study of the exhibition showed that it was very successful for visitors from a variety of backgrounds, degrees of interest, and knowledge levels. The overwhelming majority of visitors felt that the controversial aspects of the exhibition were presented sensitively on an acceptable level. Of the three educational goals, the affective goal of the exhibition, via personal stories, was most frequently considered its main message.[10]

Visitor perceptions of the exhibition were stable over time: a follow-up study three months later showed that two-thirds of the visitors said they had thought about the exhibition since their visit. For example, one woman in her 20s said: "I found myself still thinking about some of the things in the exhibit even weeks later. For example, I hate to say it, but the dice exhibit (about the probability of getting AIDS) really made me think about who I go out with these days." An analysis of these findings concluded that the exhibition's success results from the visitors being able to relate to its main ideas and phenomena on three levels: (1) on their own terms, depending on their individual levels of interest, knowledge, and experience; (2) via an accessible, safe, and fun social environment; and (3) via a rich real-world context where they could directly experience, in multiple ways, the science and technology behind these ideas and phenomena.[11]

The health messages of "What About AIDS?", if presented alone, could have been uncomfortable for science museum visitors. But by using a combination of interactive elements, along with clear, simple text and graphics, based on cognitive, affective, and behavioral goals, the traveling exhibition created a neutral place for families and young people to feel comfortable learning and talking about these sensitive health messages.

The Consortium's other activities

In addition to developing the traveling exhibition, the Consortium developed staff training materials, teacher education materials, school assembly programs and many new interactive exhibits that were made available to science museums throughout the USA.[2] The Consortium was also the catalyst for the Chicago Museum of Science and Industry's 4,000 sq. ft. permanent exhibition, "AIDS: The War Within" that was funded by a Chicago pharmaceutical firm and opened in 1995. During its eight-year lifespan, the exhibition was visited by millions of visitors. Designed with the same three educational goals as the traveling version, it contained different exhibit components.[12]

In 1994 the National AIDS Exhibit Consortium evolved into a new organization, the National Health Science Consortium, designed to develop and implement national educational programs in the health sciences and biomedical fields. The new organization was short-lived: its first and only exhibition was on Women's Health, funded by the National Institutes of Health (NIH). In 1996 the NIH released its Science Education Partnership Award (SEPA), with the goal of developing "partnerships between public and private sector organizations, scientists, and educators … to improve student understanding of the health sciences in pre-K–12 education, and increase the public's understanding of science."[13] This move, welcomed by the members of the Consortium, made funding available to all science museums for future health exhibitions and programs.

Guidelines for the present and future

Based on the above review of the Consortium and its work, what guidelines may be gleaned for use by professionals working in the field of ISL, particularly when addressing emerging pandemics and other urgent public health issues? We suggest several:

> *When dealing with sensitive and controversial issues, collaboration with diverse partners is essential.* In the Consortium, collaboration occurred on several levels: on the level of the eight different science museums that cooperated to produce effective health exhibitions and health programs that reached millions of people; on the level of the collaborative work of the Franklin Institute – both inside the museum's different departments as well as outside the museum with representatives of diverse professional and community groups; and on the level of the 26 museums the traveling exhibition; the preparatory workshops provided the foundation for this dual inside-outside collaboration and led to transformative change

in the museums. On all three levels, the guiding principle was for museum staff to be open to diverse perspectives, to work through and resolve difficult issues collaboratively, and to compromise but without losing their own points of view. This focus on collaboration has been developed and reinforced over time within the wider ISL community.

Both cognitive and affective goals are essential to promote behavioral goals relating to heath. The Consortium's emphasis on the cognitive goals in this triad was supported by front-end evaluation that showed that most potential visitors lacked knowledge about the science of AIDS (e.g., about viruses and the immune system) in order to support the public health messages they had learned from other sources. The exhibition evaluation showed that the affective goal was equally important and that developing compassion toward AIDS sufferers helped visitors connect the science of AIDS with recommended healthy behaviors. Clearly the development of empathy can lead to motivated learning. But current research about the challenges of convincing members of the public to change their health behaviors suggests that the development of empathy is tricky. For example, a public health study found that graphically intense and dramatic pro-vaccine messages can backfire, leading to increased misperceptions and reduced intentions to vaccinate.[14]

Provide multiple choices for visitors, both relating to their learning modalities and to the subject matter. In the exhibition, visitors could choose from a variety of different activities, such as engaging in interactive activities, watching a video, interacting with computer programs, listening to audiotapes, and reading text. They could choose to learn via straightforward presentations of facts and concepts, epidemiological charts and graphs, and/or personal stories of sufferers, and they could express their own stories and provide feedback on the exhibit to the other visitors and museum staff. This wide menu of choices made it possible for each visitor to find something to which they could relate, at their own levels of knowledge and interest.[11]

Social science research can inform our efforts to help people make wise health decisions. One of the ways health exhibitions such as "What About AIDS" differ from classic science museum exhibitions is that the former involve human decision-making. In the 1990s science museum professionals were often "uncomfortable with issues-related programming, preferring 'concrete' science that did not venture into politics. The changing nature of points of view on the subject also evoked some concern, as did the fact that the topic ... raises questions that cannot be answered definitively."[15] Since that time, the US National Academies of Science, Engineering, and Medicine have worked to raise awareness of social science research related to "The Science of Science Communication"; relevant videos and publications are widely available.[16] In addition, social science research has addressed issues such as the lack of trust in science and governmental agencies, such as the Department of Health and Human Services, CDC, NIH, and state public health departments. This research points to a need to develop "depoliticized outreach programs targeted at the most socially disadvantaged groups, and to design vaccination strategies conceived with people from different social and racial backgrounds to enable them to make fully informed choices."[17]

Actively develop and maintain access to institutional history. This essay has presented a brief history and some lessons of a science museum Consortium and its traveling exhibition that, over time, could have been lost to the collective memory of the ISL community. As science museum professionals, we do not do a service to our institutions or our field if we forget the past and/or fail to institute ways to record and remember it. Shared collective memory of our exhibitions – based on written documentation and studies – can provide salient lessons, useful models and perhaps even inspiration for future efforts.

The world's recent encounter with COVID-19, starting in 2019, challenged the ISL community's capacity to respond to a different global pandemic that spread more rapidly and more widely than the HIV/AIDS pandemic, massively disrupting health, economic, social, and educational systems throughout the world. It is beyond the purpose of this chapter to present and evaluate the scope and quality of the community's response to this challenge, but we hope that future studies will do so, adding to the knowledge base that will help us best prepare for and respond to society's future pandemics and its other public health needs.

Acknowledgements

The authors met online via Zoom during June–August of 2022. During this time, we found and discussed articles, reports, and other documents – many unpublished – that form the backbone of this study. We are grateful to the Archives Department at the Boston Museum of Science for providing many of these resources. Four of the authors were members of the Consortium and played key roles in its work: Martin Weiss was the Chief Scientist at the New York Hall of Science, Roberta Cooks was Senior Exhibit Developer at the Franklin Institute, Larry Bell was Vice-President for Exhibits at the Boston Museum of Science, and Wendy Pollock was coordinator of ASTC's Traveling Exhibitions Program and managed the exhibition's three-year tour. The online meetings were organized by Sherman Rosenfeld, who wrote the chapter with the assistance of the other authors. The authors would like to acknowledge the important input of Barry Aprison, Project Director and Secretary-Treasurer of the Consortium; John Falk, director of the Consortium's evaluation efforts; and Lynn Dierking.

Notes

1 Shilts, R. (1987). *And the Band Played on: Politics, People, and the AIDS Epidemic.* New York, NY: St. Martin's Press.
2 Aprison, B. (1994). *Final Project Report: National AIDS Exhibit Consortium.* Unpublished document. 38 pages and 12 appendices. Aprison, B. (1993). The National AIDS Exhibit Consortium. *Curator,* Vol. 36, No. 2, 88–93. The ASTC was instrumental in forming the Consortium.
3 SLi was incorporated as the Institute for Learning Innovation (ILI) in 1998. Both organizations were directed by Dr. John Falk. The formative evaluation included studies of the interactive computer exhibits (the AIDS Kiosk) in nine different venues, such as health clinics, shopping malls, and university settings.
4 Falk, J.H.,& Holland, D.G. (1993). *Remedial Evaluation Research Results: "What About AIDS?" Traveling Exhibit. National AIDS Exhibits Consortium and the Franklin Institute.* Science Learning, inc. Unpublished manuscript.
5 Falk, J.H., &Weiss, M. (1994). Utilizing museums to promote public understanding of science: Early adolescent misconceptions about AIDS prevention. In Bitgood, S. (ed.) *Proceedings of 1992 Annual Visitor Studies Conference.* Jacksonville, AL: Center for Social Design.
6 Janofsky, M. (1993). Exhibit Views AIDS Frankly for the Young. *New York Times,* August 8, 23.
7 Cooks, R. (1998). Is There a Way to Make Controversial Exhibits that Work? *The Journal of Museum Education, 23* (3): 18–20.
8 Witting, S. (1995). Tackling AIDS Prompts Self-Examination, Change. *ASTC Newsletter, 23* (1):, 1, 12, 24. The workshops were also funded by CDC and the Metropolitan Life Foundation.
9 Pollock, W. "What About AIDS? 1994 Tour Summary and Highlights" and "What About AIDS? 1995 Tour Summary and Highlights." Unpublished documents. These reports provide in-depth descriptions of how 12 host science museums utilized the exhibition to "involve their communities in a variety of ways to present related programming"; Witting, S. (1995);. Cooks, R. (1998).
10 Holland, D.G.,& Falk, J.H. (1994). *What About AIDS? Traveling Exhibition: Summative Evaluation.* Science Learning, inc. Unpublished manuscript.
11 Dierking, L. (2005). Museums, Affect and Cognition: The View from Another Window. In Alsop, S. (ed.). *Beyond Cartesian Dualism: Encountering Affect in the Teacher and Learning of Science,* 111–122, Dordrecht, The Netherlands: Springer.
12 See Holden, C. (1995). Education Booster for AIDS. *Science, 268* (5207): 35. The exhibit won an honorable mention at the 1996 8th Annual Exhibit Competition program, given by the American Association of Museums.
13 Science Education Partnership Awards (SEPA): History and Objectives. https://nihsepa.org/about/history-and-objectives/.
14 Nyhan B., Reifler J., Richey S., Freed G.L. (2015). Effective messages in vaccine promotion: a randomized trial. *Pediatrics,* 133 (4): e835-42. https://doi.org/10.1542/peds.2013-2365. Epub 2014 Mar 3. PMID: 24590751.
15 Quote from Minda Borun, former Director of Research and Evaluation at the Franklin Institute. In Mintz, A. (1995). *Communicating Controversy: Science Museums and Issues Education.* Washington, DC: Association of Science-Technology Centers.
16 See http://www.nasonline.org/programs/nas colloquia/completed_colloquia/science-communication.html.
17 Bajos, N., Spire, A., Silberzan, L., Sireyjol A., Jusot F., Meyer L., Franck J., Warszawski J. (2022). When Lack of Trust in the Government and in Scientists Reinforces Social Inequalities in Vaccination Against COVID-19. *Frontiers in Public Health,* Vol.10. https://www.frontiersin.org/articles/10.3389/fpubh.2022.908152.

3

DESIGNING FOR DIVERSITY

Judy Diamond, Marianne Achiam, and Devra Hock

Science is ultimately about making sense of the world. Whether this sense-making occurs in natural environments, laboratories, or designed environments such as science centers and museum exhibitions, science learning happens as an interaction of human thoughts with natural phenomena, objects, signs, tools, and artifacts. When we think about designing science learning environments, we attend to what the visitor brings to the encounter, what the designed environment offers, and the diversity of ways creative cognitive activity can happen in the space between them.

At face value, this conception of science learning is depicted as an experience that is open and available to all. Recent research, however, has revealed a different picture, suggesting that many exhibits do not offer unbiased and equitable science learning opportunities for visitors. Encounters with objects, signs, tools, and artifacts in exhibits lead some visitors to feel left out or excluded because they are not familiar with the language of explanatory labels or the expectations of how to act in response to interactive opportunities. Visitors sometimes feel marginalized by how an exhibit prompts others in the museum to behave and the spontaneous social groups that can form around exhibits.[1] In this essay, we consider what it means to design for a broader diversity of science learners, discussing visitors, exhibits, and the possible interactions between them. Then we present two examples of exhibitions that have used a variety of tools and techniques to encourage broad science learning in all visitors.

Visitors come to museums with a wealth of individual experiences. As they enter, adults and children have established attitudes and associations with the museum environment and the objects and ideas within. They may have visited that type of museum before or are familiar with a similar image or concept

DOI: 10.4324/9781003145387-4

from media or toys. They might have touched on the topic in school, read something about it, or heard about it from friends. Regardless of the source, those connections influence how visitors interact with their surroundings. More generally, visitors' own history, culture, and beliefs shape their sensory and cognitive preferences. We don't just input stimuli; we *select* certain stimuli and ignore others. And the selection processes occur at every level: what we sense, what is coded in memory, and what is available for retrieval. Ultimately, these active selection processes make every individual a unique processer of information and experience. Familiarity with a type of object or tool influences how visitors interact with it. For example, a visitor familiar with scientific research or a student who has been given access to scientific equipment in school might engage naturally with interactive science exhibits. In contrast, someone with no prior experience of scientific apparatus might be baffled by the invitation to manipulate objects, and they may feel uncertain and excluded by that environment.

Many natural history museums create their object-rich galleries with the expectation that visitors will conform to a cultural norm of quiet and subdued behavior, regardless of whether this may marginalize visitors who express interest and pleasure by active movements and vocalizations. Archer and colleagues suggest that for some young adult visitors, engaging with exhibitions may contradict their sense of selves, and they struggle to find acceptable and available ways to interact with the exhibition.[2] Modern museum exhibits are meant for all to enjoy and learn from, but cultural and socioeconomic barriers often prevent visitors from experiencing what an exhibition offers.

Exhibits

Many science exhibits engage members of the public with aspects of scientific processes, attempting to establish links to the underlying scientific disciplines. Natural history exhibits may utilize objects from the museum's collections, and their displays of evolutionary sequences or transitional fossils sometimes intentionally illustrate reasoning characteristic of the historical sciences. Science center exhibitions are often deeply influenced by the discovery pedagogy pioneered by Frank Oppenheimer, the founder of the Exploratorium. In the earliest days, Oppenheimer realized that it would not be enough just to have exhibits that prompted visitors to manipulate and observe experimental apparatus. He realized that artists were effective exhibit designers, and he conceived of the museum as a place of art, science, and perception, rather than one that tried to recreate the historical process of scientific investigation.[3] Science and technology museums, planetariums, zoos, botanical gardens, and aquarium exhibitions all struggle with how to represent the scientific disciplines that underpin their processes and collections, while at the same time, engaging their public freely in exploration and play.

At first glance, a close connection between science exhibition design and scientific disciplines seems constructive, since science claims to produce impartial, value-neutral accounts of nature. But an increasing body of research demonstrates how science can often exclude gender, ethnic, class, and ability minorities, and even more so when these characteristics intersect.[4] If science exhibitions uncritically adopt the knowledge, values, and practices of the scientific disciplines they represent, they risk adopting the exclusive mechanisms of those disciplines as well.

The community of museum professionals has begun to recognize the inherent biases within scientific investigation. It is nonetheless difficult for designers to redirect and reinvent the process of exhibition design, and often the power structure in their institutions is focused on maintaining historical scientific values. For instance, Feinstein and Meshoulam studied science centers and museums across the US and found numerous examples of institutional norms that perpetuated a white, male, and middle-class culture. In another example, Robinson showed how conventional institutional practices in the National Museum of Australia elevated established Eurocentric museum approaches, thereby overriding the perspectives of the Indigenous people who had been specifically invited to contribute to the development of an exhibition.[5]

Interactions

Science learning happens as the human mind is shaped by its interactions with phenomena, objects, tools. Those interactions also exist within a socially mediated environment. Effectively, cognition is distributed across the space of internal cognitive processes, external artifacts, and social relationships. It is influenced by the so-called affordance space, constituted by the visitor's abilities in relation to the opportunities for action made possible by the environment.

Don Norman's work influenced the idea that affordance space applies to various designed objects; for instance, how the design of an exhibit shapes how visitors interact with it. Behind glass, a specimen invites visual inspection only. The shape and color of handles and buttons define how they may be used – to lift or push. The designed elements of each object have a direct, but imperceptible, conversation with the visitor's intuition: this is for touching, this should be examined closely, this is to be looked at only, this can be pushed, and this will operate with a wave of the hand. However, the notion of affordance does not in itself explain why one visitor is able to interact comfortably with a hands-on exhibit, but another is put off or even alienated by it.[6]

The Gestalt psychologist Kurt Lewin emphasized that rather than considering the properties of an object, we should think about the features of the situation in which the object is encountered. In other words, objects (or exhibits) suggest different actions to us because of the different ways we relate to them.

Emily Dawson offers a striking example of this: A certain African bird species, displayed at a science museum in London, prompted Sierra Leonean visitors to perform a ceremonial dance involving hunting and eating the birds. Dawson's reported example exemplifies that the process of learning is highly personal and strongly influenced by an individual's prior knowledge and experiences. Prior experiences and knowledge frameworks of visitors will impact what they seek out and retain from museum exhibitions and explain how the same exhibition can prompt different actions and responses in various groups or individuals.[7]

The affordance space that emerges between visitor and exhibit can be deeply personal, varying from visitor to visitor, even while the exhibit remains the same. Rather than presenting a challenge to the notion of objectivity in science, the personalization of exhibit engagement can be embraced, so that exhibitions are specifically designed to allow for varied and personal expressions. Science learning is achieved only when individuals decide to incorporate new information and experiences into their mental constructs, infusing the new into what they already know and accept. In this way, exhibits designed to allow individual interpretation and expression present a pathway to durable science learning in ways that conventional exhibits may not.

Examples of exhibitions prompting science learning for a broad audience come from efforts to present the scientific evidence for evolution. Scientists tend to have a unified voice in advocating for the universality of evolutionary theory as an explanatory system for the diversity of life on Earth. But visitors experience museum evolution exhibits through a wide range of preconceptions. These can range from vague memories about islands and finches, to recollections about contradictions and incongruities, and even a sense of threat or alarm that evolution seeks to negate cherished beliefs.

Two National Science Foundation (NSF)-funded museum exhibits found pathways to presenting ideas about evolution to diverse visitors in ways that encouraged exploration and reconciliation with prior beliefs. In each of these projects, the exhibits offered interactive experiences that invited social play and spontaneous conversations. Studies of how visitors explored the exhibits revealed how the experience influenced their reasoning about evolution. In Diamond's *Explore Evolution* project, visitors explore how different scientists found evidence for evolution in creatures as diverse as viruses, diatoms, ants, flies, finches, people, and whales. As a result of their experience, visitors made small but significant transitions toward thinking more like evolutionary biologists. Surprisingly, many visitors understood and accepted the scientific explanations for the diversity of life, while also maintaining creationist or other religious beliefs. Many religious groups visited the exhibits, and evangelical Christians explained that they brought young people to the gallery to allow them to formulate their own sense of how to maintain their religious beliefs alongside the scientific ideas presented. Visitors to a museum exhibit may not change their underlying beliefs. But these studies demonstrated that exhibits can improve people's understanding of scientific explanations.[8]

The NSF-funded Life on Earth tabletop digital exhibit, designed by Chia Shen and her colleagues, afforded visitors the opportunity to explore how living organisms are related to one another, and it reinforced the concept that all life derives from a common ancestor. Visitors to the exhibit developed an intuitive sense about evolutionary relationships by zooming across tens of thousands of organisms, selecting specific species to see how they relate to one another, and then exploring the common pathways by which they connect on a tree of life. Although text was provided about the organisms, the primary engagement was tactile and visual, appealing to the vast numbers of visitors who come to museums to engage visually and physically. Through responsive graphics and a playful interface, this big data approach to presenting evolution demonstrated how museums can effectively engage diverse visitors with otherwise difficult subject matter.[9]

Museums have only begun to explore how to engage their visitors with scientific processes in ways that encourage individual expressions and individual differences. One can imagine that exhibit labels could speak to diverse audiences without implying that there is only one way to view the world. One pathway for investigation might be what psychologists call Theory of Mind, the ability of people to look at the world from another's perspective. Theory of Mind suggests that it is possible to know something and still understand the perspective of others who lack that knowledge.[10] Museums could incorporate Theory of Mind into how exhibit labels are developed, explicitly recognizing that visitors may have underlying beliefs that are quite variable and not always open to change. A developer writing exhibit labels may not know *what* visitors think about a topic. But it is possible to present scientific content while accepting that people have differing perspectives and that scientific explanations can exist alongside cultural and religious beliefs. While continuing to present scientifically correct information, it is possible to appreciate the diversity of ways that any one person may approach a concept. Our research has shown that museum visitors often hold contradictory belief systems, like evolution and creationism, and apply each belief system to a different context. For example, people are comfortable giving scientific explanations in academic settings while putting forth spiritual accounts in religious settings. Clear and simple writing, with well-formed explanations and multiple connections to familiar things, provides options for visitors to consider how their own ideas might relate to the new ones being presented in the exhibit.

McKenna-Cress and Kamien describe how tangible interactions that visitors have in museum spaces – how visitors interact with exhibits, physically, intellectually, and emotionally – can lead to the intellectual impact that sparks lasting impressions. Recognition of this dynamic milieu gives museums the opportunity to celebrate the diversity of their visitors, while embracing the idea that not everyone needs to think in the same way.[11] It is time for museum professionals to acknowledge not only the systemic biases that exist within science but also those within museum exhibitions themselves. Acknowledging such deficiencies enables museums to create exhibits that include perspectives and contributions from groups historically under-represented across the sciences and within museums.

Notes

1 Dawson, E. (2014). "Not designed for us": How informal science learning environments socially exclude low-income, minority ethnic groups. *Science Education, 98* (6): 981–1008. https://doi.org/10.1002/sce.21133. Nicolaisen, L. B., & Achiam, M. (2020). The implied visitor in a planetarium exhibition. *Museum Management and Curatorship, 35* (2), 143–159.

2 Archer, L., Dawson, E., Seakins, A., DeWitt, J., Godec, S., & Whitby, C. (2016). "I'm being a man here": Urban boys' performances of masculinity and engagement with science during a science museum visit. *Journal of the Learning Sciences, 25* (3): 438–485. https://doi.org/10.1080/10508406.2016.1187147. Dawson, E., Archer, L., Seakins, A., Godec, S., DeWitt, J., King, H., Mau, A., & Nomikou, E. (2019). Selfies at the science museum: exploring girls' identity performances in a science learning space. *Gender and Education, 32* (5): 1–18. https://doi.org/10.1080/095402 53.2018.1557322.

3 Oppenheimer, F. (1972). The Exploratorium: A playful museum combines perception and art in science education. *American Journal of Physics, 40* (7), 978–984.

4 Machin, R. (2008). Gender representation in the natural history galleries at the Manchester Museum. *Museum & Society, 6* (1): 54–67. See also Haraway, D. 1989. *Primate Visions: Gender, Race, and Nature in the World of Modern Science.* New York and London: Routledge. Achiam, M., & Marandino, M. (2014). A framework for understanding the conditions of science representation and dissemination in museums. *Museum Management and Curatorship, 29* (1): 66–82. https://doi.org/ 10.1080/09647775.2013.869855.

5 Feinstein, N. W., & Meshoulam, D. (2014). Science for what public? Addressing equity in American science museums and science centers. *Journal of Research in Science Teaching, 51* (3), 368–394. https://doi.org/10.1002/tea.21130. Robinson, H. (2017). Is cultural democracy possible in a museum? Critical reflections on Indigenous engagement in the development of the exhibition encounters: Revealing stories of aboriginal and Torres Strait Islander Objects from the British Museum. *International Journal of Heritage Studies, 23* (9): 860–874. https://doi.org/10.1080/ 13527258.2017.130093110.

6 Achiam, M., May, M., & Marandino, M. (2014). Affordances and distributed cognition in museum exhibitions. *Museum Management and Curatorship, 29* (5), 461–481. https://doi.org/10.1080/09647775.2014.957479. Norman, D. A. (2013). *The Design of Everyday Things.* New York: Doubleday.

7 Lewin, K. (1917/1983). Kriegslandschaft [War landscape]. *Zeitschrift für angewandte Psychologie, 12,* 440–447. Dawson, E. (2014). "Not designed for us": How informal science learning environments socially exclude low-income, minority ethnic groups. *Science Education, 98* (6): 981–1008. https://doi.org/10.1002/sce.21133.

8 Diamond, J., & E. M. Evans. (2007). Museums teach evolution. *Evolution, 61* (6), 1500–1506. Spiegel, A. N., Evans, E. M., Frazier, B., Hazel, A., Tare, M., Gram, W., & Diamond, J. (2012). Changing museum visitors' conceptions of evolution. *Evolution: Education and Outreach, 5* (1), 43–61.

9 Horn, M. S., Phillips, B. C., Evans, E. M., Block, F., Diamond, J., & Shen, C. (2016). Visualizing biological data in museums: Visitor learning with an interactive tree of life exhibit. *Journal of Research in Science Teaching, 53* (6), 895–918.
10 Mitchell, P. (2011). Acquiring a theory of mind. In *An Introduction to Developmental Psychology*, 2nd Edition, edited by A. Slater, & G. Bremner, 381–406. Hoboken, NJ: John Wiley & Sons.
11 McKenna-Cress, P., & J. Kamien. (2013). *Creating Exhibitions*. Hoboken, NJ: John Wiley & Sons. Diamond, J. (2019). Inclusion and relevance in natural history museums. In *Science Museums in Transition: Unheard Voices*, edited by H. McLaughlin, & J. Diamond, 1–6. London: Routledge.

4

ORIGIN OF INTERACTIVE SCIENCE CENTRES AND THE STATUS OF SCIENCE CENTRES IN THE MIDDLE EAST

Mike Bruton and Fatema Jasim

Interactive science centres and museums tend to develop and flourish when science and technology are strongly developed in a given culture. Many museums in the developed world were launched during the first and second industrial revolutions in Europe from the mid-1700s to the early 1900s. Science centres in the modern sense (with the majority of their displays interactive, extensive, people-centred activities, and few if any specimen or artifact displays or collections) followed during the third industrial revolution from the mid-1960s onwards, with the Exploratorium in San Francisco and the Ontario Science Centre in Toronto in the vanguard.

The above scenario, however, ignores the fact that Europeans were not the first or only culture to foment industrial revolutions that promoted the development of museums and the proliferation of interactive methods of teaching and learning. Every period of history has produced brilliant men and women who have made significant advances in science and technology when their socio-economic circumstances created the right environment for them to carry out research, develop uses for their knowledge, and pass their wisdom on to others (young and old) in effective ways.

Early development of interactive centres of learning in the Islamic world

The Islamic scholar, Ibn Khaldun, was one of the first sociologists to chronicle the rise and fall of nations and to contribute to an understanding of why civilizations go through cycles of dominance and decline.[1] The so-called 'Dark Ages' in Europe, which were characterized by significant economic, intellectual and cultural decline, lasted from the fall of Rome in 395 CE to the

DOI: 10.4324/9781003145387-5

beginning of the Renaissance in 1300 and largely coincided with the 'Golden Age of Islamic Science' (ca 800–1500 CE) when science and technology, as well as interactive learning institutions, flourished in the Islamic world. Of course, there is no such thing as 'Islamic science', just as there is no Chinese, European or African science. Science is a universal concept to which all cultures contribute over time, so we are referring to Islamic *contributions* to science.

During the 700-year period from about 800 CE onwards a unique suite of socio-economic circumstances made it possible for Muslim scholars to flourish. These circumstances were different from those pertaining in other major cultures, such as China, India, Europe, Africa, and the America, at that time. During this period Islam became heir to the intellectual heritage of prior civilizations and a haven in which a wide range of intellectual disciplines and traditions found a new lease of life. It was therefore far more than a bridge over which ideas from antiquity passed to mediaeval Europe as claimed by many Western scholars. In the early Houses of Wisdom, Muslim scholars working in teams translated Persian, Syriac, Indian, and especially Greek texts into Arabic, including those by Euclid, Aristotle, Hippocrates, Galen, Ptolemy, and Archimedes.[2–4]

Furthermore, the introduction of the hypothetico-deductive scientific method, and open-minded experimentation, characterized the science carried out by Muslim scholars, from the chemist Jabir ibn Hayyan in the 8/9th century to the engineer and physicist, Ibn al-Haytham, in the 10/11th century, and beyond. In his research on optics, Ibn al-Haytham insisted that the wisdom of the Greeks could not be accepted at face value and had to be tested through experimentation. By putting his concepts on the theory of vision to various tests, he introduced the scientific method of proof centuries before Roger Bacon or Isaac Newton. Muslim scholars also introduced the practice of peer review to evaluate their work and citations to confirm their source material.

Other reasons why Muslim scholars flourished at this time included the introduction of the Arabic numerals 1 to 9, the concept of zero and the decimal point into mathematics for the first time by Muslim scholars, the vast extent of the Islamic empire that enabled their explorers to be exposed to a variety of new cultures and ideas[1,4–7] and a common language, Arabic, that gave ordinary Muslims access to scholarly knowledge, as compared to Latin which was used primarily by academics and the clergy.[1,8]

Innovations included improvements to the design of mechanical devices inherited from the Phoenicians, Greeks and Romans but also many new inventions.[1,4,9–19] These inventions included the development of interactive displays, which they called 'trick devices', as well as hydraulically powered robots that elucidated scientific principles and showcased the latest technology. These devices were designed to satisfy the curiosity of scholars, young and old, and provide 'aesthetic pleasure' within the courtly circles that commissioned them.[1,3,4,10,14,16,19] Models of many of the trick devices made by Muslim scholars can be seen in museums in Turkey, Saudi Arabia, Dubai and Sharjah, in the Ibn Battuta Mall in Dubai, and in the travelling exhibitions '1001 Islamic Inventions' and 'Sultans of Science'.

Although they did not use the term 'science centres', Muslim scholars did develop interactive displays, hands-on teaching methods and science demonstrations in these early educational institutions. For instance, in the House of Wisdom in Baghdad in the 9th century, the three Banu Musa brothers developed a variety of 'trick devices' (*jihaz khudea*) in which they made clever use of air and water pressure, floats and siphons to baffle, entertain and educate their audiences. In 850 CE they published the *Book of Ingenious Devices* (كتاب الحيل; *Kitab al-Hiyal*; literally *The Book of Tricks*).[1,3,11] Other examples of early Muslim automatons and interactive devices included a flock of metal birds that sang automatically while sitting in a tree in the palace of Caliph al-Mamun in Baghdad in 827 CE and a tree with birds that flapped their wings and sang in the garden of the Abbasid caliph, al-Muktadir, in Baghdad in 915 CE.[1] One of the most remarkable early Muslim engineers, al-Jazari, who served the Urtuq Kings of Diyarbakir (now south-east Turkey) from 1174 to about 1210, made giant automated water clocks that showcased the advanced state of Islamic engineering at the time and served to inform scholars and laypeople about the potential of robots.[1] In Cairo Ibn al-Haytham invented the camera obscura while researching the properties of light and the functioning of the human eye.[1,20]

One way to define a science centre is, 'A safe place for dangerous ideas'[21], but a more conventional definition is 'An educational institution that uses the most effective methods to teach science, technology, mathematics, and engineering, with a primary emphasis on interactive displays'. If that is the case, then the House of Wisdom in Baghdad, operational from *ca* 750 to 1258 CE, should qualify as a precursor to modern interactive science centres, with the other hands-on learning institutions developed during the Golden Age of Islamic Science also contributing significantly in this regard. The Renaissance, followed by the three industrial revolutions in Europe and North America, would later produce new generations of scientists and science communicators who built on the foundations laid by Islamic scholars.

Science centres in the Middle East

There are about 85 science centres in the Middle East, distributed as follows: Abu Dhabi (5), Ajman (1), Bahrain (2), Cyprus (4), Dubai (5), Iran (5), Iraq (3), Israel (7), Jordan (6), Kuwait (1), Lebanon (6), Oman (7), Palestinian Authority (1), Qatar (4), Ras Al Khaimahi (3), Saudi Arabia (12), Sharjah (3), and Turkey (10). Egypt is excluded as it is discussed in the chapter in this book on African science centres.

Abu Dhabi. The Children's Museum in the Louvre Abu Dhabi caters for kids aged 6–12 years and has exquisite interactive displays that encourage an appreciation of art, science, technology, emotions, and mindfulness. Fayoonah Science Lab caters for children aged 5–12 years with workshops on physics, chemistry, team building and sports. The Abu Dhabi History Museum and Aquarium includes many realistic dioramas and a maritime museum. Located in the desert, the Al Sadeem Astronomy Observatory gives visitors the opportunity to view the night sky away from the city lights. At the Little World Discovery Center, children role play as scientists and carry out experiments under supervision.

Ajman. The Brilliant Minds Center offers courses for young children on numeracy, literacy, phonics, grammar, writing, mathematics, coding, robotics, and 3D printing.

Bahrain. The Bahrain Science Centre was established in 2012 to offer informal learning opportunities on human physiology and anatomy, food science, exercise science, astronomy, physics, chemistry, engineering, geology, and zoology. In addition, monthly themed exhibitions and science shows focused attention on topical STEM issues. In 2016 it was transferred from the Ministry of Social Development to the Ministry of Youth & Sports Affairs and renamed the Bahrain Science Center for the SDGs (the 17 Sustainable Development Goals of the United Nations)[22] and now has a range of interactive displays on the UN's SDGs. The Sharifa Alawadhi Youth and Children Club (19) has acquired most of the interactive displays from the original Bahrain Science Centre and has become a science centre in its own right.

Cyprus. The Kition Planetarium and Observatory in Kiti, which opened in 2008, offers a wide range of activities related to astronomy and space science. Mutlu Çocuklar Bilim Parkı in Gerolakkos is a STEAM education centre for children, with interactive displays and experiences on biology, geography, robotics, and other fields. The Science & Space Cafe Nicosia in Egkomi is an informal STEM education facility and café. The Akrotiri Environmental Education Centre offers exhibits and information related to the natural and cultural environment of the Akrotiri Peninsula in Cyprus.

Dubai. The History of Cinema Museum displays over 350 digital clips and interactive models on the history of cinema that were collected by Akram Miknas, a Bahraini businessman with a passion for photography. Infinity des Lumières is an immersive digital art and science experience whose displays change regularly. OliOli is a world-class children's play museum designed for kids aged 1–11 years. The 3D World Selfie Museum invites visitors, young and old, to use their creativity and imagination and take selfies against the background of artworks, giant animal models and staged effects. The Museum of Illusions in Dubai is a small, family-friendly facility with numerous optical illusions, brain teasers, and hands-on puzzles and experiences that has a unique aesthetic.

Iran. The Iran Science Park in Book Garden, Tehran, is a well-developed children's science museum with a wide array of interactive displays and science and technology demonstrations. The Sciences & Astronomy Center of Tehran, which covers the physical sciences and astronomy, offers interactive displays, a 14-inch telescope as well as scale models of inventions made by historic Muslim scholars. The Iranian National Museum of Science and Technology in Tehran has a variety of static and interactive displays on early technologies, optics, medicine and astronomy. The Kids Science Hall in the Tehran Book Garden is an extensive, kids-friendly experience with caricatures of famous characters from children's books as well as interactive displays. The Museum of Science and Nature at Shahid Chamran University in Ahvaz, which opened in 2003, has an extensive array of scale models on human physiology, anatomy, pathology, parasitology, embryology, and genetics; and the largest functional model of the human heart, 46 000 times the normal size, which demonstrates over 600 types of cardiovascular diseases.

Iraq. The Illusion Museum Erbil in Irbil and the Illusion Museum Duhok in Dihok have an amazing range of optical illusions and interactive displays on optics and light. Kid's Fun is a colourful, educational indoor playground in Dihok with giant games and puzzles.

Israel. Israel's largest science museum is the Madatech National Science, Technology and Space Museum in Haifa, which opened in 1983 in the Technion, the original site of the Israel Institute of Technology. This museum offers a wide range of interactive displays on the basic sciences (mathematics, physics, chemistry, and biology) and applied sciences (aviation, road safety, green living) as well as interactive galleries on imagination, magic, puzzles and games, and Cinematrix, a 3D theatre. The outdoor Noble Energy Science Park includes a variety of challenging physical displays including the famous 'Boyo' human yo-yo.

The Bloomfield Science Museum, established in 1992, is a children's science and technology museum that offers interactive displays, workshops, and performances. Within the Weizmann Institute of Science in Rehovot, the Clore Garden of Science is an open-air science museum where visitors can touch, experiment, and play with exhibits that demonstrate the scientific principles behind natural phenomena such as rainbows, gravity, and solar energy. The Carasso Science Park in Beersheba is a family-friendly science museum that offers a wide range of static and interactive exhibits and outdoor displays for the whole family. Havayeda Teva in Bat Yam is an interactive science centre where children can learn about scientific concepts using games, displays and models. The Science Museum in Hadera is an interactive centre for children of all ages that makes physics, chemistry, astronomy, and other topics accessible through hands-on engagement.

Jordan. The goal of the Jordan Star for Space Science Centre in Amman is to popularise space science, astronomy, and aviation, and to revive interest in one of the most important ancient Arab sciences, astronomy. In addition to a wide range of interactive displays the centre offers educational programmes, workshops, conferences, and summer camps that specialize in space science. Ealam Tak, the Regional Center for Space Science and Technology Education for Western Asia, is also located in Amman. The Children's Museum Jordan in Amman is an interactive science centre for young visitors that was launched by Queen Rania Al Abdullah in 2007. It is part of Al Hussein Public Parks and is a spacious, well organized and highly innovative facility with both indoor and outdoor interactive displays. The Mind Lab in Amman has many displays and activities that help to develop the mental capacity of children.

The Haya Cultural Center in Amman, established in 1976, has educational play areas and a planetarium as well as workshops and studios where art, dance, and cooking classes are held. The Jordan Museum in Amman, built in 2014, is the largest museum in Jordan and hosts the country's most important archaeological findings. It is a museum in transition from static exhibitions to challenging interactive displays and touch screens that explore the history, traditions, culture, and language of Jordan and the Arab world.

Kuwait. The Kuwait Foundation for the Advancement of Sciences (KFAS), a private non-profit organization, was established in 1976. Its mission is to create a thriving culture of science, technology and innovation for a sustainable Kuwait. The charter of the KFAS requires local share-holding companies to contribute 5% of their annual net profits to fund the foundation; this amount has now been reduced to 1%. Several ambitious projects have been funded by the KFAS including The Scientific Centre Kuwait (TSCK) in Kuwait City, inaugurated in 2000, which is now a major edutainment destination that promotes a passion for science and technology and encourages environmental awareness. The TSCK includes an Imax Theatre, laser planetarium, interactive science centre, maritime museum, terrariums, and an aquarium, the latter named after the founding director, Dr Mijbil Almutawa. The Imax Theatre, which was upgraded with the latest technology in 2017, was the first in the Middle East. The Dhow Harbour Maritime Museum includes an original wooden dhow from pre-oil days in which an audio-visual presentation on the history of Kuwait can be viewed.

Lebanon. The Science Center in Dbayeh, Beirut, is a hands-on, interactive facility packed with innovative games that encourage informal learning on a wide variety of science and technology topics. Planet Discovery in the Beirut Souk offers interactive displays on acoustics, food science and biology as well as educational games, puppet shows, and science shows for young children. Geek Express in Beirut provides live and online coding and robotics courses for kids and teens from the Middle East and North Africa. Young visitors are taught the essentials of computer science while they programme their own video games, apps, and websites. The Cranium STEAM Centre in Kasrouane offers exciting workshops for children on web development, coding, robotics, and educational games. The Hall of Fame Museum in Mazraat El Ras has models of many famous people that move, talk and interact with visitors. Planet Discovery in the Beirut Souk offers interactive displays on acoustics, food science, and biology as well as educational games, puppet shows and science shows for young children.

Oman. The Oman Children's Museum, founded in 1990, is located in Qurum Nature Park in Muscat. It is one of the oldest children's museums in the Middle East and offers a wide range of interactive displays on general science. PDO Knowledge World is the new name for the Oil & Gas Exhibition Center that was originally established in 1995 in Muscat by Petroleum Development Oman. The museum offers an interactive journey that explores the history of the discovery, extraction and use of fossil fuels in Oman as well as other human interactions with the environment. The Museum of History and Science at the German University of Technology in Oman links the Golden Age of Islamic Science to current university studies through interactive displays on the contributions of Arabic scholars to different scientific disciplines.

Nutty Scientists Oman in Seeb is a fun, interactive science centre for young children that offers a variety of hands-on workshops and excellent informal tuition on science and technology. Engineering for Kids in Muscat provides after-school engineering classes where kids are taught how to build robots, create video games and learn about the role that engineers play in modern society. Little Gym Muscat is based on a philosophy of allowing kids to experience success in a fun, caring environment that is focused on individual accomplishment rather than competition. The centre offers age-appropriate, curriculum-based gymnastics and movement programmes for children aged 10 months to 12 years that promote coordination, balance, rhythm, and flexibility as well as build strength and improve fitness. The new Liwa Center for Science and Innovation in Muscat, established in 2020, promises to be one of the foremost facilities for promoting the implementation of Oman Vision 2040 which aims to keep pace with regional and global changes, generate and seize opportunities to foster economic competitiveness and social well-being and stimulate technological and economic growth.

Palestinian authority. AlNayzak Science House is an interactive science centre for children in an historic stone building in the old city of Birzeit in Ramallah. It is a non-curricular educational institution that helps visitors, especially school learners and teachers, to learn about science and technology in a practical, fun and interactive way.

Qatar. The 3-2-1 Qatar Olympic and Sports Museum in Doha celebrates the history of sports, especially the Olympic Games, and includes interactive galleries where visitors can test their mental and physical abilities. Kidzania Doha in Aspire Park is a fun place where kids role-play as doctors, firefighters, pilots, TV presenters, lifesavers and flight attendants and discover their interests. The Museum of Illusions in Doha offers an interactive experience to children and adults that amuses, amazes and informs visitors and encourages them to be creative and learn about vision, perception and the functioning of the human brain. KidzMondo Doha is an international edutainment concept that offers an indoor theme park in the form of a kid-sized city where children experience age-appropriate activities in a dynamic environment that imparts knowledge through playful learning.

Ras Al Khaimah. The National Museum of Ras Al Khaimah served as the residence of the ruling Quwasim family until 1964 and is now a public museum and science centre that is a Pandora Box of historical wonders from the emirate. The Ras al Khaimah Science Club offers informal science and technology tuition to youths, and RAK Nature's Treasures, founded in 2014, comprises an interactive natural history and agriculture museum, petting zoo and vegetable farm with large collections of gemstones, fossils, plants, seashells and mounted insects on display.

Saudi Arabia. SciTech in Al Khobar has interactive exhibitions on archaeology, natural history, chemistry, physics, mechanics, and space science as well as an astronomical observatory, IMAX theatre and aquaria. The King Salman Science Oasis in Riyadh has a wide range of high-end interactive displays on science and technology. The Museum of the History of Technology in Riyadh offers many static displays with a few interactives. Other informal science education facilities in Riyadh include the Mishkat Interactive Center and the Museum of Illusions.

The Zamil Science Oasis in Unayzah is an ultra-modern science centre with an amazing array of interactive displays including a flight simulator. The Museum of Science and Technology in Islam (MOSTI) at the King Abdullah University of Science & Technology (KAUST) in Thuwal is a spectacular, high-tech celebration of the achievements of early Islamic engineers and scholars. Interactive science education facilities in Jeddah include the Jeddah Science Center and Fakieh Planetarium. The Energy Exhibit in Dhahran was established by Saudi Aramco and has a wide range of displays and experiences related to oil as a source of energy. The King Abdulaziz Center for World Culture, also established by Saudi Aramco in Dhahran, is an expansive interactive science, arts and culture museum that is locally known as *Ithra* ('richness' in Arabic). It was inaugurated by King Salman bin Abdulaziz in December 2017. The NEOM Experience Centre in Sharma is the forerunner of a series of ambitious tourist and family edutainment experiences in and around the new city of Neom in the north of Saudi Arabia on the Red Sea coast that are part of Vision 2030 of the Kingdom of Saudi Arabia.

Sharjah. The Sharjah Science Museum, a family-friendly science museum which opened in 1996 in Al Albar, is widely rated as one of the best children's science centres in the UAE. The Sharjah Discovery Centre in the Al Sharjah National Park in Gharayen is a colourful, open space that is also very child-friendly and features a wide variety of hands-on exhibits and educational activities for young children. Another museum, of art and science, was established in Sharjah in 2017 to provide educational and training programmes for young and old.

Syria. As far as we can ascertain, there are no functional science centres or parks in Syria. In 2007 Damascus University announced the development of a Technological Pole, and the Hadiqat taknulujiya al-ma'lumat ('IT Plaza'), a science and technology park, was due to be established in 2008 in the Western Mezze, but their completion has been delayed by the civil war.

Turkey. The METU Science and Technology Museum, on the campus of the Middle East Technical University (METU) in Ankara, offers a wide range of interactive displays and experiences related to ancient and modern technological tools. Konya Science Center is a large, interactive science museum that includes working replicas of al-Jazari's Elephant Water Clock and other ancient Islamic inventions. Kocaeli Bilim Merkezi is a science centre in Sikapark, Izmit, which has many interactive displays and hosts Technofest, a large aviation, space and technology festival. The M.S.Ö. Air & Space Museum in Eskisehir has interactives on aeronautics and static displays of flying memorabilia and airworthy vintage aircraft, including a fully restored 1940 DC-3 'Turkish Delight' Dakota.

The Feza Gürsey Science Center in Altınpark, Ankara, which opened in 1993, is named after a famous Turkish mathematician and physicist. The original interactive exhibits were made by the Ontario Science Centre in Toronto in line with the curriculum requirements of secondary schools in Turkey. Other science centres in Turkey include Bursa Bilim Ve Teknoloji Merkezi in Bursa and the Gaziantep Planetarium and Science Center, where a science theatre production is performed by robots. The Sancaktepe and Uskudar science centres in Istanbul are large facilities with innovative interactive displays and many hands-on visitor experiences. The Miniskop Science Workshop in Istanbul is an informal educational facility for young children with many opportunities for educational play and experiential learning on science and technology.

Several countries and emirates in the Middle East have contracted franchised informal science education organizations, such as MAD Science, Brilliant Minds, Curiosity Lab, Science Xplorers, Mind Lab, Museum of Illusions, KidsMondo, Kidzania and Science Made Fun, to offer science displays, workshops, camps, and other activities to their citizens.

Discussion

There has been a rapid development of interactive science centres and museums in the Middle East with most of the 94 facilities discussed here being established since 2005. Ambitious national goals and generous budgets have resulted in some of these facilities competing with the world's best in terms of levels of sophistication and innovation in their architecture, interactive displays and educational programmes. Dubai, Abu Dhabi, Qatar, and Neom (in Saudi Arabia) stand out as candidates to become some of the 'smartest' cities in the world, with sophisticated science centres and museums to match. Furthermore, the high levels of technology uptake in many Middle East countries have meant that young people there are attracted to, and comfortable with, the modern technologies, such as digital simulations, augmented reality, artificial intelligence, gamification and immersive experiences, that are now commonly used in interactive science centres.

A characteristic of science centres and museums in the Middle East is that they cater for the needs of all age, ethnic, and socio-economic groups, not just an educated elite. In many cases the only places where young people of different ethnicities and religious beliefs meet are in science centres and other informal community facilities, such as youth centres. Furthermore, although most informal educational facilities in Middle East countries are concentrated around the large cities, in some they are widely distributed in peri-urban and rural areas where they reach historically disadvantaged youths and families. It is important that all Middle Eastern countries develop interactive science centres and museums collaboratively through their regional network, NAMES (North Africa and Middle East Science Centre Network).

Many science centres and museums in the Middle East have achieved an excellent balance between displays and activities that showcase their proud scientific heritage while also exposing visitors to the latest trends in science and technology, especially in the fields of artificial intelligence, robotics, wireless telecommunication and computer programming. Several Middle East science centres and museums play a vital role in highlighting the significant contributions made by early Muslim scholars during the Golden Age of Islamic Science whereas others focus attention on the achievements of modern scholars from the Middle East. This dual focus should be encouraged as many modern 'screenagers' are losing touch with their technological heritage.

Unfortunately, the interactive displays in many of the Middle East science centres mentioned above are poorly maintained. A typical pattern is that newly established facilities implement strict maintenance procedures, but their intensity soon wanes, which results in many visitor complaints and negative TripAdvisor scores. Furthermore, new institutions tend to offer a wide range of people-centred educational activities, such as science shows and technology demonstrations, but these decrease in frequency and quality over time. Another characteristic is that the entrance fees to Middle East science centres vary widely, from free to exorbitant, depending partly on whether the facility is in the public or private sector. Also, a great deal still needs to be done in terms of achieving consistently high levels of staff training, exhibit interpretation, visitor management and educational delivery. Overall, Middle East countries can be proud of the way in which they have developed their science centres and museums, given the obvious caveat that there is always room for improvement.

Acknowledgements

The initial research that the first author conducted on Islamic contributions to science and technology was carried out while he was employed by MTE Studios in South Africa, Dubai, Bahrain and Saudi Arabia. In this capacity he developed the content for a wide range of interactive displays on Islamic contributions to science and technology in several science centres and museums. He is grateful to the CEO of MTE Studios, Ludo Verheyen, for offering him these inspiring opportunities. Since then, he has continued to conduct research and is grateful to his many correspondents around the world for their assistance and guidance, in particular, Anwar Mall, Mushtaq Mahmood and the late Anwar Goolam. Fatema Jassim, who was a staff member at the Bahrain Science Centre from 2012 to 2015, when the first author was director, is grateful to the many people who have encouraged her studies and involvement in science centres in the Middle East.

Notes

1 Al-Hassani, S.T.S. (ed.) 2006. *1001 Inventions. Muslim Heritage in Our World*. Manchester: Foundation for Science, Technology and Civilization.

2 Hayes, J.R. (ed.) 1978. *The Genius of Arab Civilization, Source of Renaissance*. Boston: MIT Press.

3 Al-Khalili, J. 2010. *Pathfinders. The Golden Age of Arabic Science*. London: Allen Lane.

4 Lyons, J. 2009. *The House of Wisdom. How the Arabs Transformed Western Civilization*. London: Bloomsbury.

5 Mackintosh-Smith, T. 2002. *Travels with a Tangerine. A Journey in the Footnotes of Ibn Battutah*. London: Picador.

6 Gibb, H.A.R. 2006. *The Travels of Ibn Battuta*. New Delhi: Goodward Books.

7 Goolam. N. 2017. *The Story of the Transmission of Knowledge from the Islamic World to Europe*. Wandsbeek: Reach Publishers.

8 Freely, J. 2009. *Aladdin's Lamp. How Greek Science Came to Europe through the Islamic World*. New York: Alfred A. Knopf.

9 Watt, W.M. 1972. *The Influence of Islam on Mediaeval Europe*. Edinburgh: Edinburgh University Press.

10 Nasr, S.H. 1976. *Islamic Science. An Illustrated Study*. Westerham: World of Islam Festival Publishing Company.

11 Hill, D.R. 1993. *Islamic Science and Engineering*. Edinburgh: Edinburgh University Press.

12 Rashed. R. 1996. *The Encyclopaedia of the History of Arab Science*. London: Routledge.

13 Saud, M. 2003. *Islam and Evolution of Science*. New Delhi: Adam Publishers.

14 Sezgin. F. 2006. *Scientific Excellence in Islamic Civilisation. Islamic Science Ahead of Its Time*. Frankfurt: Institute for the History of Arabic-Islamic Science.

15 Ilyas, M. and Ilyas, B.S. 2008. *Glimpses of Muslim Genius in Science*. Kuala Lumpur: A.S. Nordeen.

16 Morgan, M.H. 2008. *Lost History. The Enduring Legacy of Muslim Scientists, Thinkers, and Artists.* Washington: National Geographic.
17 Morgan, M.H. 2009. *Arab Science. A Journey of Innovation.* Doha: Qatar Foundation for Education, Science and Community Development.
18 Djaziri, S.E. 2010. *Arabia. Cradle of Civilisation.* Manchester: British Museum Heritage Centre.
19 Bruton, M.N. 2023. *Discovering the Sultans of Science.* Cape Town: Footprint Press.
20 Bruton. M.N. 2021. *Curious Notions. Reflections of an Imagineer.* Cape Town: Footprint Press.
21 Bruton, M.N. 2001. Second Science Centre World Conference, Heureka, Vantaa, Finland.
22 UN [United Nations] General Assembly. 2015. Transforming Our World: The 2030 Agenda for Sustainable Development. A/RES/70/1. https://sdgs.un.org/2030agenda.

5

HELPING YOUNG CHILDREN MAKE CONCEPTUAL CONNECTIONS IN A SCIENCE MUSEUM

Benjamin D. Jee, Florencia K. Anggoro, and Mary Grace Harris

Museums of science and natural history enable young children to directly interact with scientific artifacts and models, promoting new ways of thinking about the world.[1,2] As cognitive psychologists with an interest in children's learning, museums offer an exciting context in which to explore young children's acquisition of scientific ideas. There are a number of ways to test the educational effectiveness of museum exhibits and to measure children's thinking and learning in informal environments.[3,4] A great deal can be learned by observing children's free engagement with an exhibit, including their conversations with accompanying adults. Understanding and improving these early experiences may be highly consequential. The science knowledge that children acquire *before* entering school predicts their science achievement in later years.[5]

Of course, informal interactions in a museum do not inevitably lead to scientific understanding. One challenge is that many children (and adults for that matter) enter the museum with preconceived notions about how the world works. These preconceptions, while intuitively compelling, can run afoul of scientifically accepted ideas in astronomy, biology, chemistry, geology, and physics.[6] In astronomy, for example, young children tend to confuse the apparent motion of the Sun with actual motion and thus attribute the day/night cycle to the Sun moving up, down, or around the Earth.[7] When children are presented with a scientific explanation for day and night (Earth's axial rotation), their intuitive beliefs can impede their learning by blending with the scientific model, by filling in gaps in their understanding, and by interfering with the recall of scientific ideas.[7–10]

Research on cognition and science learning has found that nonscientific intuitions are remarkably stubborn in the face of instruction.[11–13] In a striking example, a majority of Harvard graduates were found to express nonscientific

DOI: 10.4324/9781003145387-6

intuitions when asked to explain the cause of the seasons.[14] Rather than replacing nonscientific intuitions, science learning may involve the assembly of competing sets of ideas that coexist with the intuitive preconceptions.[15,16] To help children understand core models in science, such as the heliocentric solar system, it is important to show how these models are related to observable phenomena, such as sunrise and sunset.[17,18] These connections reinforce the explanatory power of the scientific model and may discourage intuition-based reasoning.

We have recently explored an instructional approach that emphasizes the connections between scientific models and real-world observable phenomena. Our approach was based on cognitive research and theory on analogical thinking and relational reasoning, which speaks to the process of cognitively aligning sets of ideas.[19–21] In school-based research, we tested whether third graders' causal understanding of the day/night cycle was improved through instruction in which modeled events – Earth's rotation – were explicitly connected to events observable from Earth's surface – sunrise, sunset, etc.[17] In our first study, one group of participants viewed synced-up displays of the model and the sky as an instructor indicated and explained the relevant correspondences. We called this the *relational scaffolding* condition, because the instruction involved guided supports for learning the relationship between the model's behavior and the observable phenomena. We found that participants in the relational scaffolding condition learned significantly more than those who received comparable model-based instruction without relational scaffolding. Our second study found that when the observable and modeled events were shown *sequentially* (i.e., not explicitly connected) rather than simultaneously, low-knowledge (but not high-knowledge) students learned significantly less. Thus, the explicit visual alignment between a scientific model and relevant observable phenomena can be especially helpful to students with little scientific understanding. This finding fits a larger pattern of results revealing that explicit comparison leads to greater learning than other forms of studying multiple representations.[22–25]

Our goal was to test a similar relational scaffolding approach in the context of the museum. Of course, the informal learning setting precludes explicit instruction. Because visitors often come to museums in multi-generational groups, we hoped to capitalize on parent-child dynamics to promote exploration and connection of scientific ideas to everyday life.[26] Our challenge was to create a standalone exhibit that would encourage young visitors and their accompanying caregivers to make the sorts of connections relevant to grasping the explanatory power of the scientific model without compromising the informal nature of the museum experience. We collaborated with designers at the EcoTarium, the museum of science and natural history in Worcester, MA, to create an interactive exhibit, pictured in Figure 5.1. Crucially, the exhibit displays a rotating model of Earth as well as a representation of the sky, including a familiar skyline in Worcester. There was also a model Sun, an actual light source that illuminated the side of model Earth that faced toward

FIGURE 5.1 The day/night cycle exhibit created for the study.

it. A small Lego man was affixed to the Earth model at roughly the location of Worcester. The visitor could use a crank to turn the Earth model around on its axis. As the crank was turned, the display of the sky would change in sync. When the location of the Lego man was first exposed to the "Sun's" light, the Sun would appear in the east of the sky display. When the Lego man faced the Sun directly, the Sun would appear at its highest point in the middle of the sky. As the Lego man was turned away from the Sun, the Sun would set in the sky. A second light source was installed behind the sky display to illuminate the sky when the Sun was visible, creating an eye-catching contrast to the darker night sky. A sign about the sky display indicated that the display showed what the Lego man would see in the sky from his location in Worcester. A third component of the display was a small clock with an hour hand that also turned as the Earth was rotated. This, too, was synced with the other displays, such that the timing of sunrise and sunset were both plausible and consistent.

In the absence of a trained instructor guiding the child through the relevant connections between elements of the display – the rotating Earth, the Sun in the sky, and the time on the clock – we incorporated various forms of relational scaffolding into the exhibit itself. We used a picture of the local skyline to import meaning into the display of the sky. Superficially similar objects are easier to compare and can prompt learners to make connections.[23,24,27] The surface similarity between the sky display and a known location could help the visitor make connections to their own prior observations of sunrise, sunset, etc. We also considered the spatial layout of the exhibit. We wanted the Earth model and the sky display to be in close proximity, so that visitors could look back and forth between them more easily. Closeness in space, and the absence of distracting visual clutter, makes it easier to perform these visual comparisons.[28,29] We also considered the spatial properties of the control mechanism of the exhibit, the rotating crank. As opposed to a button or switch, the crank rotated around, providing an intuitive link to the idea of a rotating Earth.[30] The crank also turned only one way, which meant that the Earth model rotated only in the eastward direction, as is the case in reality. Finally, we incorporated scaffolding in the form of language, including the title of the exhibit and built-in captions, which emphasized the cause of the day/night cycle. Even simple signs can help adult caregivers appreciate the educational value of a museum exhibit.[31] The causal emphasis of the title could promote an adult's use of causal language, which is especially important when children first engage with an exhibit.[1]

In summer 2021 we conducted a pilot study to explore children's and adults' interactions with our exhibit. We set up the exhibit in a relatively secluded area of the EcoTarium and, over the course of several days, invited visitors with young children to participate in our research. We set up video recording equipment and, after obtaining parental consent, asked the children to try out the exhibit, usually along with the adult caregiver. We recorded the speech and nonverbal behavior that occurred during each interaction, which typically lasted a few minutes (see example visitor interaction in Figure 5.2).

FIGURE 5.2 Museum visitors interacting with the exhibit.

Our research assistants transcribed the approximately one dozen visitor inter-actions that took place during the piloting period. Our main goal was to identify evidence that visitors were using the exhibit to make relevant connections – including the key causal link between Earth rotation and the Sun's apparent motion, as well as other connections between aspects of the exhibit and visitors' everyday experiences. We discovered a number of examples of such conceptual connection-making involving adults and children. For example, we captured the following interaction between a two-year-old boy and his mother:

> The boy turns his attention back to the Earth model, asking for a turn at the crank. He uses the crank to turn the Earth model, while appearing to look between the sky and the clock feature. The participant's mother then turns his attention back to the Earth model, asking him if the Lego figure is in the sunlight or the shade, to which he correctly answers, "the shade." ... The mother proceeds to prompt the participant, asking "Can you make it daytime again?" to which the participant turns the Earth model slowly, making sure the Lego is exposed to the light from the model Sun and out of the shaded area.

Here we see the mother orchestrating her child's use of the exhibit to clarify the cause-and-effect relationship between Earth rotation and the day/night cycle. The child is actively engaged in the exhibit and, in response to his mother's question, demonstrates an awareness of the link between exposing a location on the Earth model (that of the Lego man) to sunlight and daytime.

Adults also encouraged children to make connections between the exhibit and familiar real-world space, as in the following example involving a five-year-old girl and her mother:

> The mother says, "Watch, where are we? We are right there" and points to the Lego person on the Earth model. The girl looks between the Earth model and sky display as her mother points. She continues to crank the Earth model and points to the sunlight. The girl watches the Earth model spin as she turns the crank. The mother directs the girl's attention to the Earth model and says, "See, it's sunset over here—it's nighttime—and over here it's morning time." The girl asks, "Where do we live?" and the mother answers and points to the Lego person on the Earth model.

We also observed that the adults used the exhibit to confront common intui-tive preconceptions, including the idea that the Sun moves around the Earth, for example:

> A five-year-old boy watches the Earth model as his brother turns the crank and then looks at the sky display when it becomes illuminated.

The mother points to the Earth model and says, "See how the Sun never goes up and down." The boy looks at his mother while referencing the Earth model and says, "It goes round and round." ...The mother points to the clock and says, "See, this is what time it is." The boy watches the clock as he slowly turns the Earth model. The mother asks, "Is the sun actually going down?" The boy answers "No." When the sky goes dark as the boy turns the crank, the mother points to the Sun and says, "Look, the sun is still there." The boy and his mother talk about how the Sun "stays up and never goes down." The boy keeps turning the Earth model and watches as the sky lights up again. The mother asks, "Is the sun moving or is the Earth moving?" The boy responds "Earth."

Altogether, we found promising evidence that museum visitors capitalized on the relational scaffolding within our day/night cycle exhibit. With the help of a parent, children made causal connections between Earth motion and events in the sky, recognized links between represented and real-world space, and were dissuaded from scientifically inaccurate ideas about Sun motion. The exhibit provided children of various ages with opportunities to increase their knowledge – from a basic recognition of Earth's rotation to a more sophisticated understanding of the link between rotation and familiar observable phenomena. Our pilot work suggests that young children can be guided toward scientifically relevant ideas through the relational scaffolding – visual similarities, synced-up displays, causal language, etc. – embedded in an interactive exhibit. This fits with prior research in which children's relational thinking is leveraged to promote conceptual learning in a museum context.[23,32] We are currently conducting an experiment on the museum floor to test our hypothesis that relational scaffolding produced these effects. We will compare visitors' behavior at the exhibit when relational scaffolding is present (as in the pilot study) or absent – i.e., when the model Earth is shown alone, a typical "model only" display. If the alignment between the model Earth and the synced-up display of the sky enhances children's understanding of the day/night cycle, then we would expect that the sorts of conceptual connections that we observed during our pilot study would be less likely to occur when only the Earth model is included in the exhibit.

As well as more rigorous experimental tests of the day/night exhibit, we are interested in exploring relational scaffolding in museum exhibits on other topics, such as the particle model in chemistry. Like the solar system model in astronomy, the particle model is central within its domain, is one of the first scientific models that students are exposed to, and is deeply counterintuitive.[33,34] Instruction that emphasizes the connections between macroscopic and submicroscopic events is found to enhance student learning of particle-based explanations.[35] Thus, a museum exhibit that shows macroscopic events, such as melting, alongside the corresponding particle-level events (increased particle motion and separation) could help young visitors grasp these conceptual connections and provide support for parents in facilitating children's learning.

Nonscientific intuitions about the world take hold in early childhood and are notoriously stubborn.[6] Yet young children are capable of understanding scientific explanations when adequate cognitive supports are provided.[17,36,37] The informal learning context of the museum is an early opportunity to support this understanding. We hope to learn more about the methods and materials that help children and their caregivers make scientifically relevant conceptual connections and, ideally, to spark their future interest in science.

Notes

1 Callanan, M. A., Legare, C. H., Sobel, D. M., Jaeger, G. J., Letourneau, S., McHugh, S. R., et al. (2019). Exploration, explanation, and parent–child interaction in museums. *Monographs of the Society for Research in Child Development, 85*(1), 7–137.
2 Hurst, M. A., Polinsky, N., Haden, C. A., Levine, S. C., & Uttal, D. H. (2019). Leveraging research on informal learning to inform policy on promoting early STEM. *Social Policy Report, 32*(3), 1–33.
3 Diamond, J., Horn, M., & Uttal, D. H. (2016). *Practical evaluation guide: Tools for museums and other informal educational settings.* Lanham, MD: Rowman & Littlefield.
4 Sobel, D. M., & Lipson, J. L. (2016). *Cognitive development in museum settings: Relating research and practice,* New York, NY: Taylor & Francis.
5 Morgan, P. L., Farkas, G., Hillemeier, M. M., & Maczuga, S. (2016). Science achievement gaps begin very early, persist, and are largely explained by modifiable factors. *Educational Researcher, 45*(1), 18–35.
6 Shtulman, A. (2017). *Scienceblind: Why our intuitive theories about the world are so often wrong.* New York, NY: Hachette Book Group.
7 Vosniadou, S., & Brewer, W. F. (1994). Mental models of the day/night cycle. *Cognitive Science, 18*(1), 123–183.
8 Plummer, J. D., Bower, C. A., & Liben, L. S. (2016). The role of perspective taking in how children connect reference frames when explaining astronomical phenomena. *International Journal of Science Education, 38*(3), 345–365.
9 Vosniadou, S., & Skopeliti, I. (2017). Is it the Earth that turns or the Sun that goes behind the mountains? Students' misconceptions about the day/night cycle after reading a science text. *International Journal of Science Education, 39*(15), 2027–2051. https://doi.org/10.1080/09500693.2017.1361557.
10 Shtulman, A., & Harrington, K. (2016). Tensions between science and intuition across the lifespan. *Topics in Cognitive Science, 8,* 118–137.
11 Coley, J. D., & Tanner, K. (2015). Relations between intuitive biological thinking and biological misconceptions in biology majors and nonmajors. *CBE Life Sciences Education, 14*(1). https://doi.org/10.1187/cbe.14-06-0094.
12 McCloskey, M., Washburn, A., & Felch, L. (1983). Intuitive physics: The straight-down belief and its origin. *Journal of Experimental Psychology: Learning, Memory, and Cognition, 9*(4), 636–649. https://doi.org/10.1037/0278-7393.9.4.636.
13 Sadler, P. M., Coyle, H., Miller, J. L., Cook-Smith, N., Dussault, M., & Gould, R. R. (2010). The astronomy and space science concept inventory: Development and validation of assessment instruments aligned with the K-12 national science standards. *Astronomy Education Review, 8*(1). https://doi.org/10.3847/AER2009024010111.

14 Schneps, M. H., & Sadler, P. M. (1989). *A private universe*. Astronomical Society of the Pacific.

15 Shtulman, A., & Legare, C. (2019). Competing explanations of competing explanations: Accounting for conflict between scientific and folk explanations. *Topics in Cognitive Science, 12*(4). https://doi.org/10.1111/tops.12483.

16 Shtulman, A., & Lombrozo, T. (2016). Bundles of contradiction: A coexistence view of conceptual change. In D. Barner, & A. S. Baron (Eds.), *Core knowledge and conceptual change* (pp. 53–71). Oxford University Press. https://doi.org/10.1093/acprof:oso/9780190467630.003.0004.

17 Jee, B. D., & Anggoro, F. K. (2019). Relational scaffolding enhances children's understanding of scientific models. *Psychological Science, 30*(9), 1287–1302.

18 Jee, B. D., & Anggoro, F. K. (2021). Designing exhibits to support relational learning in a science museum. *Frontiers in Psychology, 12*, 636030.

19 Gentner, D., & Hoyos, C. (2017). Analogy and abstraction. *Topics in Cognitive Science, 9*(3), 672–693.

20 Goldwater, M. B., & Schalk, L. (2016). Relational categories as a bridge between cognitive and educational research. *Psychological Bulletin, 142*(7), 729.

21 Jee, B. D., Uttal, D. H., Gentner, D., Manduca, C., Shipley, T., Sageman, B., & Tikoff, B. (2010). pAnalogical thinking in geoscience education. *Journal of Geoscience Education, 58*(1), 2–13.

22 Alfieri, L., Nokes-Malach, T. J., & Schunn, C. D. (2013). Learning through case comparisons: A meta-analytic review. *Educational Psychologist, 48*(2), 87–113.

23 Gentner, D., Levine, S.C., Ping, R., Isaia, A., Dhillon, S., Bradley, C., & Honke, G. (2016). Rapid learning in a children's museum via analogical comparison. *Cognitive Science, 40*, 224–240. https://doi.org/10.1111/cogs.12248.

24 Jee, B. D., Uttal, D. H., Gentner, D., Manduca, C., & Shipley, T. (2013). Finding faults: Analogical comparison supports spatial concept learning in geoscience. *Cognitive Processing, 14*(2), 175–187.

25 Hansen, J., & Richland, L. E. (2020). Teaching and learning science through multiple representations: intuitions and executive functions. *CBE—Life Sciences Education, 19*(4), ar61.

26 Rigney, J. C., & Callanan, M. A. (2011). Patterns in parent-child conversations about animals at a marine science center. *Cognitive Development, 26*(2), 155–171.

27 Thompson, C. A., & Opfer, J. E. (2010). How 15 Hundred Is like 15 cherries: Effect of progressive alignment on representational changes in numerical cognition. *Child Development, 81*(6), 1768–1786.

28 Matlen, B. J., Gentner, D., & Franconeri, S. (2020). Spatial alignment facilitates visual comparison. *Journal of Experimental Psychology: Human Perception and Performance, 46*(5), 443–457. https://doi.org/10.1037/xhp0000726.

29 Mayer, R. E., & Moreno, R. (2003). Nine ways to reduce cognitive load in multimedia learning. *Educational Psychologist, 38*(1), 43–52. https://doi.org/10.1207/S15326985EP3801_6.

30 Allen, S. (2004). Designs for learning: Studying science museum exhibits that do more than entertain. *Science Education, 88*, S17–S33. https://doi.org/10.1002/sce.20016.

31 Song, L., Golinkoff, R. M., Stuehling, A., Resnick, I., Mahajan, N., Hirsh-Pasek, K., & Thompson, N. (2017). Parents' and experts' awareness of learning opportunities in children's museum exhibits. *Journal of Applied Developmental Psychology, 49*, 39–45.

32 Callanan, M., Martin, J., & Luce, M. (2016). Two decades of families learning in a children's museum: A partnership of research and exhibit development. In D. Sobel, & J. Jipson (Eds.), *Cognitive development in museum settings: Relating research and practice* (pp. 15–35). New York, NY: Taylor & Francis.

33 Harrison, A. G., & Treagust, D. F. (2002). The particulate nature of matter: Challenges in understanding the submicroscopic world. In *Chemical education: Towards research-based practice* (pp. 189–212). Dordrecht: Springer.

34 Wiser, M., & Smith, C. L. (2008). Teaching about matter in grades K-8: When should the atomic-molecular theory be introduced? In S. Vosniadou (Ed.), *International handbook of research on conceptual change* (pp. 233–267). Hillsdale, NJ: Erlbaum.

35 Samarapungavan, A., Bryan, L., & Wills, J. (2017). Second graders' emerging particle models of matter in the context of learning through model-based inquiry. *Journal of Research in Science Teaching, 54*(8), 988–1023.

36 Kelemen, D. (2019). The magic of mechanism: Explanation-based instruction on counterintuitive concepts in early childhood. *Perspectives on Psychological Science, 14*(4), 510–522.

37 Shtulman, A., & Walker, C. (2020). Developing an understanding of science. *Annual Review of Developmental Psychology, 2*, 111–132.

6

HOW PLANETARIUMS OFFER DIVERSE OPPORTUNITIES TO ENGAGE A BROAD AUDIENCE IN STEM

Julia Plummer

Planetariums offer a distinct opportunity to engage a broad audience in STEM in ways that inspire and go beyond what most traditional educational platforms afford. Each planetarium is unique in its design, though most include a dome-shaped ceiling, a projector to show images and/or video onto the dome of celestial objects and events, and some manner of seating. Planetariums primarily serve to educate audiences – from preschool to adult – in astronomy and space exploration.[1] However, surveys of planetariums suggest two slightly different goals, each emphasized more or less by individual professionals: (a) educating for increased STEM knowledge and (b) educating for increased interest in STEM, especially to foster learning in and beyond the planetarium learning environment.[2,3] This duality reflects the nature of the planetarium as both formal and informal.[4] While both goals are cultivated in informal and formal settings, formal environments tend to emphasize developing conceptual knowledge while informal settings tend to place more emphasis on engendering interest, emotion, and engagement.[5] Many planetariums exist in museums, science centers, or standalone/portable venues; this situation allows for audiences that are motivated to visit based on free-choice and as a once-off or infrequent occurrence, which may recommend to the planetarium a goal of affective gains over conceptual gains. On the other hand, many other planetariums are housed in K-12 schools and universities; further, planetariums in museums and science centers often support the curricula of local schools through programs targeting local standards. Thus, other planetariums have the potential for more extended learning opportunities or connected learning, thus shifting the focus to supporting conceptual goals.

DOI: 10.4324/9781003145387-7

How do planetariums support learning for diverse audiences?

Planetarium programs have the capacity to improve learning outcomes even after a brief visit for audiences. For example, studies have found significant improvement in children's (aged 6–9 years) knowledge of the patterns of apparent motion of the Sun, Moon, and stars after attending a single planetarium program.[6,7] Recent studies also point to ways current planetarium education supports the affective domain. For example, interest in astronomy often drives visits to planetariums; Petrie[8] found that parents who took their preschool-age children to the planetarium were motivated to do so by their own interest in astronomy (75%) or their children's interest in astronomy (49%). In a case study of 12- to-14-year-old students who visited a planetarium and science center, students' memories from their visit related to their initial interests in astronomy.[9] Students' affective learning in the planetarium was strongly associated with enjoyment, which could relate the ability of the context to engender in visitors an interest in themes presented.[9]

Research on informal learning environments, including the planetarium, emphasizes how these venues provide opportunities for learning that goes above and beyond what audiences might find in other settings. Rather than considering the planetarium a "better classroom," it can best be understood as providing new opportunities for learning that complement the classroom. And for informal, free-choice audiences, the planetarium provides new ways of supporting existing interests or expanding interests not otherwise explored in the home or other settings. Below, I discuss themes drawn from recent research in the literature which highlight the planetarium's potential to enhance or expand learning, with a particular focus on K-12 audiences, in ways that make particular use of the unique learning environment.

Supporting classroom education

Class field trips to planetariums support teachers' work to educate K-12 students in astronomy and other STEM fields. Syntheses of research on informal learning suggest that the most successful way to use a field trip in K-12 education occurs when teachers include the following preparations: (a) provide orientation for students about the location, agenda, and learning goals; (b) engage students in pre-visit activities aligned with learning goals; (c) plan field trip activities that support classroom curriculum while also utilizing what is unique about the setting; and (d) engage students in post-visit classroom activities that reinforce learning goals and provide opportunities for reflection.[10]

In other words, field trips are most successful in supporting student learning when they are relevant and integrated into curricular goals and activities in the classroom. However, in a study of elementary teachers who bring their students to the planetarium on fieldtrips, Schwarz, Ghent, and Plummer[11] found that most did not follow these methods to integrate the visit into their curriculum. Most teachers had already finished their space science unit and were using the field trip as final wrap-up. On the other hand, the majority planned to include some form of follow-up after the visit, such as class discussions or written assignments. Schwarz and colleagues suggest teachers may not be aware of the planetarium's capabilities and how these might be used to support their curricular goals. The teachers also expressed interest in receiving curricular materials from the planetarium suggesting they are open to increased integration between their lessons and the planetarium visit.

Additional studies provide evidence of how integration between planetarium visits and classroom lessons may support student learning. Schmoll[12] investigated the use of the SMILES framework[13] which (1) integrates school and museum learning, (2) provides conditions for self-directed learning, and (3) facilitates learning strategies appropriate to the setting. While the fifth-grade students and teacher in the study made positive use of a visit to the planetarium, Schmoll offers recommendations to improve integration between classroom and planetarium. First, students should be given the opportunity to collect data in the planetarium program and explicitly review the data in the classroom. Second, self-directed learning should be facilitated in the classroom, through choice and social learning during projects, rather than the planetarium where such capacity is limited. Third, planetarium programs should minimize the number of topics addressed and to provide an outline or overview at the beginning of programs to help students' track shifts in topics covered in a program.

Plummer, Kocareli, and Slagle[7] compared combinations of classroom and planetarium visits to third-grade students' attainment of celestial motion explanations. The study's goal was to examine the contribution of differing instructional designs, including a planetarium program focused on the apparent motions of the Sun, Moon, and stars; a traditional classroom lesson teaching primarily rotation and orbital motions; and revised classroom lesson which used space-based celestial motions to explain the apparent motions observed in the planetarium. By comparing learning gains assessed through pre- and post-interviews, Plummer and colleagues found that students who first went to the planetarium program then engaged in a lesson which made use of the motions learned in the planetarium demonstrated the most improvement. The findings suggest that the planetarium can enhance learning opportunities by engaging students in observations not easily replicated in the classroom when teachers are prepared to build on those observations in the classroom in ways that help make sense using modeling and explanation building.

Using the immersive capacity of the learning environment

The planetarium community actively develops new techniques and technologies to support a growing interest in the immersive capabilities of the planetarium. The immersive fulldome environment uses projection systems to display an image on the inside of a dome surface with the goal of completely filling each audience member's field of view (FOV).[14] Immersion describes "the objective, quantifiable features of the display that result from the particular software and hardware, and the extent to which they are comparable to the level of sensory input that would be received in the real world"[14] (p. 562). The immersive capacity of the fulldome environment may improve learning, including explanations addressing how immersion supports memory, spatial learning, and data comprehension.[14]

Memory. An established model for working memory assumes the use of a central executive, with limited storage, assisted by two separate storage and maintenance processes for visual and auditory information and an episodic buffer, a limited capacity storage that integrates information into episodes.[15] An outcome of this model is cognitive load theory[16] which suggests that because working memory is limited, it can become overloaded when information demand is high. As a consequence, demands on working memory can be reduced – and consequently opportunities for learning increased – when the same information is presented to the visual and auditory modalities. Sumner, Reiff, and Weber[17] investigated student learning (grades 3–12) after participating in a fulldome, immersive planetarium program. All grade levels showed statistically significant increases in their knowledge of basic Earth science concepts. They found greater improvement on items covered in the program across more than one modality. Cognitive load theory suggests that when content was spread across visual and auditory modalities, students faced reduced cognitive load and increased their opportunities for recall of those concepts.

Data comprehension. The fulldome, immersive environment may allow a learner to visualize complex data sets and identify trends in data.[18] The Worldwide Telescope (WWT) offers one potential solution to bringing astronomical data sets to the immersive, fulldome environment as a scientific data visualization platform. The WWT interface "acts as a virtual sky, allowing users to explore all-sky surveys across the electromagnetic spectrum, to overlay data from NASA's Great Observatories, and to import their own imagery and tabular data" (p. 1).[19] In addition to the fulldome's capacity to support scientific work with large-scale data sets and complex analyses, the immersive properties of the planetarium also contribute to younger students' opportunities to make their own observations of astronomical phenomena. As noted earlier, a field trip to a planetarium can generate data for students to be used in deeper sense-making activities when back in their classrooms.[7,12,20] One commonality across these studies was that the students were all observing celestial motion phenomena – patterns of three-dimensional motion. The spatiality of their observations was likely supported by the immersive quality of the program, as discussed in the next section.

Spatial learning. The immersive properties of the fulldome planetarium have the potential to support spatial learning.[13,21] Given the spatial nature of astronomy,[22] facilitating spatial thinking – knowledge of spatial relationships, use of spatial representations, and reasoning for spatial problem-solving[23] – will assist visitor learning in the planetarium. One primary benefit of the immersive planetarium is the ability to visualize spatial relationships, especially the three-dimensional nature of astronomical phenomena, for audiences more accurately than on a flatscreen.[21] The fulldome display fills the visitor's FOV, an important consideration as limiting FOV reduces physical and cognitive performance.[24] The large FOV may allow users to pick up on spatial cues that facilitate the formation of coherent cognitive maps of the 3D astronomical phenomena.[13] Shelton and McNamara[25] found that participants were better able to align new spatial structures in a reference system when that reference system aligns with their own local reference system. In other words, in designing experiences in the immersive environment of the planetarium, we need to consider the audience's perspective – their egocentric reference frame.

In recent years, a few planetarium studies have considered how these immersive properties of planetarium programs support the children's spatial learning. Chastenay[26] investigated the use of immersion in teaching middle school students (aged 12–14 years) the reason for the lunar phases. The fulldome planetarium experience allowed learners to have both the space-based and Earth-based reference frames in order to facilitate their understanding of the phenomenon. Most participants' explanations improved and they were able to apply this knowledge to predict similar phenomena such as phases seen on Mercury and of the Earth from the Moon. Türk and Kalkan[27] compared learning outcomes from seventh-grade students (aged 12–13) who attended a planetarium program to those who studied the same content in the classroom. They found that "students in the planetarium assisted group were more successful in comprehending subjects that require 3D thinking, a reference system, changing the time and observation of periodic motion" than those who only studied these concepts in the classroom (p. 1). Across these studies, the use of the fulldome planetarium may have facilitated how children interpreted the reference system alignment.[25]

Embodied cognition and embodied design

Immersive learning considers how the visual environment shapes the learner's experience in the planetarium while embodied cognition adds one's own body to the equation, suggesting that the body is a tool for sense-making. Embodied cognition explains human conceptual knowledge and cognitive processes in terms of how the human body interacts with the environment.[28] We learn through sensorimotor input and thus our cognitive models we create of the world are grounded in body-based actions and interactions. Embodied design proposes learning environments purposefully designed to utilize this model of learning to support student learning.

Research in the planetarium utilizing an embodied design approach has considered ways to integrate audience members' physical motions into the program to promote embodied sense-making. Across a series of studies, Plummer and colleagues [6,7,19] have designed planetarium programs in which the planetarium educator prompts early elementary students to mimic the motion of the Sun, Moon, and stars as they appear to move across the sky to facilitate their embodied learning of apparent celestial motion. Each study found significant improvement in children's descriptions of these movements, as assessed through interviews accompanied by gestures or drawings. Children benefited from the use of embodied actions in the planetarium environment that were congruent with the concepts being learned.[29]

Researchers have also investigated the use of embodied design to help students learn to explain the apparent movement of celestial objects. Plummer, Kocareli, and Slagle[7] interviewed third graders before and after a visit to the planetarium, where they used embodied actions to mimic the motions of the Sun, Moon, and stars, and classroom lessons, where they used their own bodies to mimic the motion of the Earth to explain why these celestial objects actually move. Students who participated in both embodied design lessons (planetarium and classroom) made significantly greater improvement than students who only studied the motion of the Earth, without the use of their bodies, and students who only attended the planetarium.

Learning through storytelling

The use of narrative in science communication has the potential to increase audience comprehension, interest, and engagement.[30,31] Narratives help audiences learn by emphasizing how ideas interrelate to demonstrate coherence.[32] Storytelling is widely used in the planetarium, both in live shows and pre-recorded shows, to connect audiences to science.[33] Yu[33] describes how a story-based narrative format can be used to construct astronomical stories for the planetarium in order to transform unfamiliar and abstract concepts into memorable structures. For example, "[s]tories of individual struggle in science—which could involve the difficulties of understanding nature, or conflict with others who have similar or antagonistic goals—draw the listener in by personalizing a scientist's conflict, to make it understandable and relatable" (p. 89).

Stories can improve elementary students' attitudes toward science. Meyers[34] used quantitative methods to investigate how pairing folk tales with scientific explanations might increase attitudes toward science, among the third- to sixth-grade student participants (N = 3,500). Initially, results suggested students did not improve their attitudes toward science, but after revisions to the science narrative and program designs, students showed significant improvement in their science attitudes although these improvements were small. However, Meyers' study does not delve into the reasons behind why students' attitudes may have improved or what elements of the storytelling may have been most beneficial. While storytelling is widespread in the planetarium field, research studies that explore the nuances of how variations in narratives and storytelling practices are few. Very few studies have investigated the role of stories or storytelling in children's learning or engagement in the planetarium. A recent exploratory study[35] points to the potential role that story *characters* may play in engaging younger children with science phenomena in lessons that integrate experiences across classroom and planetarium. More research is needed that examines and expands on similar points of leverage within narrative frameworks.

Future directions for the planetarium community

The planetarium community embraces a broad mission to inspire and educate audiences of all ages in space science and beyond. The themes explore in this essay areas of research highlighting how planetariums engage and educate diverse audiences. Yet, more research is needed that explores broader contexts and a more diverse audience base to extend our understanding of how to make the planetarium accessible to all visitors. For example, accessible design for visitors with visual impairments, hearing impairment, and mobility disabilities has been installed at Tecnópolis in Argentina using a multi-sensorial approach to communicating the Universe.[36] Other planetariums have made efforts to bring astronomy to visitors with visual impairments: tactile materials for astronomy have been developed, such as books like Touch the Stars;[37,38] tactile planetarium domes have been developed in Brazil for visitors with visual impairment to explore stars, constellations, and their names;[39] and a planetarium program was developed by a group in Spain using sound coming from different directions and haptic hemispheres to teach the constellations.[40] However, more research is needed to explore how these innovations support engagement and learning in historically underserved populations to broaden the evidence base for strategies that improve access to astronomy. Such research has begun, such as with recent studies on accommodations for members of the deaf or hard-of-hearing communities. Hintz and colleagues[41] found promising results with the use of head-mounted displays (HMD) to deliver an American Sign Language "sound track" to learners in the planetarium show. Using a series of studies to investigate the relationship between the HMD design and audience learning outcomes, Jones and Lawler[42] suggest that HMD may better support learning than an interpreter at the front of the audience, if the relative brightness of the American Sign Language video matches the star brightness. Similar use of design-based research might benefit other areas of the planetarium field in developing accessible content.

The nature of the planetarium as a naturalistic environment may constrain the kinds of research that can be conducted. The most commonly used research method for research on the planetarium learning environment is quantitative.[3] However, many of these studies have been of single-group design which, while providing some insight for the community, limits the extent to which we can connect what was learned to the design of the learning environment. Future research using controls or comparison groups to strengthen the validity of findings will help us answer questions about the impact of new planetarium designs on how children learn or when children. In addition, a shift toward more use of mixed and qualitative methods has taken place over the last few decades.[43] For example, Plummer, Kocareli, and Slagle[7] combined qualitative methods (pre/post interviews) with the use of comparison groups to investigate how the planetarium environment can be successfully combined with classroom instruction to further support elementary students' spatial reasoning in astronomy. Studies that consider rigorous methodologies, such as combining the use of qualitative methods to make sense of how visitors make sense of the learning environment with the use of comparison-based studies, offer the field opportunities to gain new understanding of how and why visitors learn in the planetarium.

Notes

1 Petersen, M. C. (2019). State of the (Full)Dome 2019. Retrieved on June 21, 2021, from https://www.lochnessproductions.com/reference/attendance/attendance.html.
2 Small, K. J., & Plummer, J. D. (2010). Survey of the goals and beliefs of planetarium professionals regarding program design. *Astronomy Education Review, 9*(1), 010112-1–010112-10. http://dx.doi.org/10.3847/AER2010016.
3 Slater, T. F., & Tatge, C. B. (2017). *Research on teaching astronomy in the planetarium.* Springer International Publishing.
4 Plummer, J. D., Schmoll, S., Yu, K. C., & Ghent, C. (2015). A guide to conducting educational research in the planetarium. *Planetarian, 44*(2), 8–24.
5 National Research Council. (2009). *Learning science in informal environments: People, places, and pursuits.* National Academies Press.
6 Plummer, J. D. (2009). Early elementary students' development of astronomy concepts in the planetarium. *Journal of Research in Science Teaching, 46*(2), 192–209.

7   Plummer, J. D., Kocareli, A., & Slagle, C. (2014). Learning to explain astronomy across moving frames of reference: Exploring the role of classroom and planetarium-based instructional contexts. *International Journal of Science Education, 36*, 1083–1106.

8   Petrie, K. B. (2013). Early childhood learning in preschool planetarium programs (Master's thesis, University of Washington).

9   Lelliott, A. D. (2007). *Learning about Astronomy: a case study exploring how grade 7 and 8 students experience sites of informal learning in South Africa* (Doctoral dissertation, University of the Witwatersrand, Johannesburg).

10 DeWitt, J., & Storksdieck, M. (2008). A short review of school field trips: Key findings from the past and implications for the future. *Visitor Studies, 11*(2), 181–197.

11 Schwarz, K., Ghent, C., & Plummer, J. D. (2019). Why do they come? The motivation behind field trips to the planetarium. *Planetarian, 48*(1), 20–22.

12 Schmoll, S. E. (2013). *Toward a Framework for Integrating Planetarium and Classroom Learning* (Doctoral dissertation, University of Michigan).

13 Griffin, J. M. (1998). *School-museum integrated learning experiences in science: A learning journey* (Doctoral dissertation, University of Technology, Sydney).

14 Schnall, S., Hedge, C., & Weaver, R. (2012). The immersive virtual environment of the digital fulldome: Considerations of relevant psychological processes. *International Journal of Human-Computer Studies, 70*(8), 561–575.

15 Baddeley, A. (2003). Working memory: Looking back and looking forward. *Nature Reviews Neuroscience, 4*(10), 829–839.

16 Paas, F., & Sweller, J. (2012). An evolutionary upgrade of cognitive load theory: Using the human motor system and collaboration to support the learning of complex cognitive tasks. *Educational Psychology Review, 24*(1), 27–45.

17 Sumners, C., Reiff, P., & Weber, W. (2008). Learning in an immersive digital theater. *Advances in Space Research, 42*(11), 1848–1854.

18 Raja, D., Bowman, D., Lucas, J., & North, C. (2004, May). Exploring the benefits of immersion in abstract information visualization. In *Proceedings of the 8th International Immersive Projection Technology Workshop* (pp. 61–69).

19 Rosenfield, P., Fay, J., Gilchrist, R. K., Cui, C., Weigel, A. D., Robitaille, T., ... & Goodman, A. (2018). AAS WorldWide telescope: A seamless, cross-platform data visualization engine for astronomy research, education, and democratizing data. *The Astrophysical Journal Supplement Series, 236*(1), 22.

20 Plummer, J. D., & Small, K. J. (2018). Using a planetarium fieldtrip to engage young children in three-dimensional learning through representations, patterns, and lunar phenomena. *International Journal of Science Education, Part B, 8*(3), 193–212.

21 Yu, K. C. (2005). Digital fulldomes: The future of virtual astronomy education. *Planetarian, 34*(3), 6–11.

22 Plummer, J. D. (2014). Spatial thinking as the dimension of progress in an astronomy learning progression. *Studies in Science Education, 50*, 1–45.

23 National Research Council. (2006). *Learning to think spatially.* Washington, DC: National Academy Press.

24 Alfano, P. L., & Michel, G. F. (1990). Restricting the field of view: Perceptual and performance effects. *Perceptual and Motor Skills, 70*(1), 35–45.

25 Shelton, A. L., & McNamara, T. P. (2001). Systems of spatial reference in human memory. *Cognitive Psychology, 43*(4), 274–310.

26 Chastenay, P. (2016). From geocentrism to allocentrism: Teaching the phases of the moon in a digital fulldome planetarium. *Research in Science Education, 46*(1), 43–77.

27 Türk, C., & Kalkan, H. (2015). The effect of planetariums on teaching specific astronomy concepts. *Journal of Science Education and Technology, 24*(1), 1–15.

28 Wilson, M. (2002). Six views of embodied cognition. *Psychonomic Bulletin & Review, 9*(4), 625–636.

29 DeSutter, D., & Stieff, M. (2017). Teaching students to think spatially through embodied actions: Design principles for learning environments in science, technology, engineering, and mathematics. *Cognitive Research: Principles and Implications, 2*(1), 1–20.

30 Dahlstrom, M. F. (2014). Using narratives and storytelling to communicate science with nonexpert audiences. *Proceedings of the National Academy of Sciences, 111*(Supplement 4), 13614–13620.

31 Norris, S. P., Guilbert, S. M., Smith, M. L., Hakimelahi, S., & Phillips, L. M. (2005). A theoretical framework for narrative explanation in science. *Science Education, 89*(4), 535–563.

32 Millar, R., & Osborne, J. (Eds.) (1998). *Beyond 2000: Science education for the future. London: King's College London*, School of Education.

33 Yu, K. C., Petersen, C. C., LeBlanc, D., & Wyatt, R. (2020). Science storytelling for the planetarium. In D. W. Smith & W. Kemp (Eds.), *Proceedings of the International Planetarium Society 2020 Virtual Conference*, 88–92.

34 Meyers, M. B. (2005). *Telling the stars: A quantitative approach to assessing the use of folk tales in science education* (Doctoral dissertation, East Tennessee State University).

35 Allen, A., Bass, K., & Jarvis, N. (2021). An Integrated Approach to Early Elementary Earth and Space Science: Project PLANET (Unpublished summative report). Rockman et al, San Francisco.

36 García, B., Maya, J., Mancilla, A., Alvarez, S. P., Videla, M., Yelós, D., & Cancio, A. (2012). Touch the sky with your hands: A special planetarium for blind, deaf, and motor disabled. *Proceedings of the International Astronomical Union, 10*(H16), 554–554.

37 Grice, N. (2019). *Touch the Stars*, 5th Edition. National Braille Press.

38 Grice, N., Beck-Winchatz, B., Wentworth, B., & Winchatz, M. R. (2004, December). Touch the universe: A NASA braille book of astronomy. In C. Narasimhan, B. Beck-Winchatz, I. Hawkins & C. Runyon (Eds.) *NASA Office of Space Science Education and Public Outreach Conference (319, 301)*, Astronomical Society of the Pacific, San Francisco.

39 Carvalho, C. L., & de Aquino, H. A. (2015). The Universe at the Fingertips of the Visually Impaired: Building a tactile planetarium. *A Global Audience for the New Race to the Moon*, 36.

40 Ortiz-Gil, A., Ballesteros Rosello, F., Blay, P., Espinos, H., Fernandez-Soto, A., Gallego Calvente, A. T., ... & Navarro, J. (2017). *Astronomy for Inclusion* (特集 ユニバーサルデザイン天文教育研究会). 天文教育, *29*(1), 4–9. https://ci.nii. ac.jp/naid/40021246353/.

41 Hintz, E. G., Jones, M. D., Lawler, M. J., Bench, N., & Mangrubang, F. (2015). Adoption of ASL classifiers as delivered by head-mounted displays in a planetarium show. *Journal of Astronomy & Earth Sciences Education, 2*(1), 1–16.

42 Jones, M. D., & Lawler, M. J. (2019, October). Delivering sign language in a live planetarium show using head-mounted displays and infrared light. In *The 21st International ACM SIGACCESS Conference on Computers and Accessibility* (396–401). Association for Computing Machinery, New York, NY.

43 Slater, S. J., Tatge, C. B., Bretones, P. S., Slater, T. F., Schleigh, S. P., McKinnon, D., & Heyer, I. (2016). iSTAR First light: Characterizing astronomy education research dissertations in the iSTAR database. *Journal of Astronomy & Earth Sciences Education, 3*(2), 125–140.

7

NOW IT'S AFRICA'S TURN

The Status and Potential of Science Centres and Informal Science Education Initiatives on the 'Bright Continent'

Mike Bruton, Julie Cleverdon, Knowledge Chikundi, and Kenneth Monjero

Africa is a vast continent comprising 48 mainland countries and six island nations. Estimates of the number of languages natively spoken in Africa range from 1,250 to over 3,000.[1] This cultural and linguistic diversity, combined with high population densities, a relatively high percentage of people living in rural areas and poorly developed formal educational institutions, highlights the importance of informal science education facilities, especially science centres, that offer relatively language-free interactive learning experiences in both urban and rural settings.

Africa's rapidly increasing population, low pre-primary and primary school attendance, a severe shortage of teachers, low literacy and techno-literacy rates, and a lack of institutional capacity at the tertiary level lead to educational imbalances and emphasize the urgent need to develop a stronger network of informal and non-formal educational facilities that support the formal education sector. In particular, interactive science centres offer pooled educational resources and expertise that can assist under-resourced schools to not only teach science and mathematics but, probably more importantly, instill an interest in and curiosity about science in young people, female and male, of all cultures. Notwithstanding this potential contribution, no mention is made in recent reviews of the challenges and opportunities facing formal education in Africa of the important role those informal educational facilities can play in supporting formal education.[2,3]

This essay summarizes the status of science centres and informal science initiatives in Africa and recommends ways to unleash their potential to support and expand science learning opportunities in formal education.

DOI: 10.4324/9781003145387-8

The status of science centres in Africa

Depending on how you define them, and it is often difficult to distinguish between science centres and museums, there are 49 science centres in Africa, 47 on the mainland and two on adjacent islands (Mauritius and Réunion), representing about 5% of science centres worldwide.[4–6] Furthermore, science centres in Africa are very unevenly distributed, with about 17% occurring in North Africa (Tunisia, Morocco, and Egypt), only 2% in West Africa, 6% in East Africa (Kenya, Rwanda, Uganda, and Ethiopia), and 73% in South Africa, Zimbabwe, and Bostwana, with 2% in Mauritius and Réunion.

Many science centres in Africa are small by international standards and serve mainly as STEM education centres for the youth with only a few, such as those in Egypt, Ethiopia, Tunisia, Mauritius, Réunion, Morocco and South Africa, resembling the large, publicly accessible, museum-like venues that cater for both youths and adults that are widespread in the developed world. The worldwide average for science centres per capita is about one for every 10 million people. The USA and Canada have 12 times the world average while South Africa has seven times this average, and the rest of Africa only 7%. To put this disparity into perspective, the USA has 180 times as many science centres per capita as most of Africa. Of the 54 countries and territories in Africa, only 11 have science centres, with South Africa dominating the scene with 33, the third largest number worldwide per capita after the USA and Canada. Thus, the vast majority of African people do not have access to science centres and their associated outreach activities.

Tunisia. There are two science centres in Tunisia (Tunis Science City established in 2002; Monastir Sciences Palace in 2007) and one each in Botswana (at the Botswana International University for Science & Technology, BIUST), Egypt (Planetarium Science Centre in Alexandria), Ghana (Planetarium Science Centre in Accra), Kenya (Science Centre Kenya in Nairobi), Mauritius (Rajiv Gandhi Science Centre in Bell Village near Jamestown), Réunion (Museum Stella Matutina in Saint-Leu), Rwanda (Museum of the Environment, near Lake Kivu), Uganda (Source of the Nile Science Center in Kampala), and Zimbabwe (Discovereum Children's Museum in Harare), with several others in different stages of development but not as yet functional at the time of going to press.

Ghana. The Planetarium Science Centre in Ghana, the only science centre in West Africa, is a typical example of the can-do spirit of a small number of passionate people. It was founded in 2009 by Dr Jacob Ashong and his wife, Jane, using his pension funds (and his wife's savings!). Before he retired Dr Ashong was a researcher and lecturer in the Biochemistry Department at the University of Ghana. He is also a director of the company, African Gifted Foundation Ghana, which established the African Science Academy in the country. The Planetarium Science Centre is located in Cantonments, Accra, in a simple thatched building with a laptop being used for projections of constellations and celestial objects and the guests seated on fold-back chairs. It is run by Sarah Abotsi-Mastee and other volunteers and offers science demonstrations as well as planetarium shows.

Ethiopia. Of the five science centres in Ethiopia, the Addis Ababa Science Museum, established in 2005, is the largest followed by the Foka Science & Engineering Center, Kallamino STEM Center, Gondar University Science Center, and the Bahir Dar University STEM Incubation Center. The Discovereum Children's Science Centre was established by J.N. Maraire and A.N. Chiurain in Belgravia, a northern suburb of Harare, in Zimbabwe. It is sponsored by INZI ('Inspired Zimbabwe'), and its goal is to bring hope and create stronger, more vibrant communities through creative, interactive projects. This science centre closed in 2018 but may reopen.

Kenya. Science Centre Kenya (SCK) is based at the Kenya Agricultural & Livestock Organization in Nairobi. It is a fully interactive science centre that emphasizes a hands-on approach to teaching science and technology. Their vision is 'A brighter future for Kenya's youth through STEM' and they contribute strongly to the motto of Kenya, 'Harambee', which means 'let's all pull together'. Kenya, where the pioneering *M-Pesa* cell phone money transaction system was first launched in 2007, is arguably Africa's most innovative nation after South Africa, which is regarded as the continent's chief 'rain maker'.[6] SCK has also developed a strong outreach programme through the boisterous efforts of the famous 'Dr Fun' (Kenneth Monjero) who participated in *National Science Week* and performed 'Boom Science with Covid' science shows at *SciFest Africa* in South Africa in 2022. Monjero, who is a qualified plant pathologist, has widened the scope of his science centre's educational programmes to include agriculture and is a maestro at performing science shows with minimal apparatus.

Uganda. The Source of the Nile Science Centre (SNSC) in Kampala, Uganda, was established by Martin Kafeero, after he had attended a science centre capacity building leadership training course at the Australian National University's Center for the Public Awareness of Science (CPAS) in Canberra. Previously he had worked with the Initiative for Science & Technology-Tanzania (InSciT-Tanzania), an informal science learning organization that he had founded in Tanzania in 2013. According to Kafeero, Uganda, like many African countries, lacks creative problem solvers in STEM fields due to the failure of its educational ministries to develop skills-based learning opportunities. The SNSC was therefore established to help meet Uganda's skills challenge and to become a hub for innovative educational initiatives.

The vision of the SNSC is to help create a nation in which all people value STEM as a source of informed decision-making and a means to empower individuals, enrich lives, and help people to discover the pleasure and relevance of science and technology in their everyday lives. Since its establishment in July 2017 the SNSC has reached students throughout Uganda as well as in

Kigali, Rwanda. In future, Kafeero envisions the SNSC becoming a 'wonder' of informal science learning in East Africa with cutting-edge technology, innovative education initiatives, and a planetarium. He also plans to develop a 'Ghetto Science Outreach' programme that takes science to street kids in Kampala as well as programmes that promote the United Nation's Sustainable Development Goals.

Rwanda. The Museum of the Environment near Lake Kivu in Rwanda is an interactive two-storey museum with a traditional herbal medicine garden on the rooftop. It showcases renewable and non-renewable sources of energy and is designed as an educational centre for local visitors. Its goal is to help people understand and safeguard their environment and to promote integrated and sustainable development. It is one of the eight museums that make up the Institute of National Museums of Rwanda.

South Africa. South Africa has 700% of the world average of science centres per capita whereas the rest of Africa has only 7%. The first science centre in South Africa (and Africa) was established as the 'Exploratorium' (renamed the Sci-Enza Discovery Centre in 2005) on the campus of the University of Pretoria in 1977. It was founded by Professor Lötz Strauss of the Physics Department as an open laboratory that gave students the opportunity to 'play' with scientific apparatus in an informal setting.

The second oldest science centre in South Africa is the Unizulu Science Centre which opened in 1986 in Richards Bay and receives about 30,000 learners annually. This science centre works in a region with over 500 high schools, many of which remain virtually unchanged since apartheid and suffer from a severe lack of infrastructure and resources, including lack of access to qualified educators. Under the leadership of Dr Derek Fish, Unizulu has become nationally and internationally recognized for its efforts to uplift science education in disadvantaged regions within Kwa-Zulu Natal. (See Figure 7.1.)

The largest science centre in Africa is the Sci-Bono Discovery Centre which was established in 2008 in the historic Electric Workshop building in Newtown, Johannesburg. This science centre is primarily supported by the Gauteng Department of Education and is responsible for implementing the Gauteng Mathematics, Science and Technology Education Strategy. This world-class science centre receives over 300,000 visitors per year and is now one of the most popular leisure and educational destinations in South Africa's most populous province.

Science centres in South Africa established in association with universities include the UniZulu Science Centre (established in 1986; University of KwaZulu-Natal), Boyden Observatory near Bloemfontein (1976; University of the Free State), and the FOSST Discovery Centre (2008) at the University of Fort Hare in Alice.

The Boyden Observatory was established in 1927 at Maselspoort near Bloemfontein and a science centre is in the process of being developed there.

FIGURE 7.1 UniZulu Science Centre, Richards Bay, KwaZulu-Natal, South Africa.

South Africa's first digital planetarium was opened in the original dome of the Lamont-Hussey Observatory on Naval Hill in Bloemfontein. The second digital planetarium in South Africa, and the most advanced in Africa, was established at the Iziko South African Museum in Cape Town in 2018. Independently established science centres in South Africa include the Cape Town Science Centre (CTSC) (established in 2000), Nelson Mandela Bay Science & Technology Centre in Uitenhage (2013), and the two science centres sponsored by the multinational steel company, Arcelor Mittal, in Sebokeng and Saldanha (2006 and 2010) adjacent to their major operations in Vanderbijlpark (Gauteng) and Saldanha (Western Cape), respectively.

Several science centres in South Africa act as interactive visitor and/or information centres for major science and technology facilities, such as the Koeberg Nuclear Power Station near Cape Town, the National Accelerator Centre (NAC) in Faure near Stellenbosch (both in the Western Cape), the South African Nuclear Energy Corporation (NECSA) in Pelindaba, North-West province and the Hartebeesthoek Radio Astronomy Observatory (HartRAO), located west of Johannesburg in Gauteng. An interactive science centre is currently being planned for Carnarvon in the Northern Cape at the main field site of the South African Radio Astronomy Observatory (SARAO), which includes the multinational Square Kilometre Array (SKA) project.

Status of other informal science education facilities and events

Important milestones in the development of informal science education events, services and facilities in South Africa include the hosting of the first regional science centre conference in Africa at the UniZulu Science Centre in Richards Bay, KwaZulu-Natal, in 1995 and the launch of the first science centre network in Africa, the Southern African Association of Science & Technology Centres (SAASTEC), in 1996.[5] A second science centre network that serves the needs of science centres in North Africa and the Middle East, NAMES (North Africa and Middle East Science Centers Network) was launched in January 2006 and serves the science centres and science museums in this region. There is regrettably no science centre network for West or East Africa, nor is there a continent-wide umbrella body.

In 1996 the first national conference on science communication in Africa, 'Promoting Public Understanding of Science and Technology in Southern Africa', was held at the University of the Western Cape in Cape Town. The year 1996 also marked the launch of two science festivals in South Africa, *Sasol SciFest* in Sasolburg, where the giant Sasol oil refinery is based, and *SciFest Africa* in the university city of Makhanda, both of which are still operational. *Scifest Africa* comprises a National Science Festival that is held in Makhanda (previously Grahamstown) each year; regional and national outreach programmes are implemented throughout the year. The project is supported by South Africa's National Department of Science and Innovation and other sponsors. Science festivals are also held in Ethiopia, Kenya, Botswana, Namibia, and Tunisia.

In 2011 the Science Centre World Congress was held in Africa for the first time at the International Convention Centre in Cape Town. It was attended by 430 delegates from 29 countries and issued the highly regarded 'Cape Town Declaration' (2011) on the status and aspirations of science centres worldwide. In 1999 the South African government initiated annual National Science Weeks, which are still ongoing, and, in 2000, a government-led flagship science promotion project was launched in South Africa. Science weeks, as well as an annual 'Innovation Week', are also held in Kenya.

STEM clubs

The CTSC in collaboration with the Western Cape Education Department in South Africa has identified STEM Clubs as a strategic extracurricular support programme to promote STEM engagement, literacy and career development in the Western Cape. These clubs help to improve learner's STEM skills and knowledge so that they can navigate their way through the world they live in and successfully access the opportunities available to them.

STEM Clubs develop aptitude and interest in STEM subjects from an early age and sustain that interest through the learner's school career and into adulthood. They have been shown to improve a learner's performance in STEM subjects and to help learners develop a positive perception about science.[7] The STEM Clubs developed by the CTSC offer learners the opportunity to explore STEM topics in an informal setting and allow them to experiment, ask questions and experience the challenges that scientific research offers. They also promote heuristic thinking, help link abstract concepts learned in class to the real world and allow learners to be the authors of their own discoveries and experiments.

Africa Code Week

Africa Code Week is an important continent-wide initiative by the multinational computer company, SAP, and is supported by UNESCO, Irish Aid, the Association for the Development of Education in Africa (ADEA), Jokkolabs, and the Camden Trust. This initiative has been widely implemented by the CTSC. Its aim is to introduce coding and digital literacy to 8- to 16-year-olds, shape the next generation of experts in technology use and shed light on how technology can help to solve real-life problems. The programme was initiated in 2015 when 88,763 youths and 2,088 teachers were introduced to coding skills in 17 countries. The numbers increased to 3.85 million youths and 39,000 teachers in 37 countries in 2019. By 2021 the project had reached youth and teachers in 41 African countries with 48% participation by girls.

The mission of the CTSC, which was established by Mike Bruton in 1996, is 'to make a hands-on contribution to Africa's future by strengthening its science and technology culture through interactive methods of teaching, to excite the youth and general public about science and technology, and to offer educational services to learners and educators across Africa'. The CTSC began working across Africa in 2015 with the initiation of Africa Code Week. In 2015 it conceptualized and piloted the first 'Train the Trainer' workshops in Cape Town and then advised the funder (SAP) that the multipliers of education in Africa are teachers and that the resources to implement such a programme are more likely to be found in schools than in individual homes.

EduConservation

In addition to a focus on coding, the CTSC also led the EduConservation initiative in Africa and has worked closely with education departments in African countries to produce conservation-enriched curriculum material for use in schools, with a teacher-centric, Africa-centric approach. The EduConservation project was initiated by Sabine Plattner African Charities (SPAC) in partnership with Leadership for Conservation in Africa (LCA) after the 2014 World Parks Congress held in Sydney, Australia. Sabine Plattner, a keynote speaker at the congress, highlighted the need for conservationists to support the enrichment of school curricula with educational content aimed at the protection of nature. She initiated EduConservation as a curriculum enrichment programme that pilots ways to enrich the curricula of African countries with Afro-centric conservation content. EduConservation has thus far been launched in the Republic of Congo, Senegal, Morocco and South Africa with Gabon and Namibia to follow as part of the next phase.

EduConservation primarily aims to impact youth during their school-going years and ensure that children's positive attitude towards conservation is nurtured and carried through into adulthood. This is achieved by developing educational content that is relevant and has a natural fit with the existing curriculum. The project has developed and distributed an activity booklet that covers nine core topics related to environmental conservation that was developed in collaboration with local Ministries of Education for secondary schools. Over 20,000 activity booklets have been distributed to secondary schools in the pilot countries with teacher training being provided to about 300 teachers. The CTSC is implementing the SPAC EduConservation project across Africa.

International collaboration in science awareness

The Science Circus of Australia has played a strong role in promoting science awareness in Africa. The project is an initiative of the Australian National Centre for the Public Awareness of Science (CPAS) at the Australian National University (ANU) in Canberra and is run in partnership with Questacon, the National Science and Technology Centre. The Science Circus has travelled to Africa on several occasions and has been a regular participant in *SciFest Africa*, the National Science Festival in South Africa. CPAS staff have trained hundreds of African teachers and left them with locally obtained equipment to run their own inspiring science programmes. In 2018, the Science Circus toured included southern and eastern Africa. Walker presented a series of science shows entitled 'Dr. Graham's Blow Up Science' at *SciFest Africa 2022* in South Africa. Another regular contributor to *SciFest Africa* over several years has been Dr Stephen Ashworth of the University of East Anglia who has presented 'Kitchen Science' shows.

One of the most indefatigable partners in Africa is Knowledge Chikundi of Zimbabwe, often referred to as a 'one-man NGO'. Starting with a single school in 2013, Chikundi organized graduate students to assist primary and high school students to complete science fair projects; in 2013 and 2014, pass rates in science at that school improved dramatically. Building on that success, the science fair platform was expanded to the rest of Zimbabwe, and, in 2016, Zimbabwe sent its national science fair winner to the Intel Science and Engineering Fair in Arizona, USA, for the first time.

Chikundi noticed that a full-scale science fair project has too high an entry bar for many students, so he started the 'Science Busker's Festival', modelled on a concept from Singapore and with the help from CPAS and the Australian Embassy in Zimbabwe: participating students prepare a presentation on a science phenomenon and then perform for an audience. The inaugural festival took place in Harare in March 2017. The Science Buskers, now a programme of the Science House for African Youth (SHAY) Institute and Zimbabwe Science Fair, champions science that is dedicated to expanding effective STEM education, scientific curiosity and scientific research and innovation in Zimbabwe and southern Africa. Today the Africa Science Buskers Festival, sponsored by the Broadcom Foundation, is the continent's largest science communication competition for young scientists, engineers and science communicators in primary and high school.

Gender equality programmes

The CTSC serves as the fund coordinator for the EQUALS Global Partnership for Gender Equality in the Digital Age and is responsible for the overall management of the fund. The EQUALS Global Partnership is a committed group of corporate leaders, governments, businesses, not-for-profit organizations, academic institutions, NGOs, and community groups that is dedicated to promoting gender balance in the technology sector by championing equality of access, gender-sensitive skills development and career opportunities for women and men alike.[8]

The core focus areas of EQUALS include skills development. The #eSkills-4Girls Fund was established to provide digital skills training for women and girls. This fund recognizes that the barriers to gender equality are numerous and diverse with some deeply rooted in the past while others are unique to the digital world. The hurdles that curtail women's ability to fully benefit from the digital transformation include access and affordability, inherent biases and socio-cultural norms, and lack of educational opportunities and skills. Training is an important factor in overcoming some of these barriers. The #eSkills4Girls Fund is supported by BMZ (Federal Ministry for Economic Cooperation and Development in Germany) and is implemented throughout Africa by science centres such as the CTSC.

Factors limiting science centre development in Africa and recommendations for their development

There have been appeals from many quarters for the international science centre community to contribute to the development of a stronger science centre culture and network in Africa. Africa cannot do it alone and needs international support, expertise, and experience in order to develop a critical mass of science centres which will then generate their own momentum.

African governments should follow the examples of South Africa, Tunisia, Egypt, Mauritius and Réunion in promoting science centre (and science festival) development from national and/or provincial coffers. The Covid pandemic and the environmental crisis should surely have convinced politicians by now that there is an urgent need to educate people about the scientific basis of these crises and the means of resolving them through science and technology. Furthermore, the excellent record that science centres have achieved internationally should make it clear by now that they are one of the most cost-effective ways of demystifying and teaching science, strengthening a science culture and complementing the formal educational sector. Their 'hands-on, minds-on, hearts-on' approach, which engages visitors physically, mentally and emotionally, and their relatively language-free exhibits are well suited to multicultural and multilingual audiences, and they cater to visitors from a wide range of ages and socio-economic backgrounds.[4,6]

Africa suffers from a serious lack of people who have the skills to conceptualize, design, build and operate science centres. Furthermore, many science centre managers in Africa lack business skills and the ability to draw up and implement robust business plans and adhere to strict accounting procedures. These managers also need to pay more attention to measuring and evaluating their impact on their visitor base. The science centre umbrella bodies, and the Masters courses now offered at Stellenbosch University, are helping to overcome these obstacles. There is, however, no science centre umbrella body that promotes the development of science centres in Africa outside southern Africa (covered by SAASTEC) and North Africa (by NAMES).

Discussion

Science centres were founded on the principle that interactive methods are the most effective ways of demystifying and teaching science and technology and that all people, regardless of their culture or socio-economic status, deserve to have access to scientific information that profoundly affects their lives.[5] Ideally, science centres need to play a key role in changing the mindset of learners, empowering teachers and celebrating a nation's achievements in science and technology. They should also serve as forums for debate on STEM topics, not just technical issues but also social, moral and ethical issues. In order to gain support, science centres also need to demonstrate to funders and policy makers that they are not just glorified games arcades or kiddies' play areas, but serious educational facilities.

From the science centre perspective, South Africa is the most advanced country in Africa. Yet the authors of a major review of science communication conclude that despite the initial years of democracy and the optimistic view that prosperity would 'spill over into a continent-wide African Renaissance ...

given the historical disconnects between science and the majority of South African citizens, along with huge socioeconomic disparities, the challenges of creating a scientifically literate society were vastly underestimated.'[5] These authors further acknowledge that science communication policies and strategies remain largely aspirational, with limited expertise and fragmented capacity to implement the government's ambitious plans; they conclude that 'twenty-six years since the advent of democracy (in South Africa), it remains questionable (as to) what progress we have made in developing an appropriate science communication infrastructure that adequately responds to local needs and would be able to deliver a truly science-engaged knowledge society.'[5]

Today, in most of Africa, science engagement activities are almost entirely located in urban areas and are predominantly targeted at children of school-going age. At their present level of development, they cannot hope to reach the hundreds of millions of people spread over the vast continent and heighten the level of science awareness on the continent. Africa needs support from the rest of the world to develop its science centre network as everyone will benefit if African youth are able to reach their full potential.[9,10]

We need to beneficiate Africa's demographic dividend through science centres. The continent is characterized by significant technology leapfrogs and rapid technology uptake. Generation Alpha (Gen-A; children born between 2010 and 2015) will soon replace Generation Z as the new kids on the block. Gen A is the first batch of children born to millennials, and they are likely to be even more tech-savvy, game-changing and challenging than their parents, having been brought up with screens as their babysitters and toys. As several authors have argued, this is Africa's century and Africa has the prospect of being the 'bright continent' if it develops its human potential and natural resources to their full potential.[6,10] Its lingering inferiority complex, stoked by colonialism, is rapidly disappearing. Africa has bred a highly determined and resilient class of entrepreneurial people, female and male, who are prepared to take on the world despite the odds stacked against them. It is important to note that creativity is often sparked in highly constrained, resource-strapped environments and that deprivation often stimulates innovation.[6,10]

The *modus operandi* of interactive science centres equips them to face these challenges. We believe they science centres and their associated non-formal science and technology initiatives have the potential to play a major role in the African renaissance, but only if they are developed in such a way that they can reach out to most of the people in Africa.

Acknowledgements

The authors are grateful to Marina Joubert, Shadrack Mkansi, and Samridhi Sharma for providing valuable information.

Notes

1 Kaya, H. 2020. African multilingualism in teaching and learning African indigenous languages and home-grown philosophies for domestication of the Sustainable Development Goals (SDGs). *DSI-NRF Centre in Indigenous Knowledge newsletter,* June 2020, 1-8.
2 Africa-America Institute. 2015. *State of Education in Africa Report 2015. A report card on the progress, opportunities and challenges confronting the African education sector.* Africa-America Institute, New York.
3 Zeleza, P. 2021. Quality higher education 'indispensable' for Africa's future. *University World News, Africa edition,* July 2021, 1–4.
4 Trautmann, C. & K. Monjero. 2019. Science centers in Africa. *Informal Learning Review* 154 (January/February), 3–10.
5 Joubert, M & S. Mkansi. 2020. South Africa. Science communication throughout turbulent times. In: *Communicating Science: A Global Perspective* (edited by T. Gascoigne, B. Schiele, J. Leach, M. Riedlinger, B. V. Lewenstein, L. Massarani & P. Broks). Australian National University, Canberra, Australia.
6 Bruton, M. N. 2022. *Harambee: The Spirit of Innovation in Africa.* Human Sciences Research Council, Cape Town,.
7 Chittum, J. R., B. D. Jones, S. Akalin, & A. B. Schram. 2017. The effects of an afterschool STEM program on students' motivation and engagement. *International Journal of STEM Education 4* (1), 1–16.
8 https://equals.lhs.berkeley.edu/aboutEQ.html
9 Olopade, D. 2014. *The Bright Continent: Breaking rules and making change in modern Africa.* Duckworth Overlook, London.
10 Mielly, M. 2003. The aesthetics of necessity: An interview with Werewere Liking. *World Literature Today 2* (3), 52–56.

8

TEACHERS AS DESIGNERS

Learning by Exhibit Prototyping

Ilona Iłowiecka-Tańska, Katarzyna Potęga vel Żabik, and Aneta Gop

Each year, millions of visitors come to science centers all over the world, where they learn about natural phenomena by interacting with exhibits. Informal science education has become a significant feature of the STEM learning landscape in terms of research, activity, training, and international scope.[1] The educational impact of this hands-on form of learning is increasingly well documented in an extensive literature, which has thus inspired efforts to incorporate this approach into school curricula.[2]

Since 2018, the Copernicus Science Centre in Warsaw, Poland, has been organizing Summer Prototyping Schools, workshop courses for teachers on creating prototypes of learning aids, following what we call the exhibit pedagogy" approach. The key principle of this method is that the teachers learn through active interaction with a material object (i.e., an exhibit) that acts as a mediator to guide the learning process. Through the experience of active exploration, the teachers develop concepts of a given phenomenon, on a level of complexity adequate for their current development and experience.[3]

Since 2019, Summer Schools of Prototyping have themselves been the focus of a research program, in which we have studied the development of teachers' competences during such prototyping exercises. In this chapter, we describe the advantages of drawing teachers into the prototyping process. We also share what we have learned by studying the experiences of teachers participating in the workshops, as expressed in their structured diary entries and survey responses. In this paper, we present a four-stage model of the prototyping process: attempted transposition (typically unsuccessful), deeper exploration of the phenomena itself (inspired by this lack of success), re-transposition, and re-exploration. The most important outcome of this process is a profound experience of personal development, which we claim ultimately gives

DOI: 10.4324/9781003145387-9

teacher-participants a stronger basis for working with their science students in the spirit of "exhibit pedagogy".

Bringing teachers into the design process

The Copernicus Science Centre is among a group of museums that pursue the idea of an exhibition space serving as a "laboratory for all".[4] In such an exhibition space, each visitor can adopt the role of a researcher who, using experimental apparatus (a particular exhibit), learns about how a given natural phenomenon works, driven by one's own curiosity and need to know. In this approach, the museum exhibits are actually research stations. Working with them allows the teachers to create and study various natural phenomena: waves, electric charges, vibrations, the actions of forces, etc. The notion of a "laboratory for all" entails that the exhibits enable anyone, regardless of their competence, to experience the excitement, curiosity, and discovery that has inspired the development of science itself.[5] The idea of an exhibition space serving as an accessible lab entails a certain pedagogical premise for the exhibits: they are objects designed to allow the visitor to discover, for themselves, certain natural phenomena and the mechanisms via which they can be induced and modified.[6]

Although these hands-on exhibits are primarily found in exhibition spaces (museums, science centers, etc.), there are many reasons why they can and should be used in schools,[7] and just as many reasons why these exhibits should be created by teachers.[8] First among these reasons are the advantages to be gained from incorporating "exhibit pedagogy" in school education. Research provides evidence of how visitors find significance in their experiences. Under this paradigm, the process of conceptualization is envisioned as a negotiation between naïve (embodied) and cultural (verbalized, represented) ways of perceiving phenomena.[9] This negotiation between the action and ability to develop its representations can be seen as a dialogue that celebrates both perspectives while seeking reconciliation.

A second rationale for getting teachers involved in the process of creating exhibits is to make sure the ideas generated in the science center are more coherent with current teaching goals and with what might be called the "local culture" of education. At the Copernicus Science Centre, while creating exhibits, educational kits, and lesson plans for more than ten years, educators have recognized the value of including end users into the design process. The educational solutions created in such a co-creative process are ultimately more relevant to the needs of the audience, including very demanding educators and students.

For both of these reasons, involving teachers in the processes of creating aids that bring "exhibit pedagogy" into the schools seems to be an optimal move. In theory, the idea essentially has no weaknesses or limitations. It supports the introduction of a new model of learning and teaching into schools and links together the worlds of formal and informal education, as well as cultivates and harnesses the creative potential of teachers. Let us take a look at how this practice is implemented.

Summer Prototyping Schools at Copernicus

The Summer Prototyping Schools are week-long workshop courses for teachers that deal with the process of creating educational solutions. The courses are held twice a year. During the workshops, groups of about 20 teachers are tasked with creating prototypes of new experimentation stations or a measuring apparatus.

Acting in freely formed teams, teachers work on developing specific objects (learning tools), having the necessary materials and basic workshop tools at their disposal. This work is done in keeping with the "rapid prototyping" method, which encourages a quick transition from the concept phase to the first version of the prototype. The teacher-designers make modifications under the influence of their own analysis (to what extent the object manages to manifest their intentions) and feedback from colleagues, experts, and potential users invited *ad hoc*. During the workshops, the form of the exhibit prototypes and the topics they address are in principle open, as long as they fulfill the basic condition, i.e., the exhibit will allow a student-user to independently discover a phenomenon or process occurring in nature. Each workshop culminates in the testing of the objects that are developed in the exhibition, with the participation of visitors. The entire process is moderated by experienced educators and exhibit developers from Copernicus.

Studying participants' experiences

Since 2019, researchers at Copernicus have been studying the changes of attitude in the Summer Prototyping School teacher-designers, seeking to determine the direction, area, and intensity of those changes. Below we analyze data collected during four sessions of the School, held over two years (2019–2020). The basic tool for obtaining data is diary-keeping, a method that provides narrative data, structured sequentially.[10] A semi-open form of diaries is applied, which provides an opportunity for in-depth meaning analysis.[11] Diary entries were elicited from each participant, for every day of the five-day workshop course, based on answers to the following four sentence-completion questions: (1) "I enjoyed..."; (2) "The most difficult thing for me today was..."; (3) "I think..."; (4) "Today I learned..." The participants were asked to complete these sentence fragments each day after class. As a complementary source to such diary data, survey questionnaires were sent to the participants following each workshop session. Lastly, photographs and self-descriptions of prototypes created by the teacher-participants of the School were utilized as another source of data.

Prototyping is a process during which certain assumptions are made that reveal what the participants consider to be crucial for a "good" exhibit (also called a learning tool). Collaborative prototyping by educators therefore shows how they understand the essence of teaching a particular subject, be it biology, physics, chemistry, or geography. The starting point for designing an exhibit is the teachers-designers' knowledge about the phenomenon on which the user's cognition and exploration are to be focused. This process has been designed as the transfer – transposition – of knowledge available in the scientific literature and other sources (scientific theories, processes, and concepts) to a material representation of this knowledge in the exhibit.[12]

According to the model, during the Summer Prototyping Schools, a teacher with knowledge of a phenomenon transfers this knowledge into an exhibit, intended for use as a learning aid by his or her students. The program of the School assumes that the area requiring the most intensive work during the workshops is developing the pedagogical principles of the objects created, or in other words, creating the concept of the object–student interaction. It seems likely that when teachers, representing a school system with dominant transmission methods, are confronted with "exhibit pedagogy", this particular issue should be the focus of dialogue. However, analysis of the teachers' diary entries calls this assumption into question (Figure 8.1).[13]

The diary entries kept by participants were classified into five different areas of personal development: personality and temperament; social skills; manual and technical skills; pedagogical area; and knowledge and understanding.

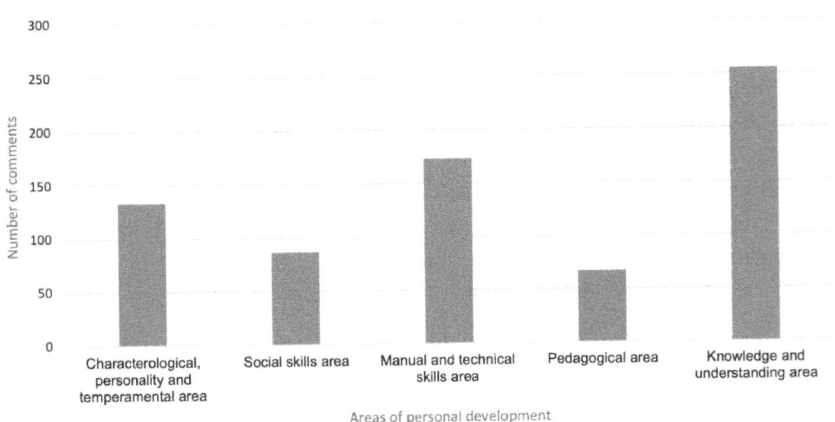

FIGURE 8.1 Areas of personal development noted by Summer Prototyping School participants (N = 546 comments for 64 teacher-designers). Each entry could be assigned to more than one category.

Each entry could be assigned to more than one category. The results show the field of knowledge and understanding to be the area of development most frequently indicated by the teachers. Moreover, the area also remained the one most intensively developed across each of the several days of prototyping work.

During prototyping, teachers are confronted with a situation where the clear-cut theory of textbooks clashes with the ambiguity of real-life practice. The teachers' diary entries described their experiences as, for example, "gaining new knowledge about the flow of electricity", "discovering all the phenomena occurring during an experiment with a cup, water and paper", "knowing when pressure decreases and when it increases", "knowing exactly what pressure is", and "understanding the flow of electricity and the behavior of electrons in a magnetic field". On the second day of the workshop there is a clear increase in the perceived development of teachers' own knowledge and understanding. The second day typically consists of the first focused explorations of phenomena that have been selected by the participants as topics for further work. By conducting experiments that ultimately help to define the main theme of the exhibit, teacher-designers encounter gaps in their existing knowledge and experience, and this experience stimulates intensive sense-making processes. The overall downward trend of diary entries about teacher knowledge and understanding in days 3 and 4 reflects the dynamics of prototyping work: from the discovery of ignorance during the first attempts to identify the essence of a given phenomenon, through a sense of dynamic knowledge acquisition during exploration and collaboration with the prototyping team, to the slow deceleration of this focus, when the exhibit takes on its final form. This three-stage process is closely related to the dynamics of the School's program.

The benefits for teacher-participants

Three particular points are worth noting here: the development of the interdisciplinary approach to science, the development of content knowledge within the process, and the development of teaching skills. The development of the interdisciplinary perspective of seeing natural phenomena is related to the team model of prototyping. The Summer Prototyping Schools are attended by teachers of science-block subjects (biology, chemistry, physics, geography, and mathematics), and so their university background may be in any of these fields. To a certain extent, the interdisciplinary nature of the School may explain the participants' sense of a sudden increase in knowledge. In working with their colleagues, participating teachers are naturally confronted with issues that stretch beyond the scope of the specific subjects they teach. This phenomenon is well demonstrated by the following example. When one team of participants, who designed a model of the heart, presented their prototype to be the broader group, one of the team members explained: *Our team includes a biologist and two physicists. We are working on an exhibit that shows how the heart works. Since it is a heart, our biologist will tell you more about it...* (statement by a teacher-participant).

The biologist talks about the model while the other team members support him by sharing technical remarks, such as adequately connecting the heart chambers' material to that of the aortas using hot glue. However, the group starts asking questions. How exactly does the electrical signal stimulate the heart contractions? How high is the pressure inside the ventricles? Who should answer these questions, the biologist or one of the physicists? While working together on the answers, the teachers begin to combine their scientific knowledge in a way that sheds new light on seemingly familiar phenomena and mechanisms. This process likely results in a sense of developing the competence of the teachers (Figure 8.2).

The development of content knowledge, which is associated with a broader, interdisciplinary perspective, involves intensive sense-making activity and is related to the development of scientific understanding.[14] Teacher-participants declare a clear sense of incremental knowledge acquisition and understanding

FIGURE 8.2 The case of an interdisciplinary approach: an exhibit prototype meant to illustrate changes in a riverbed under the influence of a changing water flow.

as they face the task of transposing theoretical concepts into material representations (exhibits). This process – which turned out to be a side effect central to participants' experience – involves teachers revising their conceptions of phenomena in experimental settings. While prototyping an exhibit, the participants are confronted with phenomena that their current knowledge does not allow them to interpret. Why is it, exactly, that when a lit candle is covered with an inverted glass and set floating in the water, it goes out, and the glass fills up to a constant level? Why is it that when a glass filled with water is covered with cardboard and quickly flipped over, the water does not spill? And so on. As teachers attempt to demonstrate a particular phenomenon and the conditions that allow it to be manipulated, they begin to notice the phenomenon's complexity. And it is precisely such complexity and ambiguity that eludes the clear-cut definitions and theories described in textbooks.[15] Prototyping thus becomes a cognitive practice.

The development of teaching skills relates to the process of shifting from pursuing a successful transposition of existing knowledge to intensive exploration of the phenomenon itself. The first step when planning a prototype is to pin down its scientific concept. Teacher-designers discuss in teams which phenomenon they plan to show in their exhibit and why. By creating a first prototype, they attempt to transpose this agreed-upon concept into a tangible object, which means striving to give the knowledge they possess into a material form, to transform that knowledge into an exhibit. The first prototype, however, usually reveals certain gaps and ambiguities in previously held concepts. The phenomenon intended to be presented often proves to be invisible, to be disturbed by unknown factors, or to proceed differently than assumed in theory. Simply put, the exhibit usually does not work. Subsequent iterations of the prototype are based primarily on exploration of the phenomenon: how can it be found in nature in its purest form, which can also be demonstrated in the form of an exhibit? What conditions of manipulation should be created so that student-users are able to discover the surprising, delightful, inspiring aspects of the phenomenon?

The prototyping process

The first prototypes are usually designed to show the students the phenomenon in a way that coincides with the textbook-given theory. The main task of the exhibit is thus to illustrate the established theory, or, to put it another way, to show that the theory manifests itself in practice. Jerome Bruner describes a sequence of representational acts (enactive, iconic, and symbolic) that parallels the process of cognitive development.[16] The *enactive* representation is embodied in actions of simulating a phenomenon in question (for example, seeing a real pig heart). The *iconic* representation, in turn, is constituted in actions of creating and operating on figural forms that are taken to stand in for that enacted performance (for example, watching how the heart works). The *symbolic* representation, in turn, reenacts the iconic activity, albeit with graphic-numerical signs of quantities (for example, seeing a schematic representation of the components of the heart).

Working on successive versions of the prototype usually has the teacher-designers returning to active exploration of the original, embodied form of the phenomenon. On the second and third day of the workshop, the participants create several successive versions of the same object, trying to find a perspective that will most successfully demonstrate the essence of the phenomenon to student-users; in this way the teachers act much like the professional design engineers working at the science center. For instance, if the topic is about air pressure, the teachers test out different variations that help to show what rising or falling air pressure means. One of the strategies teachers often employ is to draw upon phenomena that occur naturally in nature.

In this process of searching for a satisfactory way to physically represent a given phenomenon, each subsequent experiment – for example, attempting to create objects with water or with air, rapid construction of explanatory models – raises subsequent questions. Thus, such exploration consists in an intensive process of sense-making, at the core of which lies reconsidering the phenomenon itself, revising and expanding upon previously held concepts. Searching for answers, verifying hypotheses, and dispelling doubts lead to subsequent discoveries and gives participants a sense of development. In this way, the phenomenon, by its nature belonging to the world recognized by the senses, is rediscovered anew. Such discovery builds a memory trace of the experience and forms the basis for a deep understanding of the phenomenon and the mechanisms that allow it to be evoked and modified. This process is addressed in the extensive literature on the role of embodiment in learning.[17]

Teacher-designers, therefore, experiment and explore phenomena and processes while trying to work out an optimal form for their teaching aid/exhibit. The transition from searching for a successful transposition of knowledge into a material form to exploring the phenomenon itself opens up the possibility of formulating interesting premises for the exhibit. The most important outcome of this process is the experience of personal development of one's own understanding and knowledge, which ultimately gives teachers a more robust basis for working in the spirit of "exhibit pedagogy". If we attempt to simplify this complex process into a certain schematic model, it will consist of the following four stages: attempted transposition (knowledge → exhibit), exploration inspired by the outcome of this attempted transposition, re-transposition, and re-exploration. A schematic representation of this preliminarily proposed model is shown in Figure 8.3.

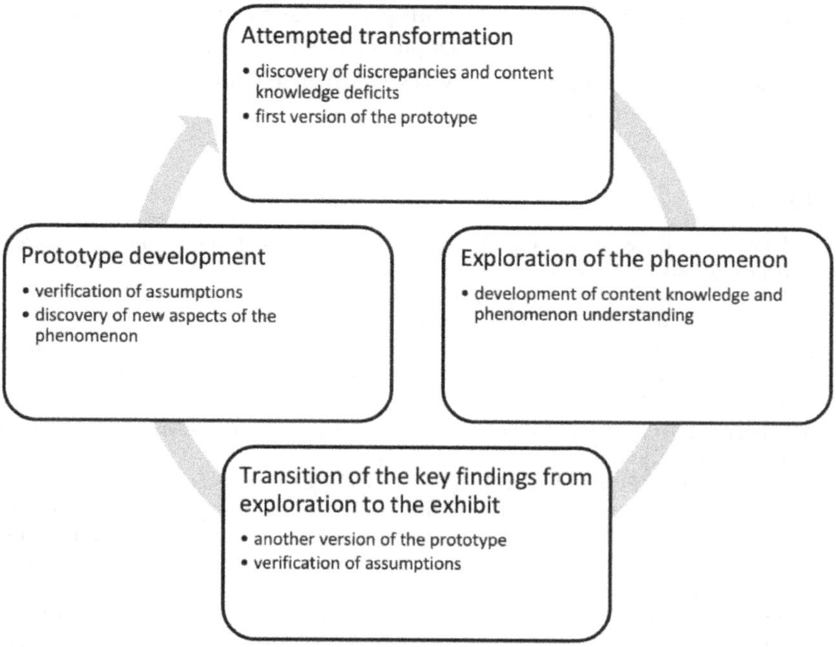

```
┌────────────────────────────────────┐
│ Attempted transformation           │
│ • discovery of discrepancies and content │
│   knowledge deficits               │
│ • first version of the prototype   │
└────────────────────────────────────┘
```

```
┌────────────────────────────┐      ┌────────────────────────────────┐
│ Prototype development      │      │ Exploration of the phenomenon  │
│ • verification of assumptions │    │ • development of content knowledge and │
│ • discovery of new aspects of the │ │   phenomenon understanding     │
│   phenomenon               │      │                                │
└────────────────────────────┘      └────────────────────────────────┘
```

```
┌────────────────────────────────────┐
│ Transition of the key findings from │
│ exploration to the exhibit          │
│ • another version of the prototype  │
│ • verification of assumptions       │
└────────────────────────────────────┘
```

FIGURE 8.3 Model of prototype development, from the perspective of teachers' cognitive processes.

Conclusions

Bringing teachers into the design process of developing exhibits has a number of advantages. One of the most important advantages is making the ideas generated in the science center more coherent with current teaching goals, ensuring that the educational solutions co-created are ultimately more relevant to the students' needs, and more broadly, incorporating "exhibit pedagogy" more extensively in the system of school education.

From the museum perspective, prototyping of exhibits is a cognitive exercise that enriches the designers' knowledge and understanding of the phenomena being addressed. The iterative process, whereby teachers as designers try to give a material form to their concept of a phenomenon and are then forced to reconsider that concept, leads them to discover deeper layers of the phenomenon itself. As we have noted, the key notion in "exhibit pedagogy" is the personal experience of discovery, which ultimately leads to the student-user forming the conceptual foundations of the phenomenon in question. In the constructivist theory of knowledge development, the existence of such foundations is a prerequisite for moving on to subsequent phases, including making generalizations, which allow definitions, theories, and patterns to be used competently as tools for describing and learning about the world.

The most important outcome of this process for teacher-designers is a profound experience of personal development of their own understanding and knowledge, which in the final analysis gives them a stronger basis for working in the spirit of "exhibit pedagogy".

Notes

1 Davidsson, E., & Jakobsson, A. (Eds.) (2012). *Understanding interactions at science centers and museums*. New York: Science & Business Media.
2 Falk, J. et al. (2012). Mapping the informal science education landscape: An exploratory study. *Public Understanding of Science, 21*(7), 865–874.
3 Sinatra, M. et al. (2014). Confidence in prior knowledge, self-efficacy, interest and prior knowledge: Influences on conceptual change. *Contemporary Educational Psychology, 39*(2), 164–174.
4 Oppenheimer F. (1968). Rationale for a science museum, *Curator: The Museum Journal, 1*(3), 206–209.
5 Oppenheimer, F. (1986). *Working Prototypes: Exhibit Design at the Exploratorium*. San Francisco: The Exploratorium. Renner, N. (2013). *Free to explore a museum: Embodied inquiry and multimodal expression of meaning*. ProQuest Dissertations Publishing. Allen, S. (2004). Designs for learning: Studying science museums exhibits that do more than entertain. *Science Education, 88*(Supplement 1), S17–S33.
6 Kullman, K. (2018). Perennial Prototypes: Designing science exhibits with John Dewey. In: Vermaas, Pieter E. and Vial, Stéphane (Eds.) *Advancements in the Philosophy of Design*. Design Research Foundations. New York: Springer, pp. 185–199.
7 Fiesser, L. (1994). PHÄNOMENTA — durch Physik Denken lernen? *Physikalische Blätter, 50*, 678–680.
8 Iłowiecka-Tańska, I., Gop, A., & Jaskulska, S. (2021). Designing teaching aids: The transformative value of the process. Teachers' narratives of their own learning during onsite and remote prototyping summer schools. *Studia Edukacyjne, 62*, 2021, Poznań 2021, 169–186.
9 Bruner, J. S. (1966). *Toward a Theory of Instruction*, Cambridge: Harvard University Press. Davidsson, E., & Jakobsson, A. (Eds.). (2012). *Understanding Interactions at Science Centers and Museums*. New York: Springer Science & Business Media.
10 Hyers, L. (2018). *Diary methods: Understanding Qualitative Research*. Oxford: Oxford University Press.
11 Harvey L. (2011). Intimate reflections: Private diaries in qualitative research, *Qualitative Research, 11*(6), 664–682.
12 Mortensen, M. F. (2010). Museographic transposition: The development of a museum exhibit on animal adaptations to darkness. *Éducation et didactique, 4*(1), 115–138.
13 From: Iłowiecka-Tańska et al. (2021). Fig. 1.
14 Lederman, N. G. (1992). Students' and teachers' conceptions of the nature of science: A review of the research. *Journal of Research in Science Teaching, 29*(4), 331–359.
15 Lederman, N. G. (1992)
16 Bruner, J. S. (1966). *Toward a Theory of Instruction*. Cambridge: Harvard University Press. Davidsson, E., & Jakobsson, A. (Eds.). (2012). *Understanding Interactions at Science Centers and Museums*. New York: Springer Science & Business Media.
17 Nemirovsky, R., Kelton, M. L., & Rhodehamel, B. (2013). Playing mathematical instruments: Emerging perceptuomotor integration with an interactive mathematics exhibit. *Journal for Research in Mathematics Education, 44*(2), 372–415.

PART II
Engaged with Earth Stewardship

9

SALMON CAMP

The Design of an Immersive Science Education Program for Native American Youth

Jeffry Gottfried

The original Salmon Camp blueprint

The Oregon Museum of Science and Industry (OMSI) and the American Indian Science and Engineering Society (AISES) collaborated in creating Salmon Camp, a unique traveling camp for Native American students following the route of migrating young salmon from Cascade Mountain spawning streams to the Pacific Ocean. Along the way, Salmon Campers experienced and studied the varied freshwater environments through which salmon travel. They were hosted by tribal elders, participated in salmon ceremonies, and collected environmental data under the supervision of fish biologists, geologists, foresters and range managers. Other salmon-related groups that interact with Salmon Camp include tribal and non-tribal fishermen, irrigators, dam operators, and the Columbia River Intertribal Fish Commission (CRITFC).

Salmon Camp has been continuously engaging Native American youth in the Columbia River watershed since the early 1990s. How has this camp survived and thrived over the years so that Salmon Campers, now parents, have sent their kids to Salmon Camp? How is it that this improbable intersection of traditional indigenous knowledge and tradition, Western science, career exploration, place-based learning, and outdoor adventure came to be? What generalizable lessons were learned that may be useful to informal science educators who endeavor to create successful programs that target underserved students? On the way to answering these questions, I'll describe Salmon Camp's first iteration, how it originated and how it evolved.

On the first day of Salmon Camp in 1992, 25 Native American middle school students boarded a bus at OMSI in Portland, along with their instructors and counselors, mostly Native American science students and graduates

DOI: 10.4324/9781003145387-11

and OMSI Science Camp staff. They began a five-day adventure of cultural, ecological, and geographic experiences, all in some way tied to the theme of the life cycle of Pacific salmon. As they left Portland and drove east out of the Willamette Valley, the Salmon Campers started their drive up and over the Cascade Mountains. During this time, their instructors challenged the students to notice how and when the dominant species of trees changed as the bus gained in elevation, climbed over the pass, and descended into Warm Springs Indian Reservation located in the high desert in the rain shadow of the Cascades. Instructors took advantage of the time on the bus for science, as well as Native cultural leaning. Many former Salmon Campers remarked that the journey itself was an eye-opener for them, having never previously traveled so extensively in Oregon and experienced its diverse landscapes and geography.

Salmon Campers were treated as honored guests by the Confederated Tribes of Warm Springs at their Hee-Hee Butte Ceremonial Grounds, not just because the participants and staff included local tribal members, but also because the meeting was seen as an intertribal gathering. (As an example of this diversity, the Portland Public Schools include native students from approximately 100 different tribes.) We were given full access to the Hee-Hee Butte Longhouse and ceremonial grounds along the Warm Springs River, where we set up our tents and shelters on land that was otherwise reserved for tribal members only. For the next two days, Salmon Campers and staff figuratively and literally immersed themselves in the study of the ecology of the Warm Springs River, a spawning stream for Chinook salmon that had made their journey from points west in the Pacific Ocean to the mouth of the Columbia River and then about 250 miles upstream and over above rapids and waterfalls of the Deschutes River to their spawning beds in the Warm Springs River.

While at Warm Springs, Salmon Campers collected and learned to identify aquatic insects and other macro-invertebrates. They also gained an appreciation of the complex life cycles of these organisms that are the food-source for juvenile salmon during their maturation in freshwater. They netted samples of juvenile Chinook salmon and rainbow trout/steelhead, placed them in an aerated aquarium and learned to identify one from the other. They observed and learned about the dynamics and importance of riparian (streamside) forests and their importance to salmon. They observed traditional Native net fishermen fishing from a platform over a waterfall on the nearby Deschutes River. They observed petroglyphs and pictographs accompanied by Warm Springs cultural specialists. Finally, the campers participated in a traditional Warm Springs salmon bake, after having received detailed instructions and stern warnings from tribal elders who performed the Salmon Ceremony regarding the sacredness of a salmon bake, and the proper respect to be shown to the salmon as well as to the tribal elders.

Upon leaving Hee-Hee Butte and the Confederated Tribes of Warm Springs Reservation, Salmon Camp traveled by bus following the Deschutes River for a distance and then drove to their second camping spot, on the Washington side of the Columbia Gorge, where they were guests of the CRITFC and the Yakama Nation.

CRITFC, as the name implies, is an organization formed by all of the federally recognized "Treaty Tribes" living along the Columbia River. They include the Warm Springs Tribes, Yakama Nation, Nez Perce Tribe and Umatilla Tribes. Youth from each of these tribes and many more attended Salmon Camp over the years. CRITFC was formed in 1977 after these tribes successfully sued the federal government and re-established their rights to fish and, at least, be a partner in the management of salmon populations. As each Tribe was empowered by the courts to fish and manage fish, it became clear that singly they lacked the resources to create the infrastructure to do this important work. As a result, they did the hard work of overcoming cultural and political obstacles creating CRITFC, an inter-tribal team of scientists, technicians, law-enforcement agents, administrators, and attorneys, to insure that native voices, viewpoints, traditions, customs and especially fishing rights were respected and a part of the decision-making process. It was wonderful that CRITFC had embraced Salmon Camp and worked diligently to see to it that we were hosted at one of their Columbia River fishing sites, where we camped, observed and interviewed Indian fishermen, learned from fish biologists and patrolled the river with fisheries law-enforcement officers.

This stop on the Salmon Camp journey was an opportunity to see Native people in positions of authority in matters of scientific and natural resource management, as well as law enforcement. After one overnight and two days, the bus was off to the Confederated Tribes of Siletz Indians, along the Siletz River, about 20 miles upstream from the Pacific. Once again, Salmon Camp received an especially warm welcome from the Confederated Tribes of Siletz Indians, in large part because Dave Hatch, president of AISES in Portland and Salmon Camp cofounder, was an enrolled member of the Confederated Tribes of Siletz Indians and later went on to serve on the Tribal Council. A special opportunity offered to us at Siletz was a series of cultural/historical presentations and by Tribal Cultural Historian Robert Kenta, who shared many cultural objects with us and taught us about the life ways and histories of the many tribes that had been forced off of their lands and brought together from as far away as California, the Willamette Valley, the Columbia River and Washington State and made to live at Siletz, along with the people who originally lived there.

As visitors, new to much of this information, our Salmon Campers and staff could only have imagined what a feat of communication, cooperation, adaptation and diplomacy it must have taken for diverse peoples to have been forced to live together while their traditional lands were taken and given to white settlers. These were very tough and brutal realities that were shocking and sobering to many of our Salmon Campers and staff.

Being a few miles above tidewater on the Siletz River gave Salmon Campers opportunities to continue their study of the many environments inhabited by salmon on their migrations from Ocean to spawning stream and back. Along the lower Siletz River, they viewed the last 20 miles of river and its entry into, and mixing with, the in-coming tide. It was in this rich mix of salt and freshwater where young salmon eat profusely on aquatic insects and other invertebrates and prepare to leave the river and go to sea, where they grow rapidly into sizable adults.

Although the Salmon Camp journey had not followed the actual migration of a single salmon run at camp, it had afforded us the opportunity to experience and study each habitat type, from spawning tributaries over waterfalls and rapids to main-stem rivers to an estuary and finally the Pacific Ocean. On the last day, Salmon Campers and staff boarded an ocean-going boat, what usually served as an educational whale-watching cruise, and motored out into an estuary and then across the bar, into the open ocean, thus completing the freshwater journey of the salmon. Along our cruise, we towed plankton nets, took samples of the bay and ocean bottoms and examined and identified the species in our samples using microscopes and hand lenses.

The richness and diversity of the day-to-day experiences of Salmon Camp was magical and key to this life-changing experience. The program described above was more or less the experience of campers for about ten years. Fortunately, we had talented and knowledgeable tribal fishermen and elders, fisheries biologists, hydrologists, geologists, from CRITFC, tribes, Oregon Department of Fish and Wildlife, USDA Forest Service and others, lining up to volunteer each year, so we never lacked for talent, information or inspiration.

Salmon Camp science club

No sooner had the first session of Salmon Camp ended, than Dave Hatch and the membership of the Portland AISES Professional Chapter expressed strong interest in taking the lead in conducting a monthly Salmon Camp follow-up program, what came to be known as "Salmon Camp Science Club." Once a month on a Thursday night, cars and pick-up trucks full of extended Indian families rolled into the OMSI parking lot. Some came from the Portland metropolitan area; others had driven over 120 miles that afternoon from Warm Springs, Siletz, or Yakima to OMSI in Portland. At OMSI, families, including parents, grandparents, younger children, aunts, uncles, and cousins, had free run of the Museum's exhibit halls and theaters on this night when the museum was open late, while Salmon Campers joined camp staff and member of AISES in science activities, demonstrations and science-math-engineering career presentations given by inspiring Native role models. This monthly meeting took place until 2000. At that time, Salmon Camp Science Club evolved into a series of weekend-long field trips to Indian reservations to participate in tribal land and water conservation and restoration projects where they interacted with tribal land managers, scientists, and students like themselves.

Salmon Camp version #2

Salmon Camp, as described above, took place each summer from 1992 to 2003, with OMSI and AISES as its two driving forces. In 2003, OMSI also received National Science Foundation funding for "Salmon Camp Research Teams," a summer high school science camp for Native American high school students that was meant to teach the uses of electronic devices such as laptop computers, global positioning systems (GPS) and data loggers in research. This ambitious undertaking was designed with little input from the Indian Community. Building upon 13 years of success of Salmon Camp, OMSI decided to (1) use the name "Salmon Camp" in its proposed "Salmon Camp Research Teams," even though the two programs were very different from one another, (2) include the funding for the original middle school Salmon Camp in the proposal, thereby ending the ten-year relationship with the group of funders who had each been supporting a portion of the program and placing the future of Salmon Camp in the hands of the National Science Foundation (NSF) and (3) end its close relationship with AISES and the Indian leadership of Salmon Camp. The decision was made for OMSI to go it alone as the recipient of the NSF grant, thereby making the Salmon Camp's middle school program a much more western and less Native cultural experience with few, if any, Native staff in visible, decision-making roles.

New OMSI leadership with no direct history of having experienced Salmon Camp chose, for reasons of income and efficiency, to increase the number of campers in the middle school Salmon Camp to 30, thereby making it logistically impossible for Salmon Camp to continue to be a traveling camp that followed the migratory route of salmon. Instead, middle school Salmon Camp became a residential camp based at OMSI's Hancock Field Station, a camp that like other of OMSI's excellent camps, introduced campers to the plants, animals, geology, and paleontology of Oregon's High Desert. The only difference was that this camp was limited to Indian middle-schoolers.

Salmon Camp version #3

Grants from the National Science Foundation had provided OMSI with increased funding in the short run but they are not intended for ongoing support of excellent programs. The inevitable happened: OMSI lost its funding for both its Salmon Camp Research Teams as well as the original Middle School Salmon Camp. Fortunately, Oregon's tribes, schools, and communal institutions had grown to value and become accustomed to the availability of

Salmon Camp for their youth. The Columbia Intertribal Fish Commission and individual tribal natural resources departments had come to view Salmon Camp as a part of their educational and employment pipeline for prospective fish and wildlife managers, researchers and technicians. Influential employees and leaders of CRITFC had sent their kids to Salmon Camp and had been impressed by its influence on their kids. When word came from OMSI that NSF funding for Salmon Camp had come to an end, yet another somewhat random meeting of two individuals saved the day for Salmon Camp.

At the Closing Ceremony for what was to be the last session ever of Salmon Camp, Dan Calvert, OMSI Camps Coordinator and passionate Salmon Camp advocate, just happened to run into Chuck Hudson, a parent of a camper who was also a biologist for CRITFC, the very person to whom CRITFC had delegated as their liaison to Salmon Camp. When Dan expressed his remorse at the demise of Salmon Camp, it inspired Chuck Hudson into action. In short order, he met with leadership at CRITFC and began making plans to continue Salmon Camp under CRITFC leadership.

The new CRITFC Salmon was different in a few very significant ways. OMSI-AISES Salmon Camp was made available to all students who identified as Native American. The new CRITFC Salmon Camp became available only to middle-schoolers from those treaty tribes that were members of CRITFC. No longer could native students from tribes native to lands outside the region or even within the region (but not members) be eligible to become Salmon Campers. The format of Salmon Camp also changed once again so as to best use the resources at hand. From 2009 to the present, each of the CRITFC member tribes takes turns hosting Salmon Camp on their historical lands, focusing on their waters and salmon-management challenges and programs. The new CRITFC program was superior to the OMSI program in that a large, Native-run organization of fish biologists, engineers, computer scientists, economists, etc., was now conducting the camp and teaching those subjects that it wanted future employees to know about.

Salmon Camp's origin story

How did Salmon Camp get started? Without addressing this question I would not be doing proper service to my readers, especially those who might wish to design and implement their own culturally appropriate science education programs. As cofounder of Salmon Camp and former Vice President for Science Education at OMSI, I consider my proudest and most rewarding experience of my professional career was the creation of Salmon Camp. This creation resulted from my close, synergistic partnership with Dave Hatch, Member, Confederated Tribes of Siletz Indians and Professional Engineer (PE).

One day in 1991, shortly after arriving in Portland from Albuquerque (where I helped to create the New Mexico Museum of Natural History and had worked closely with a number of Pueblos, Navajos and Apache educators) a strikingly handsome Native American with a commanding presence walked into my office. He introduced himself as Dave Hatch, President of AISES, Portland Professional Chapter. He told me that he was delegated by his membership to meet with the new guy at OMSI and let him know that AISES thought that OMSI was not doing enough to encourage and include Indian young people in science, math, and engineering.

I was very surprised by this bold approach and heartfelt plea to improve our record. I told Dave that I had just arrived at OMSI and knew of no program or concerted effort to engage Native students in science and technology. I did take the opportunity to describe to him my work as Director of the New Mexico Rural Science Education Project in cooperating with Native tribes to engage their youth in place-based science. I said that I was interested in talking more with him and asked him what he and AISES were willing to bring to the table in order to help OMSI better serve Native students. This unscheduled, bold meeting between the two of us got the ball rolling. I could not sleep that night thinking about Dave Hatch and remembering his fiery eyes and passionate to-the-point presentation to me. I was hooked on the idea of starting an exciting new program that would involve a true partnership with Native people and tribes, as I had done in New Mexico, breaking new ground for OMSI as well as the newly recognized field of "informal science education."

As a naturalist, environmental educator, and passionate fly fisher, my return to Oregon meant to me a return to Salmon Country, the land of rivers, spectacular watersheds, and the profuse lives and ecosystems that they supported. When Dave and I met the next week, I proposed that we create "Salmon Camp" for Native middle school aged kids. We next had to fill in the many blank spots that would define what this actually meant to our organizations, to tribes and Native people. Subsequent meetings led to a proposal to the National Science Foundation who rejected it. Reading between the lines, it was clear that reviewers had no apparent appreciation of Indian science, traditional knowledge of salmon and the many species of other animals that depended upon the salmon. A month after our rejection from NSF, Dave Hatch was invited to join the Board of the Howard Vollum Fund, an educational foundation funded by the bequest the founder of Tektronix, the inventors of the oscilloscope. In support of Dave, the Vollum Fund gave a grant to Salmon Camp each year for its first five years. Soon other institutions – the Bonneville Power Administration, the Bureau of Indian Affairs, banks and other businesses – added additional funding. So financially we were in good shape, with an excellent chance of economic sustainability due to all the interest.

Both AISES and OMSI recruited staff. AISES recruited Native science and technology students from Oregon Universities, assuring that they had both science and cultural knowledge and skills. OMSI hired science educators with specific skills in first aid, child supervision, bus driving, cooking and camping logistics. This provided a very effective team to work on behalf of our campers. However, even the best laid plans sometimes can go. Despite all of attempts of AISES and OMSI to hire culturally sensitive and skillful staff, our Salon Camp staff unintentionally committed at least two notable "cultural blunders."

At the Confederated Tribes of Warm Springs one evening, three Salmon Campers, all Warm Springs tribal members, invited our three Native women staff and a few female campers to take a dip in the warm springs. This seemed like a unique opportunity, and so they walked to the springs and took a dip. What none of our staff realized was that these springs were strictly reserved for Enrolled Members of the Confederated Tribes of Warm Springs. Other bathers objected and called both the Tribal Police and the Federal Marshals who arrested our female staff. They were give summons to appear in Tribal Court on the next Monday, the start of the next session of Salmon Camp. Thankfully, our hosts in the Warm Springs Natural Resources Department intervened with the judge and got the cases dismissed on the Sunday before their scheduled court appearances.

At the Confederated Tribes of Siletz Indians, our Native staff made another cultural *faux pas*. They organized a night hike so that campers could focus on and learn about the sounds of the night. Everyone was apparently having a good time, listing to night sounds. But after camp was over, I received a call from an angry Warm Springs parent who informed me that we had taken her daughters near a cemetery at night, where they had become possessed by spirits. In order to return their daughters to a normal state, the parents, at considerable cost, had hired a medicine man to perform a ritual that would rid them of their spell.

Without Dave Hatch and AISES as "owners" of Salmon Camp, it would not have been possible for the camp to be the success that it was. AISES, the tribal coownership of Salmon Camp, and the participation of OMSI, all contributed to make the program a unique experience for Indian middle school students. AISES members from the Portland Professional Chapter came from many different tribes, and they included scientists, engineers, mathematicians, computer scientists and natural resource managers. These Tribal STEM experts were accustomed to operating and thinking in two "worlds," the world of Western science and the world of traditional native science. They were very generous in providing insider contacts with tribal scientists, governmental agencies including the CRITFC. They were also instrumental in recruiting outstanding Indian staff for camp. All Native staff at Salmon Camp independently reported that personal invitations from AISES members led them to apply for teaching positions. AISES members also volunteered to lead Salmon Camp following-up programs like Salmon Camp Science Club and Salmon Camp Science Fair.

As a complement to AISES OMSI, with its 30+ years of experience in leading field-based science camps, made other contributions: the logistical aspects of safely and professionally transporting, feeding, supervising, sheltering and teaching youth in the wilds of Oregon. These contributions were ingrained in the culture of OMSI's. Science Camps Department, led by Camp Director Joseph Jones, himself a former OMSI science camper, research team participant and instructor. Combining the collective resources of these two resource-rich organizations launched a unique and successful educational and cultural experience for Native American middle school students.

In retrospect, our initial lack of dependence upon the NSF was a blessing in that we were not constrained to a set of generic goals or "deliverables" set by folks in Washington, DC. We were free to integrate Indigenous Ways of Knowing and Western science, something that at that time was not fully appreciated by the NSF. Neither were we creating an innovative program that would be ending in a few years when the focus of NSF funding availability changed. From the start, we were in it for the long-term, providing that Salmon Camp would prove to be valued by the tribes and Native families in our region.

Dave Hatch and I remained intimately involved in Salmon Camp from 1991 to 2002 including recruiting and hiring staff, coordinating involvement from Indian tribes and organizations. The day-to-day logistics, such as vehicles, drivers, food, safety and first aid were all expertly handled by OMSI Camps Director, Joseph Jones. We each poured our own knowledge, world views, resources and connections into Salmon Camp for the first decade of its existence, at which time it was left to other people with different ideas, values, and definitions of success.

Lessons for informal science educators

Our story of the history, design, and iterations of Salmon Camp suggests lessons that may be useful to informal science educators who want to create successful programs that aimed to engage specific ethnic or cultural groups:

> *Science educators who are not personally experienced in the community they want to serve must first develop excellent relationships with that community.* This is the most important lesson of Salmon Camp. These relationships, based upon mutual respect and program coownership, represent a critical starting point. It is not sufficient to write an educationally excellent proposal and then seek the endorsement of your program from under-served communities afterwards, no matter how excellent you think the program will be.
> *Be aware of the diversity of beliefs, values, customs, taboos, and life ways that exist within each ethnic, racial, language and/or religious group.* Recall the "cultural" blunders mentioned above. Do the best that you can to staff appropriately but be prepared to address diverse needs, customs,

and values and also to learn and respond "on the spot." Be observant of your students and invite ongoing critiques, criticism, and suggestions from your students and their parents.

The design of the program should use the unique capabilities, connections, power of each partner, and the flexibility of informal science education. This flexibility can be used by focusing on what surrounds the participants, i.e., the plants, animals, weather, water, soil, rock formations, and the like. Leave lectures to the classroom. Leave lots of time for participants to interact with the immediate environment, encouraging questions and allowing students to participate actively in the discovery and interpretation of the study site. Remember that verbs (observing, discovering, questioning, drawing, interpreting, mapping, measuring, etc.) are at least as important as nouns, the topics of study.

Collaborate with partner organizations to recruit staff from the community that you seek to serve. Ideally, the day-to-day leadership of the program should be in the hands of people who come from a similar background to that of participants. Participants who look up to their teachers and identify with them are more likely to be inspired and to remain inspired long after the program is over, when they can see someone from their community as a respected professional.

Build a team of instructors who enjoy and feel comfortable working in the field, with the participants. Hire and or train instructors who are perceptive of nature and have the ability to ask challenging and perceptive questions, as opposed to simply providing answers. Reward and encourage the asking of questions. Create a standard that says the only stupid question is the one that you didn't ask.

Based on these guidelines, and a great deal of dedicated work, Salmon Camp became an immersive, culturally rich, and long-running science education experience for Native American youth. I believe that these same guidelines can help spawn similar programs designed to serve the specific needs and aspirations of other ethnic or cultural groups.

Acknowledgments

In loving memory of Dave Hatch. Dave Hatch was an enrolled Member, Confederated Tribes of Siletz Indians, Professional Engineer, and former President of the American Indian Science and Engineering Society (AISES). He was a cocreator of the original Salmon Camp at OMSI.

10

CULTIVATING SUSTAINABILITY

Youth Food Systems & STEM Learning

Eileen G. Harrington

Food encompasses several science fields. Agriculture involves agronomy, soil microbiology, animal husbandry, veterinary medicine, and plant genetics. It includes everything from engineering and computer science to chemistry, ecology, and environmental science. Food also encompasses health sciences, including nutrition and public health. Food systems intertwine with and affect human and environmental health, and they can reinforce social and economic inequities. Issues related to food systems include climate change, habitat loss, pollution, nutrition, food security, food waste, animal welfare, and workers' rights.

This essay explores the relationship of food education programs to STEM learning in informal settings. Focusing specifically on K-12 and family programs, it provides an overview of the landscape: where and how learning occurs and its benefits and challenges. The final section outlines common approaches to food system learning that help make them stronger and hopefully lead to more resilient and sustainable communities.

Food pedagogy

Food pedagogy remains a relatively new field, and researchers and practitioners are still defining exactly what it means to be food literate. Several researchers have developed various definitions and frameworks to help guide the development of learning objectives, program goals, and evaluation and assessment efforts.[1] Goldstein outlines two approaches to food literacy: the "neoliberal consciousness model," which "maintains a focus on individual consumer responsibility and technical skills to make healthy choices within the current market" and the "critical consciousness model," which "encourages systems thinking and pursues active engagement with the food system to disrupt problematic

DOI: 10.4324/9781003145387-12

practices."[2] The first often leads to more measurable outcomes, while the second is likely necessary to bring about genuine transformative social, economic, and environmental change. Sumner, a leader in the food pedagogy field, drawing on the work of critical educators, provides a definition of food literacy that aligns with Goldstein's critical consciousness model:

> Food literacy is the ability to 'read the world' in terms of food, thereby recreating it and remaking ourselves. It involves a full-cycle understanding of food—where it is grown, how it is produced, who benefits and who loses when it is purchased, who can access it (and who can't), and where it goes when we are finished with it. It includes an appreciation of the cultural significance of food, the capacity to prepare healthy meals and make healthy decisions, and the recognition of the environmental, social, economic, cultural, and political implications of those decisions.[3]

Many food education programs focus on skills and knowledge, such as how to select and prepare healthy food, without considering the broader context. The same could be said about many science programs that emphasize imparting and understanding scientific concepts and processes, while ignoring issues around race, class, and economics.[4] This approach centers on a Western view of science as being neutral and apolitical and excludes other forms of knowledge and perspectives for examining environmental or health-related issues. Concepts of food justice and food sovereignty can bring the social and economic aspects back into this learning environment, as well as Indigenous ecological knowledge.[5] This chapter focuses on STEM learning that also brings in social and economic aspects to achieve stronger and more authentic sustainable education programs.

Learning through shopping and cooking

Food systems provide fertile ground for developing science and math skills. Often chemistry and physics concepts are introduced through cooking, such as Harvard University's open online course "Science & Cooking: From Haute Cuisine to Soft Matter Science."[6] Youth can explore engineering concepts while designing solar cookers or creating biodiesel from used cooking oils.[7] Physiology and genetics concepts emerge when looking at how different components of food, such as sugar, caffeine, and salt, affect our health and mood. Scientists at the Genetics of Taste Lab located within the Denver Museum of Nature & Science engage members of the public in research on taste perceptions.[8] Microbiology bubbles up when exploring food safety, food processing, and fermentation.[9] Math underlies all of these activities and is also reinforced when calculating daily recommended consumption of calories or nutrients; examining food miles of various products; or measuring amounts for recipes.

The foundation for fostering STEM literacy starts through interactions between children and their caregivers and often involves food. In everyday experiences, such as shopping, children and their caregivers engage in math activities, such as counting, measuring, spatial thinking, and noticing patterns or shapes.[10] The Playful Learning Landscapes Action Network seeks to capitalize on this by sparking engagement and enriching child-caregiver interactions through guided play in urban public spaces.[11] One of their initiatives brings a museum environment to the supermarket, particularly in under-resourced neighborhoods, by adding signs to spark conversations. Signage was developed, sometimes in multiple languages, in collaboration with community members, making it more locally and culturally relevant.[12] Researchers found an increase in the number and quality of child-caregiver interactions in stores that had the added signage compared to those without any.[13] Many children's museums have supermarket or cooking exhibits providing opportunities to incorporate STEM learning. Often these, like the Learning Landscapes initiatives, focus more on developing math skills than science.[14] Opportunities exist, therefore, to incorporate early science concepts such as comparing, describing, and observing or delving into the origins of common foods.

Also tapping into everyday learning, researchers and museum educators involved in the NSF-funded project "Food for Thought: Igniting, Engaging, and Measuring Family STEM Learning Using a Food Lab" recognize that family cooking can foster hands-on, collaborative informal STEM learning, while also incorporating cultural practices and intergenerational sharing of knowledge.[15] This ongoing project, which started in 2019, includes collaborators from Kent State University, the Cincinnati Museum Center, La Soupe (a food justice non-profit), and the Center of Science and Industry (COSI). The team has created short videos and engagement prompts for families to use, demonstrating how to cook various foods and the science behind them. Eating is a daily activity we all engage in so through the project, they seek to make STEM learning more accessible and inclusive for families that might not normally engage in STEM activities or come to museums. Evaluation of the activities is occurring using surveys, online activities, and audio-recording families engaging in activities using an app.[16]

Another program, Kitchen Science Investigators, out of Georgia Tech also used cooking and science to allow middle school youth to see the relevance of science in their everyday lives and implement their science knowledge in novel situations.[17] Held as part of an existing afterschool program at a YWCA, various characteristics led to the program's success at helping students better understand the scientific process and develop their science self-efficacy. These included scaffolding activities, close relationships between the facilitators and participants, and the presence of various artifacts (e.g., lab equipment, various types of ingredients) to help guide the youth to draw on their past experiences and bring them into current learning, as well as allowing for spontaneous experiments.[18] This differs from the home or museum environment where all of these components likely will not be available.

Some programs incorporate traditional means of obtaining food, while also exploring science concepts and practices. One example is the Winterberry citizen science project led by University of Alaska Fairbanks researchers, which examines the impact of climate change on four species of berries that have cultural and subsistence importance throughout Alaska and that flower in the fall and winter.[19] After receiving training, youth groups in both informal and formal settings throughout Alaska adopted 20 or more plants with at least 100 berries, observing and recording the abundance and condition of the berries in fall and spring. They also measured snow pack depth each month. The multidisciplinary program team created culturally responsive and adaptable lesson plans, some with materials in Indigenous languages, so that youth group leaders could incorporate other activities along with monitoring.[20] The researchers also tested the use of storytelling throughout the program as a way to connect the science to their everyday lives, incorporate knowledge youth bring to learning, and explore the youths' roles in possible positive futures in the face of climate change. Youth gained first-hand experience with the scientific process: gathering data, analyzing their own data from other sites in "data jam sessions," and presenting the results at professional and community meetings.[21] The close involvement with scientists throughout the project helped make the youth feel like they were conducting authentic science, and data from the project have been published in a scientific journal.[22]

Learning through cultivation

A large part of food education occurs in community gardens or urban agriculture programs. Life Lab in Santa Cruz, CA, is one of the longest-running programs to promote learning in gardens. They have created activity books and curriculum guides aligned with the Next Generation Science Standards (NGSS) that cover a variety of science concepts.[23] In gardens, youth can explore a variety of STEM topics including plant and soil science, meteorology, ecology, engineering, and math.

Some youth urban agriculture programs seek to increase interest in pursuing STEM degrees and careers among students of color, such as a hydroponics program collaboratively developed and implemented by academics, community organizers, and educators.[24] The program seeks to make STEM more relevant to youth, while learning about food systems in cities. They created a modular curriculum that aligns with the NGSS so that it can easily be incorporated into what teachers are already doing in or out of the classroom and adapted to different contexts.[25] Educators can attend a summer institute that introduces the curriculum and how to set up the hydroponics system. Another initiative from this group, Urban HydroFarmers, for students of color from three Boston high schools includes STEM learning and career readiness. In a greenhouse and science lab at Boston College, students develop their botany, engineering, and computer science skills by learning about hydroponics, setting up their own system, and tweaking it to improve productivity. One cohort of students worked with middle school students with moderate special needs to assist them in constructing a hydroponic system that included a solar array to power it, reducing its environmental impact. Produce grown from the system set up at this school was donated to a nonprofit that runs a job-training program for homeless individuals to move into the food service industry.

Several positive outcomes grew out of these initiatives. They increased interest among underrepresented groups in STEM fields. For youth from recent immigrant families from agricultural regions, the program provided opportunities to share their STEM learning with other family members around familiar topics. Also, for those students with special needs, it helped them realize that they can develop and implement scientific experiments on their own. In a study of 234 elementary school students who participated in the afterschool hydroponics program, they found a significant increase in girls' self-concept of seeing themselves as scientists.[26] Hydroponics provides the opportunity to incorporate different aspects of STEM learning, while fostering science self-efficacy: "Participants engage in engineering design and problem solving through building the systems, learn chemistry and biology content to maintain the nutrient and pH levels that sustain plant growth, and use math skills and scientific practices to design, conduct, analyze, and share experiments involving their plants."[27] Finally, hydroponics represents an attractive alternative to community gardens since it can be done in a smaller, more controlled environment, does not depend on local growing seasons, and can be more closely linked to other afterschool activities at the site.

Other programs also incorporate peer teaching, combining youth development with science concepts.[28] In Crop City, a garden-based youth development program for middle school students, high school students go through a training program to lead the activities. Involvement in the training program fueled a passion in the high school students for wanting to teach other youth and solidified their own learning.[29] Youth can also become involved in other aspects of program delivery and evaluation. The high school educators for Crop City come from a related project-based internship program developed by the same organization, and the youth conducted a peer-to-peer evaluation of the internship program. Having the youth become more involved in conducting the evaluation, "…was a means to accessing the elements of the program that were important to them and to each other as participants, rather than those elements that were only known or valued by those in charge of the internship and its evaluation."[30]

Participation in youth garden programs that take a project-based approach helps develop a sense of agency among youth and counters the sense of doom that can emerge out of some food education programs. The Garden Mosaics exemplified this approach by integrating garden-based science learning with intergenerational mentoring, community-based projects, and developing cultural humility.[31] The program developed by researchers at Cornell University was implemented at various sites, mostly community gardens, throughout the US and South Africa.[32] Incorporating Participatory Rural Appraisal (PRA), techniques often used in sustainable development, youth conducted a community garden inventory and collected oral histories of elder community gardeners to gain a deeper sense of plant choices, gardening practices, and their relationship to the gardeners' cultures. For many immigrants, community gardens are a space to integrate into their new country, while also maintaining links to their heritage and preserving traditional ecological knowledge by growing particular produce, not available in local stores. Youth and gardeners also collaboratively created a community map exploring other sources besides the community garden for fresh fruits and vegetables, recreation, social and cultural engagement, and experiences with nature. Data from oral histories and community garden inventories at each site were added to databases to collect information on culturally relevant and environmentally friendly gardening practices, in a form of citizen science. When evaluating the program, researchers found that the PRA techniques promoted positive interactions among educators, youth, and gardeners and fostered learning among "participants varying in age, ethnicity, motivation, and academic background and ability."[33]

Youth also added to another citizen science project out of Cornell, Weed Watchers, as they gathered data about weeds in their community garden in order to help scientists develop eco-friendly solutions to controlling weeds in urban gardens. Another common science-related activity was soil testing in the garden since many urban gardens are located on reclaimed land that might have soil contamination and/or might not have all the nutrients needed to help plants thrive. After gathering all of their data from the various activities, youth developed action projects to address a need in their community. Projects included creating a new garden on an abandoned lot, generating compost and sharing it with other gardens, and donating produce grown in a garden to a women's shelter. Participating in the Garden Mosaics program allowed students to explore not just the science concepts related to gardening, but also learn more about the cultural, social, and economic aspects of food systems. As two Garden Mosaics leaders point out: "Thus, community gardening provides opportunities for learning that address multiple societal goals, including creating a populace that is scientifically literate, that practices resource stewardship, and that is engaged in civic life."[34]

Concluding thoughts

This chapter provided an overview of the landscape of youth food systems and STEM learning in informal settings, but it is by no means exhaustive. Time and money constraints might limit formal evaluations and/or writing reports or articles to share with the broader informal science community. A recent survey of youth urban agriculture programs in 35 of the most populated cities in the US found only 40 organizations that integrated STEM learning into their activities.[35] This survey relied more on internet searches than literature databases, while for my review I focused more on databases. Only three of the 40 organizations identified came up in my search of databases.

Because of the variety of approaches, learning outcomes, and contexts presented in this chapter, it would be difficult to state which were the most impactful. Reviewing them, though, reinforces the need for a critical food system and STEM learning approach to ensure lasting and transformative change to our food systems. Those that take a critical pedagogical approach possess five common features: experiential or embodied learning; systems thinking; being rooted in the local context; valuing different ways of knowing and developing new knowledge; and building community. Experiential or embodied learning involves getting our hands in the dirt, using all our senses to make discoveries. Systems thinking entails examining more closely the interdependencies in our food systems and becoming more aware of our individual and local connections to larger environmental, social, and economic issues. It allows us to reconnect with parts of the system that have become hidden in the conventional agro-industrial system. Being rooted in the local context allows the program to take into consideration the cultural, economic, and environmental realities of the participants. It helps foster a sense of place, allowing learners to feel a deeper connection to where they live. Valuing different ways of knowing ensures that all voices take part in the learning process, which enriches the experience for all. It also means incorporating alternative methods of analysis and evaluation, such as the arts or popular education techniques. Often the most valuable part of any program is the community that develops around it – the sharing and learning together. This also means focusing not just on individual-level behaviors but also the structures that perpetuate the inequities in our current food system.

Fostering transformative change is hard work and takes time. As demonstrated by the many initiatives outlined in this chapter, however, we are not alone in this work. Food system issues are complex and often there is not one simple solution to solving them. By exploring the various approaches, venues, and benefits of food systems learning, people can engage with those that work best for them and ultimately cultivate more sustainable food systems.

Notes

1 Azevedo Perry, E., Thomas, H., Samra, H. R., Edmonstone, S., Davidson, L., Faulkner, A., Petermann, L., Manafò, E., & Kirkpatrick, S. I. (2017). Identifying attributes of food literacy: A scoping review. *Public Health Nutrition, 20*(13), 2406–2415. https://doi.org/10.1017/S1368980017001276; Rosas, R., Pimenta, F., Leal, I., & Schwarzer, R. (2021). FOODLIT-PRO: Conceptual and empirical development

of the food literacy wheel. *International Journal of Food Sciences and Nutrition, 72*(1), 1–13. https://doi.org/10.1080/09637486.2020.1762547; Truman, E., Lane, D., & Elliott, C. (2017). Defining food literacy: A scoping review. *Appetite, 116*, 365–371. https://doi.org/10.1016/j.appet.2017.05.007; Vidgen, H. A., & Gallegos, D. (2014). Defining food literacy and its components. *Appetite, 76*, 50–59. https://doi.org/10.1016/j.appet.2014.01.010.

2 Goldstein, S. (2016). Youth and food literacy: A case study of food education at The Stop Community Food Centre. In J. Sumner (Ed.), *Learning, food, and sustainability: Sites for resistance and change* (pp. 181–200). Palgrave Macmillan, p. 183.

3 Sumner, J. (2013). Food literacy and adult education: Learning to read the world by eating. *Canadian Journal for the Study of Adult Education, 25*(2), 79–92, p. 86.

4 Crosley, K. L. (2013). Advancing the boundaries of urban environmental education through the food justice movement. *Canadian Journal of Environmental Education (CJEE), 18*(0), 46–58; Gahl Cole, A. (2007). Expanding the field: Revisiting environmental education principles through multidisciplinary frameworks. *The Journal of Environmental Education, 38*(2), 35–44.

5 Donatuto, J., Campbell, L., LeCompte, J. K., Rohlman, D., & Tadlock, S. (2020). The story of 13 Moons: Developing an environmental health and sustainability curriculum founded on Indigenous first foods and technologies. *Sustainability, 12*(21), Article 8913. https://doi.org/10.3390/su12218913; Swayze, N. (2009). Engaging Indigenous urban youth in environmental learning: The importance of place revisited. *Canadian Journal of Environmental Education, 14*(0), 59–73; Walter, P. (2012). Educational alternatives in food production, knowledge and consumption: The public pedagogies of "Growing Power" and "Tsyunhehkw." *Australian Journal of Adult Learning, 52*(3), 573–594.

6 Harvard University. (2022). *Science & cooking: From haute cuisine to soft matter science*. EdX. https://www.edx.org/course/science-cooking-from-haute-cuisine-to-soft-matter.

7 Kyari, J. (2018). Sun-powered lunch hours. *Green Teacher, 117*, 17–20; Nagchaudhuri, A., Mitra, M., & Domnique Henry, X. S. (2017, January). Experiential learning activities for K-12 outreach and undergraduate students involving production and utilization of biodiesel. *Proceedings of the ASEE Annual Conference & Exposition*. https://doi.org/10.18260/1-2--28322.

8 Denver Museum of Nature & Science. (2022). *Genetics of taste*. https://www.dmns.org/science/health-sciences/projects/genetics-of-taste/.

9 Macbeth, A. J., Zurier, H. S., Atkins, E., Nugen, S. R., & Goddard, J. M. (2021). Engaged food science: Connecting K-8 learners to food science while engaging graduate students in science communication. *Journal of Food Science Education, 20*(1), 31–47. https://doi.org/10.1111/1541-4329.12215; Verran, J., Redfern, J., Moravej, H., & Adebola, Y. (2019). Refreshing the public appetite for "good bacteria": Menus made by microbes. *Journal of Biological Education, 53*(1), 34–46. https://doi.org/10.1080/00219266.2017.1420678.

10 MacDonald, A., Fenton, A., & Davidson, C. (2018). Young children's mathematical learning opportunities in family shopping experiences. *European Early Childhood Education Research Journal, 26*(4), 481–494. https://doi.org/10.1080/1350293X.2018.1487163.

11 Playful Learning Landscapes Action Network. (2019). *About*. https://playfullearninglandscapes.com/about/.

12 Hassinger-Das, B., Zosh, J. M., Bustamante, A. S., Golinkoff, R. M., & Hirsh-Pasek, K. (2021). Translating cognitive science in the public square. *Trends in Cognitive Sciences, 25*(10), 816–818. https://doi.org/10.1016/j.tics.2021.07.001.

13 Hassinger-Das, B., Bustamante, A. S., Hirsh-Pasek, K., & Golinkoff, R. M. (2018). Learning landscapes: Playing the way to learning and engagement in public spaces. *Education Sciences, 8*(2), Article 74. https://doi.org/10.3390/educsci8020074.

14 Braham, E. J., Libertus, M. E., & McCrink, K. (2018). Children's spontaneous focus on number before and after guided parent-child interactions in a children's museum. *Developmental Psychology, 54*(8), 1492–1498. https://doi.org/10.1037/dev0000534.

15 Morris, B., Dunlosky, J., & Owens, W. (2021, May 11). *Food for thought: Food labs and family STEM learning.* STEM for All Video Showcase: COVID, Equity & Social Justice, virtual. https://stemforall2021.videohall.com/presentations/2182.

16 Morris, B. J., Link to external site, this link will open in a new window, Owens, W., Ellenbogen, K., Erduran, S., & Dunlosky, J. (2019). Measuring informal STEM learning supports across contexts and time. *International Journal of STEM Education, 6*(1), 1–12. http://dx.doi.org/10.1186/s40594-019-0195-y.

17 Clegg, T., & Kolodner, J. (2014). Scientizing and cooking: Helping middle-school learners develop scientific dispositions. *Science Education, 98*(1), 36–63. https://doi.org/10.1002/sce.21083.

18 Clegg, T., Gardner, C., & Kolodner, J. (2010). Playing with food: Moving from interests and goals into scientifically meaningful experiences. In K. Gomez, L. Lyons, & J. Radinsky (Eds.), *Learning in the disciplines: Proceedings of the 9th International Conference of the Learning Sciences* (Vol. 1, pp. 1143–1150). International Society of the Learning Sciences. https://repository.isls.org//handle/1/2657.

19 Spellman, K., Mulder, C., Sparrow, E., Stanley, S., Cost, D., Shaw, J., Villano, C., Buffington, C., & Parkinson, L. (n.d.). *Winterberry: Citizen science for understanding berries in a changing north.* Retrieved June 6, 2022, from https://sites.google.com/alaska.edu/winterberry/home.

20 Spellman, K. V., Shaw, J. D., Villano, C. P., Mulder, C. P. H., Sparrow, E. B., & Cost, D. (2019). Citizen science across ages, cultures, and learning environments. *Rural Connections, 13,* 25–28.

21 Spellman, K. V., Cost, D., & Villano, C. P. (2021). Connecting community and citizen science to stewardship action planning through scenarios storytelling. *Frontiers in Ecology and Evolution, 9,* Article 695534. https://doi.org/10.3389/fevo.2021.695534.

22 Mulder, C. P. H., Spellman, K. V., & Shaw, J. (2021). Berries in winter: A natural history of fruit retention in four species across Alaska. *Madroño, 68*(4), 487–510. https://doi.org/10.3120/0024-9637-68.4.487.

23 Life Lab Science Program. (2022). *Curriculum and Activity Guides.* https://lifelab.org/new-store/.

24 Patchen, A., Aeschlimann, A., Vera-Cruz, A., Kamath, A., Jose, D., DeLisi, J., Barnett, M., Madden, P., & Rupani, R. (2017). Seeding the future: Blending urban gardening with community outreach and STEM learning. *Connected Science Learning, 1*(3). https://www.nsta.org/connected-science-learning/connected-science-learning-may-july-2017/seeding-future; Patchen, A. K., Zhang, L., & Barnett, M. (2017). Growing plants and scientists: Fostering positive attitudes toward science among all participants in an afterschool hydroponics program. *Journal of Science Education and Technology, 26*(3), 279–294.

25 Patchen, A., Aeschlimann, A., Vera-Cruz, A., Kamath, A., Jose, D., DeLisi, J., Barnett, M., Madden, P., & Rupani, R. (2017).

26 Patchen, A. K., Zhang, L., & Barnett, M. (2017).

27 Patchen, A. K., Zhang, L., & Barnett, M. (2017), p. 282.

28 Farinde, A. A., Tempest, B., & Merriweather, L. (2014). Service learning: A bridge to engineering for underrepresented minorities. *International Journal for Service Learning in Engineering, 9,* 475–491; Wooten, M., & Corlew, J. (2021). Engaging Nashville's youth in farming, food choice, and food access issues: Two programs by a Nashville nonprofit. In I. DeCoito, A. Patchen, N. Knobloch, & L. Esters (Eds.), *Teaching and learning in urban agricultural community contexts* (pp. 57–76). Springer.

29 Wooten, M., & Corlew, J. (2021).

30 Wooten, M., & Corlew, J. (2021), p. 75.

31 Krasny, M. (2005). *Garden Mosaics program manual*. Cornell University. https://civeco.files.wordpress.com/2018/08/manual-complete.pdf.

32 Liddicoat, K. R., Simon, J. W., Krasny, M. E., & Tidball, K. G. (2007). Sharing programs across cultures: Lessons from Garden Mosaics in South Africa. *Children, Youth & Environments, 17*(4), 237–254.

33 Doyle, R., & Krasny, M. (2003). Participatory Rural Appraisal as an approach to environmental education in urban community gardens. *Environmental Education Research, 9*(1), 91–115. https://doi.org/10.1080/13504620303464, p. 111.

34 Krasny, M. E., & Tidball, K. G. (2009). Community gardens as contexts for science, stewardship, and civic action learning. *Cities and the Environment, 2*, 1–18, p. 13.

35 Gareau, T. P., & Moscovitz, A. (2021). An overview of urban agriculture youth programs in major cities of the U.S. and the integration of STEM curriculum and activities. In L. T. Esters, A. Patchen, I. DeCoito, & N. Knobloch (Eds.), *Research approaches in urban agriculture and community contexts* (pp. 165–183). Springer International Publishing.

11

WILD HEARTS

Exploring the Connection between STEM Learning and Conservation Psychology in Zoos and Aquariums

Kathayoon Khalil

To curtail rapid environmental degradation and promote the sustainable coexistence of all living things, environmental education providers offer experiential programs that connect learners to the natural world. Zoos and aquariums, one such group of providers, curate hundreds of millions of nature and animal experiences worldwide each year[1] for visitors and audiences across many dimensions of age, racial and ethnic demographics, and socioeconomic status. As informal learning institutions, zoos and aquariums offer a multitude of societal benefits, including opportunities for social bonding, pro-environmental skill and behavior development, and the chance to forge emotional connections with animals. Zoos and aquariums also serve as potential venues for science, technology, engineering, and math (STEM) education and provide opportunities for visitors to engage in learning centered on wildlife and habitat conservation, though many of the applications for this work are yet to be explored. In this essay, we explore the connection between STEM learning and the study and application of conservation psychology in the zoo and aquarium field. Through this synthesis, we find spaces in which these areas can overlap more effectively to enhance their unique objectives and further the conservation missions of these facilities.

Informal STEM education

Informal learning settings refer to those that exist outside of the formal education system, including but not limited to zoos, aquariums, museums, and science centers. In many of these settings, the learning that occurs is often self-directed and controlled by the learner. This approach can be particularly helpful in science learning and interest development. Informal STEM

DOI: 10.4324/9781003145387-13

education has been defined as, "lifelong learning in science, technology, engineering, and math that takes place across a multitude of designed settings and experiences out-side of the formal classroom".[2] Furthermore, STEM learning includes content knowledge and skill-building, scientific thinking (e.g. inquiry), fixing social and environmental problems, and engaging in collaborative experiences.[3] STEM learning in informal settings is not relegated to cultural or science museums – this type of learning can permeate day-to-day life including social encounters and household tasks.[4] Because STEM subjects are traditionally seen as being biased in their accessibility toward dominant cultures, exclusive of women and people of color, the opportunity to learn STEM subjects in informal settings is even more important; informal settings are accessible to a broader range of ethnicities, races, and socioeconomic backgrounds than traditional formal education.[5,6] Focusing on STEM subjects in informal education means potentially reaching historically underrepresented communities and enriching their learning and interest in these topics.

For many informal learning settings such as science and natural history museums, the focus has largely been on the natural sciences – particularly biological and physical – over the other three arms of STEM: technology, engineering, and mathematics.[7,8] This is largely the case in zoos and aquariums, where much of the learning and messaging centers on animals and their ecosystems. Science in zoos and aquariums often manifests in animal behavior, endocrinology, ecology, zoology, and other relevant disciplines. But while the subject of science dominates these settings, other aspects of STEM are not entirely excluded, and opportunities to better integrate technology, engineering, and mathematics are plentiful. Despite this potential, however, zoo and aquarium visitors do not often see these facilities as explicit in their STEM learning opportunities.[3] This presents a unique chance to reframe the zoo or aquarium experience and determine where and when STEM outcomes might be pursued or highlighted.

Conservation psychology

To inform this discussion on new learning opportunities, we turn to the field of conservation psychology. Conservation psychology is an interdisciplinary school of thought that integrates approaches and theories from a variety of the learning and social sciences to understand the relationships between human and nature with a specific focus on conservation. The field of conservation psychology seeks to marry the theoretical with the practical, using research to support improvements to practice and drawing from real-life examples to inform further research.[9]

In zoos and aquariums, conservation psychology provides a foundation for much of the education and social science work that occurs in and through these facilities. Conservation psychology has been applied in zoos and aquariums in myriad different research studies and evaluations, including work to understand how to foster empathy for animals[10] and feelings of connectedness to nature.[11] At a higher level, conservation psychology and the work of zoos and aquariums converge on a desire to influence conservation behavior and encourage people to consider their impact on the planet.

An initial framework for research in conservation psychology suggests several avenues for study: conservation behaviors and valuing nature at both the individual and group levels, and across theoretical, applied, and evaluative dimensions.[9] STEM education can surely fall under several of these categories, sharing with conservation psychology an emphasis on the practical applications of research and the synthesis of applied work into theory development.

Using conservation psychology for STEM education

Considering the intersection of STEM and conservation psychology provides intriguing new applications for both fields. Particularly, combining STEM education and conservation psychology gives us novel insights in how to address and generate solutions to conservation issues, especially within the context of a zoo or aquarium visit. We examine this convergence in three key areas: first, we will discuss inquiry-based education as a viable approach for informal learning. Then, we will examine empathy as a lens and a framework through which we can curate rich experiences. Finally, we will look at the generation of conservation solutions through empathy-based inquiry as a critical task of conservation education.

Inquiry-based STEM education. Zoos and aquariums have long been sites for inquiry-based education, including foundational projects like *Wild Research*.[12] Inquiry-based education asks learners to engage in a cyclical approach to questioning, observation, data collection, and reporting with the goal of generating deeper and more equitable engagement.[13] Outside of zoos and aquariums, inquiry-based education has offered an effective approach to STEM education in both informal and formal learning environments. The increased focus on STEM education in informal environments can contribute to a more holistic approach to STEM that is inclusive of a broader range of individuals, particularly those who may not interact with these subjects in the formal school system such as early childhood learners.[14] Furthermore, increased access to STEM experiences can support engagement and interest in ongoing learning.[15,16]

STEM education has emphasized problem-solving based on a scientific approach that integrates knowledge, attitudes, and skills in service of questioning and drawing conclusions supported by evidence and based in real-world problems.[17] Inquiry and STEM are tightly interrelated; inquiry has been conceptualized as the common language of the four branches of STEM,[3] and the effectiveness of inquiry-based education is higher when applied to science subjects.[18] Informal STEM education in particular focuses on including play and social learning to encourage experimentation and questioning.[18]

Peer-reviewed studies on conservation-focused STEM education are not nearly as plentiful as those relating to other social issues, and there is certainly room for growth as we consider how STEM topics can be emphasized in zoos and aquariums. There are numerous opportunities, here – technology can be paired with other interpretive and educational tools to support and heighten learning.[19] Similarly, engineering and mathematics can be specifically targeted to supplement learning about animals; for example, asking learners to think about the structural challenges to creating zoo exhibits or the math involved in calculating diets. While there are almost certainly internal examples of these practices in zoos and aquariums around the world, intentional and cohesive strategizing on these topics could help to establish best practices and contribute to our broader understanding of STEM education. Perhaps most interestingly, integrating empathy into STEM education can lead to increased interest in science among students;[20] this offers a unique opportunity to informal learning institutions like zoos and aquariums who are looking to articulate and achieve empathy outcomes while positioning themselves as STEM learning hubs.

Empathy and STEM. While inquiry offers an approach to informal education, empathy offers an affective lens through which we can consider zoo and aquarium experiences. Empathy for animals is defined as "a stimulated emotional state that relies on the ability to perceive, understand, and care about the experiences and perspectives of another person or animal".[10] Empathy provides a promising pathway to pro-environmental behaviors by acting as an intrinsic motivator for many conservation actions.[21] There are a variety of ways in which zoos and aquariums can intentionally foster empathy for animals; ultimately, many of these strategies center on perspective-taking exercises. Inviting guests to learn about technology used with animals represents one such strategy. For example, technology has been used in zoos and aquariums to enrich the animal lives and offer opportunities for visitors to empathetically connect with animals.[22] In two studies, guests were invited to observe and learn about how computers can provide cognitive enrichment for orangutans.[8,23] In these studies, visitors perceived that the animals were benefitting from the exposure to technology[8] and demonstrated empathy for the individual animals in the study by commenting on the animals' intentions and desires, as well as their similarity to humans.[14]

In these studies, the focus was primarily on the animals and not necessarily on the technology being used. The results provide intriguing potential for future work wherein visitors can engage in deeper learning about the specific technologies being used as well as the implications for animal welfare. Related to empathy, visitors and program participants can be offered opportunities to take the perspective of animals, using that perspective to further their own inquiry or to design new experiences for these individuals. Indeed, empathy is a core tenet of design thinking – an approach to engineering that centers the human experience.[24] Taking the perspective of others allows designers and engineers to create more effective and innovative solutions to unique challenges. This approach has distinct potential for zoo and aquarium educators and animal care staff to engage visitors in new ways, using empathy and design to encourage people to share the perspective of an animal and consider ways to improve or enrich that animal's life.

Conservation action and solutions. One of our central frameworks, conservation psychology, asks us to think deeply about pressing environmental issues and engage in solutions-oriented work. Conservation psychology differentiates itself from other fields of study by emphasizing the equivalent importance of research and practice, using each to inform and validate the other. By combining STEM education and conservation psychology, we can encourage innovative thinking related to conservation action and environmental problem-solving. The results of this synthesis could also lead to new approaches to animal wellness, visitor experiences, and community engagement.

As discussed, zoos and aquariums represent a distinctive confluence of experiences – the ability to foster empathy for animals, the arena to share STEM learning opportunities related to nature and wildlife, and ultimately the goal of supporting thriving human and animal populations across the world. The next evolution of conservation recognizes conservation as a people problem; but as people perpetuate conservation problems, they can also be the purveyors of conservation solutions. Conservation professionals have recognized with increasing clarity the importance of multimodal, interdisciplinary approaches to conservation work, particularly the integration of social sciences. This introduces the possibility of zoos and aquariums evolving into a role as overt conveners, presenters, and generators of conservation solutions.

While zoos and aquariums have long centered conservation and education within their missions, they have also struggled to decide how best to approach complicated conservation issues with guests, particularly when the motivations of guests may differ from the desired outcomes of the institution. Heightened conversations about the ability of zoos and aquariums to influence conservation action and support peoples' connections with the natural world lie firmly within the purview of conservation psychology but can also be informed by the approaches used in STEM education. Thus far, conservation action and behavior change in zoos and aquariums have been approached using models such as community-based social marketing,[25] the transtheoretical model of behavior change,[26] and the theory of planned behavior,[27] to name a few. The most recent addition to this suite of frameworks is the CARE model, which synthesizes the applicable characteristics of several other theories and applies them to the zoo and aquarium context.[28] Common to all these approaches is the idea that behavior change is dependent on the interplay between motivators and barriers for a particular behavior or constellation of behaviors. The successful development of conservation solutions is dependent in many cases on effectively motivating behavior change; while our understanding of the influence of a zoo or aquarium experience on behavioral outcomes is growing, there is still more to examine and clarify regarding the ideal conditions and messaging. An integration of conservation psychology and STEM enlivens this discussion by offering relevant frameworks for understanding, addressing, and solving many of these issues and by offering diverse voices to contribute to a conversation whose success relies on its ability to be inclusive and welcoming.

Within zoos and aquariums, audiences including visitors and program participants can both learn about solutions to conservation issues that have been enacted worldwide and generate innovative new approaches to addressing environmental problems. Distinct from many other learning spaces, zoos, aquariums, and some museums shape experiences around the presence of live animals. Live animals can influence these conversations in critical ways, particularly if used to increase empathy or perspective taking on behalf of the animal. Research suggests that live animals elicit caring behaviors more effectively than inanimate representations, especially among young children who are developing their sense of self and others, as well as their social skills.[29] Experiences with live animals can also increase knowledge retention and foster positive attitudes toward animals and the environment across age groups.[30] Keeping animals in managed care settings, however, does have important welfare considerations. If visitors perceive animals to be exhibiting signs of poor welfare, this can contribute to the development of negative empathy or empathic sorrow. This can become a barrier to conservation messaging or attitude development.[31] Conversely, perceiving positive welfare in animals can elicit feelings of empathic joy which can in turn inspire support of conservation work and willingness to engage in behaviors that extend the well-being of the animal.[10] Framing messaging, exhibitry, and programming to highlight animal welfare and elicit positive empathy reactions can encourage collaboration, inclusion, and optimism – all critically necessary to conservation work. Highlighting the unique contributions of the physical, natural, and social sciences through the integration of STEM and conservation psychology can turn zoos and aquariums into workshops for innovative and effective solutions, encouraging learners to look at complex problems in new ways.

Lastly, the equity considerations related to STEM education and zoos and aquariums introduce an interesting and potentially critical facet to this work. As mentioned earlier, STEM education offers the potential to engage historically underrepresented audiences in these four subject areas, which have been traditionally exclusive in their constituencies. Similarly, zoos and aquariums are centering diversity, equity, inclusion, and access work in more intentional ways through their programming, marketing, staff, and philosophical approaches to conservation. The Association of Zoos and Aquariums (AZA) requires that accredited institutions have a written diversity, equity, access, and inclusion (DEAI) program that includes visitors, staff, volunteers, and vendors.[32] Though there is currently little published data on this work, many institutions have addressed this requirement by creating internal DEAI strategies, creating DEAI outcomes, and even hiring staff dedicated to this body of work. In service of their DEAI goals, many zoos and aquariums have also engaged conservation psychology principles to meaningfully integrate communities into conservation and education work to increase sustainability and uptake of solutions.[33] For many institutions, this work is still growing and evolving, with an ever-present need for new approaches and ideas for process and implementation. By applying conservation psychology to STEM education in these settings, we can encourage a more justice-oriented approach to conservation, one that draws from the overlapping priorities of these fields to encourage meaningful collaboration.

While STEM education excels in teaching learners how to engage in the practice of science or the applications of technology, mathematics, and engineering, conservation psychology asks us to use very similar approaches for the benefit of wildlife and the environment. As zoos and aquariums continue to redefine their relevance in modern society, the nexus of these two avenues can offer a new role for these institutions to fill – as integrative, engaging co-creators of conservation solutions that are accessible and resonant with diverse audiences and communities.

Notes

1 WAZA. (2021). *World Association of Zoos and Aquariums*. Retrieved from http://www.waza.org

2 Center for the Advancement of Informal Science Education. (2021). *What is informal STEM learning?* Retrieved from https://www.informalscience.org/what-informal-stem-learning.

3 Gupta, R., Fraser, J., Rank, S., Brucker, J., & Flinner, K. (2019). Multi-site case studies about zoo and aquarium visitors' perceptions of the STEM learning ecology. *Visitor Studies, 22*(2), 127–146.

4 Allen, S., & Peterman, K. (2019). Evaluating informal STEM education: Issues and challenges in context. *New Directions for Evaluation, 2019*(161), 17–33.

5 National Research Council. (2009). Theoretical Perspectives. In *Learning Science in Informal Environments: People, Places, and Pursuits* (pp. 27–53). Washington, DC: The National Academies Press.

6 Tal, T., & Dierking, L. (2014). Learning science in everyday life. *Journal of Research in Science Teaching, 51*(3), 251–259.

7 Bybee, R. (2010). Advancing STEM Education: A 2020 vision. *Technology and Engineering Teacher, 70*(1), 30–35.

8 Basham, J., & Marino, M. (2013). Understanding STEM education and supporting students through universal design for learning. *Teaching Exceptional Children, 45*(4), 8–15.

9 Saunders, C. (2003). The emerging field of conservation psychology. *Human Ecology Review, 10*(2), 137–149.

10 Young, A., Khalil, K., & Wharton, J. (2018). Empathy for animals: A review of the existing literature. *Curator: The Museum Journal, 61*(2), 327–343.

11 Chen-Hsuan Cheng, J., & Monroe, M. (2012). Connection to nature: Children's affective attitude towards nature. *Environment and Behavior, 44*(1), 31–49.

12 Heimlich, J., Yocco, V., Myers, C., & Myers, L. (2011). *Wild Research: Summative Evaluation*. Edgewater, MD.

13 Myers, C., & Myers, L. H. (2009). Science is not a spectator sport: Three principles from 15 years of Project Dragonfly. In R. Yager (ed.), *Inquiry: The Key to Exemplary Science* (pp. 29–40). Arlington, VA: NSTA Press.

14 Hurst, M. A., Polinsky, N., Haden, C. A., Levine, S. C., & Uttal, D. H. (2019). Leveraging research on informal learning to inform policy on promoting early STEM. *Social Policy Report, 32*(3), 1–33.

15 Bell, P., Lewenstein, B., Shouse, A., & Feder, M. (2009). Learning science in informal environments: People, places, and pursuits. *Museums & Social Issues, 4*(1), 113–124. https://doi.org/10.1179/ msi.2009.4.1.113.

16 Deák, C., Kumar, B., Szabó, I., Nagy, G., & Szentesi, S. (2021). Evolution of new approaches in pedagogy and STEM with inquiry-based learning and post-pandemic scenarios. *Education Sciences, 11*(7), 319.

17 Bybee, R. (2013). *The Case for STEM Education: Challenges and Opportunities*. Arlington, VI: NSTA Press.

18 Cairns, D., & Areepattamannil, S. (2019). Exploring the relations of inquiry-based teaching to science achievement and dispositions in 54 countries. *Research in Science Education, 49*, 1–23.

19 Perdue, B., Stoinski, T., & Maple, T. (2012). Using technology to educate zoo visitors about conservation. *Visitor Studies, 15*(1), 16–27.

20 McCurdy, R. P., Nickels, M. L., & Bush, S. B. (2020). Problem-based design thinking tasks: Engaging student empathy in STEM. *The Electronic Journal for Research in Science & Mathematics Education, 24*(2), 22–55.

21 Berenguer, J. (2007). The effect of empathy in proenvironmental attitudes and behaviors. *Environment and Behavior, 39*(2), 269–83.

22 Perdue, B., Clay, A., Gaalema, D., Maple, T., & Stoinski, T. (2012). Technology at the zoo: the influence of a touchscreen computer on orangutans and zoo visitors. *Zoo Biology, 31*(1), 27–39.

23 Webber, S., Carter, M., Sherwen, S., Smith, W., Joukhadar, Z., & Vetere, F. (2017). Kinecting with orangutans: Zoo visitors' empathetic responses to animals' use of interactive technology. In *Proceedings of the 2017 CHI conference on human factors in computing systems*. Denver, Colorado. pp. 6075–6088. https://dl.acm.org/ doi/10.1145/3025453.3025729.

24 Razzouk, R., & Shute, V. (2012). What is design thinking and why is it important? *Review of Educational Research, 82*(3), 330–348.

25 McKenzie-Mohr, D. (2011). *Fostering Sustainable Behavior: An Introduction to Community-Based Social Marketing* (3rd ed.). Gabriola Island, Canada: New Society.

26 Prochaska, J. O., & Velicer, W. F. (1997). The transtheoretical model of health behavior change. *American Journal of Health Promotion, 12*(1), 38–48.

27 Ajzen, I. (1991). The theory of planned behavior. *Organizational Behavior and Human Decision Processes, 50*(2), 179–211.

28 Routman, Khalil, Schultz, & Keith (2022). Beyond inspiration: Translating zoo and aquarium experiences into conservation behavior. *Zoo Biology, 41*(5), 398–408. CARE stands for Cultivate Caring, Amplify Intent, Remove Barriers, and Expand Impact.

29 Myers Jr., O., & Saunders, C. (2002). Animals as links toward developing caring relationships with the natural world. In P. H. Kahn, Jr., & S. R. Kellert (Eds.) *Children and Nature: Psychological, Sociocultural, and Evolutionary Investigations*, 153–178. Cambridge, MA: MIT Press.

30 Povey, K. D., & Rios, J. (2002). Using interpretive animals to deliver affective messages in zoos. *Journal of Interpretation Research, 7*(2), 19–28.

31 Miller, L. J., Luebke, J. F., & Matiasek, J. (2018). Viewing African and Asian elephants at accredited zoological institutions: Conservation intent and perceptions of animal welfare. *Zoo Biology, 37*(6), 466–477.

32 Association of Zoos and Aquariums. (2022). *The Accreditation Standards and Related Policies.* https://www.aza.org/accred-materials.

33 Cranston, K. (2016). *Building & Measuring Psychological Capacity for Biodiversity Conservation [Doctoral dissertation, Antioch University].* Retrieved from OhioLINK Electronic Theses and Dissertations Center: http://rave.ohiolink.edu/etdc/view?acc_num=antioch147203418.

12

OUT-OF-SCHOOL SCIENCE EDUCATION INSTITUTIONS FOR SUSTAINABILITY

Marianne Achiam and Henry James Evans

Humanity faces a number of global problems such as climate disruption, loss of biodiversity, deforestation and not least, the ongoing COVID-19 pandemic. These problems are often described as 'wicked problems' because of their complexity. Often wicked problems have multiple stakeholders with different perspectives and thus different (and uncoordinated) suggestions for solutions, none of which offer definitive resolutions.[1] Many of these problems are the result of a materialistic-mechanistic worldview that has led to the long-term overutilisation of Earth's human and natural resources. To counteract these problems, we must therefore 'expand our thinking and fundamentally correct our current behavior'.[2]

There is a broad agreement that the transition to sustainability requires education.[3,4] However, the cross-cutting competencies and transdisciplinary perspectives that are required to make this transition are not easily translated into the practices and structures of formal school system[5] because these competencies and perspectives are radically different from those associated with the 'traditional' or established science that is reproduced in schools (Table 12.1).

In response, we argue that out-of-school science education institutions such as natural history museums, science and technology museums, science centres, planetariums, zoos, botanical gardens and aquariums have a critical and unique role to play in educating children and youth for a sustainable future. In the following sections, we discuss how out-of-school science education institutions are ideally situated to engage learners in the five key aspects of sustainability science presented in Table 12.1. We illustrate our points with examples from out-of-school science education institutions.

DOI: 10.4324/9781003145387-14

TABLE 12.1 Selected Characteristics of 'Traditional' Science and Sustainability Science[2,6,7]

'Traditional' science	*Sustainability science*
Monodisciplinary	Inter/transdisciplinary
Addresses academic peers	Involves extended peer communities
Driven by scientific curiosity	Driven by real-world problems
Unequivocal results	Ranges of options
Temporally fixed	Temporally dynamic

Sustainability science: Inter- and transdisciplinary

Sustainability is not circumscribed by any one scientific discipline, but works simultaneously between, across and beyond them.[8] In contrast, most formal science education systems are arranged by subjects that correspond to well-known research disciplines (i.e. biology, physics, chemistry, mathematics). This organisation makes it difficult for schools to realise inter- and trans-disciplinarity in lessons and programmes. However, out-of-school science education institutions are not limited by disciplinary boundaries. Instead, their systems-based perspectives mean that they can offer learners experiences with transdisciplinary ways of working in programmes, exhibitions and more. For instance, natural history museums often focus on cross-cutting content such as evolutionary relationships and biogeography,[9] while science and technology museums may present processes of human innovation and ingenuity that transcend disciplinary boundaries.[10] Finally, the focus on biomes in many modern zoos and aquaria integrates biodiversity and ecosystem perspectives.[6]

Sustainability science: Extended peer communities

As mentioned, sustainability science deals with problems at the intersection of society, environment and economy. This means that it involves a range of non-academic stakeholders[11,12] and that science is just one kind of input to the co-creation of knowledge that includes practical, lay and indigenous knowledge.[13,14] In this way, each stakeholder becomes a peer. In fact, this 'extended peer community' approach is a critical part of sustainability science.[15]

Established or 'traditional' science has typically not been inclusive of different perspectives or groups of stakeholders. Across a number of science disciplines, people of non-dominant cultures, gender identities or backgrounds experience exclusion[16–18]; unfortunately, this exclusion may also be reproduced in school science.[19,20] In this respect, too, out-of-school science education institutions may have an important role to play. They have important expertise in making complex subject matter accessible to a broad diversity of learners. And even though not all out-of-school science education institutions are as inclusive as they could be,[21] still there are many examples of these institutions engaging diverse publics in dialogue with each other and with scientists to co-create new knowledge. Initiatives such as citizen science, hands-on workshops[22,23] and makerspaces[24] are some of the ways that museums and science centres attempt to create equitable experiences with science.

Even so, achieving equity in practice is challenging. For instance, ideologies in citizen science include inclusive and participatory approaches,[25] yet many projects involve participants only in data collection, leading to questions of how participatory citizen science really is. In addition, certain groups in society are typically underrepresented in citizen science projects,[26] which are thus not reflective of the diversity of modern society. Similar challenges seem to affect makerspaces.[27]

Sustainability science: Driven by real-world problems

Sustainability science is defined by the problems that drive it, rather than by the academic disciplines it draws on.[28] Additionally, the wicked problems of sustainability are complex and messy, and not easy to define. These characteristics challenge school science, because school science tends to define science in terms of the linear, algorithmic 'scientific method'.[29–32] Here is another opportunity for out-of-school science education institutions and for their expertise communicating complex content matter through physical, multisensory and social experiences.[6] A compelling example is provided by the exhibition Klima X in the Finnish science centre Heureka. This exhibition required visitors to walk through water wearing rubber boots. A slowly melting ice block was located in the middle of the room, and a soundtrack played the sounds of thunderstorms. Visitors interacted with these media in a range of different ways, leading to measurable changes in their emotional involvement in climate change.[33] Another example is offered by the Botanical Garden of the University of Zurich. Here, visitors immersed themselves in different climate scenarios of the year 2085. These experiences allowed them to build a visceral and nuanced understanding of the implications of climate change on their regions and local areas.[34]

Sustainability science: Ranges of options

Wicked sustainability problems challenge the idea that science can provide the right or the definitive answer.[35] Because wicked problems are found at the intersection between science and society, science and technology alone cannot provide proven, unambiguous answers. Instead, responses to these problems must be evaluated against a range of often contradictory scales.[36] This means that the range of options for addressing wicked problems often involve significant trade-offs.[37]

In recent years, out-of-school science education institutions have begun to question positivist and authoritative accounts of science.[22] Instead, many of these institutions are seeking to present science in more contextualised and dialogic ways: Science as part of culture and society rather than distinct from it. This shift in perspective makes out-of-school science education institutions important places to pioneer the ambiguity and open-endedness of wicked sustainability problems. For instance, Klimatopia is a sustainability-themed exhibition at the Danish science centre Experimentarium that incorporates three different future pathways representing possible scenarios for Earth – each led by a girl from the future who has crash-landed in our time. These scenarios are based on the modelling that are part of reports by the Intergovernmental Panel on Climate Change. At the end of the exhibition, visitors respond to a behaviour-themed interactive, based on their own lifestyle choices, such as meat consumption and travel. These answers combine to highlight the future pathway that their own choices are taking the Earth towards.

Sustainability science: Considers temporal dynamics of problems

Finally, sustainability science focuses on the (uncertain) future facing humanity, meaning that it has a built-in temporal element.[38] Moreover, much of what we know about the wicked problems that face us today comes from evidence of the past,[39] for instance ice cores,[40] tree rings[41], and sediment layers[42] that reveal signs of long-term temperature variability. Even so, the time scales involved in sustainability problems are often difficult for learners to grasp because they do not fit within the timeframe of a lifetime or a generation.

With respect to the temporality of wicked problems, out-of-school science education institutions have unique perspectives to offer. For instance, the living collections of aquariums, zoos, and botanical gardens can engage learners in the conservation history of a species as well as its present status and the efforts required to preserve it for the future.[6] Natural history museums and science and technology museums have vast collections that can be thought of as memory banks to represent the world's material diversity and adaptive intelligence in the past as well as the present.[43] For example, the Danish Museum of Science & Technology is presently exhibiting innovations that have profoundly changed society over the past 150 years, from cars and airplanes to space exploration. This way of connecting the past and the future may help learners grasp the immediacy of the problems we face rather than deferring them to an undefined future time.[44]

Discussion

In this paper, we have argued that even as formal science education struggles with promoting important skills and knowledge to address wicked problems, out-of-school science education institutions offer important complementary pathways to sustainability competencies. Out-of-school science education institutions have significant expertise, experience, and institutional perspectives that can be used to engage learners in wicked sustainability problems in constructive and critical ways.[45] Although museums, science centres and related institutions have long been recognised for the contributions they can make to school science, the perspectives we suggest here mark a more progressive and radical use of out-of-school science education institutions, thus renewing their relevance to the formal education system.[46,47]

As we have illustrated in this text, some out-of-school science education institutions are already offering well-planned experiences to engage their visitors in the open-endedness of wicked problems. However, far from all out-of-school science education institutions address the global problems we face. One need only glance through the recent report from the Intergovernmental Panel on Climate Change[48] to grasp how serious a situation we find ourselves in. Why then, are not all museums and related institutions actively engaged in communicating about global wicked problems and sustainable solutions?

Sustainability is not well defined for education

One obstacle to sustainability education is the concept itself. Sustainability is a fuzzy term with a range of interpretations and understandings,[49] meaning that both inside and outside of schools, the notion remains amorphous and poorly defined for education.[50,51] Even though, as we have argued in this text, out-of-school science education institutions are well placed to disseminate sustainability, they still face the significant challenge of deconstructing and reconstructing sustainability concepts and notions to fit their own practices.[50,52–54]

In recognition, perhaps, of this challenge, special interest organisations for out-of-school science education institutions have taken steps to prompt their member institutions to address sustainability. For example, in 2020, the World Association of Zoos and Aquaria (WAZA) released its sustainability strategy, suggesting that its member institutions work externally and internally across the UN's sustainable development goals.[55] Even so, this strategy seems less than transformational, advocating for instance that its member institutions 'develop education programmes that can practically address issues of sustainability'. Another example is the International Council of Museums (ICOM), which formed a working group on sustainability in 2018 with specific attention on building sustainability into its members' practice.[50] Here, we also find mention of the UN's sustainable development goals, yet the resolution to 'incorporate sustainability into [ICOM members'] own internal and external practices and educational programming' does not provide operational directions on how to do so. Although on the one hand, these interest organisations provide top-down prompts to their member institutions, it remains up to the individual institutions to operationalise these ideas.[56] We hope this essay offers some such guidance.

The influence of funders and sponsors

Uncertain funding patterns are becoming a fact of life for many out-of-school science education institutions. This may be perceived as another challenge for institutions' efforts to address sustainability. Because governmental funding can no longer be taken for granted, these institutions must rely more and more on donations and funding from private sources. As a consequence, they may feel unable to ignore the values, ideologies and discourses of those funders.[57] This tension has been discussed in relation to for instance David Koch's financial support of the dinosaur exhibition at the Smithsonian National Museum of Natural History[58] or Shell's sponsorship of the *Our Future Planet* exhibition at the London Science Museum.[59] In these cases, the museums in question have been criticised for obscuring the science of climate change and effectively greenwashing the corporate sponsors behind their exhibitions.

Another perspective on this debate is offered by museum experts Robert Janes and Richard Sandell, who have argued that discussions of shortages have become a habit of mind in the museum community; a habit of mind that is limiting and even counterproductive to their purposes. Museum professionals, they argue, 'know intuitively that money is not the measure of their worth [...] everything that is required to fulfil the true potential of museums is there'.[43] Rather than letting the perceived shortages of money, staff, technology or support become an excuse to maintain the status quo, Janes and Sandell argue that museums should use their privileged positions to develop new stories about sustainable connections between people and planet. We would extend this argument to other types of out-of-school science education institutions as well.

Feigned neutrality

Finally, a third reason science education institutions are not addressing global wicked problems may be their perceived need to take a so-called neutral stance on socio-scientific issues such as the biodiversity crisis or climate disruption. The assumption here is that science education institutions cannot risk alienating their supporters by taking stands on social or political issues.[43] They thus place themselves on the side lines of new or controversial issues, even though appeals to neutrality become increasingly tenuous in the face of the mounting socio-scientific problems confronting us.[60]

During the preparation of this essay, a prominent museum expert shared with us a sense of disillusionment about the ability of museums to play a significant role in helping people engage with the social and environmental problems that confront humanity. She wrote:

> I've had discussions with many who have worked in museums for decades. There is definitely a general sense of frustration regarding the reluctance of museums to align with Black Lives Matter, with climate change, around white privilege, and inequity.
> (anonymised museum expert, personal communication, 21/12/2020).

We have argued elsewhere that museums are not, and have never been, neutral.[57] The problem with museums' (and other science education institutions') *feigned* neutrality is that it effectively supports the status quo, rather than questioning it. This avoidance of responsibility is not only in direct opposition to the educational mission of science education institutions, it stands in the way of the real impact these institutions could make on preparing citizens for a more sustainable future.

Final remarks

In this essay, we described and discussed the critical and unique role that can be played by natural history museums, science and technology museums, science centres, planetariums, zoos, botanical gardens and aquariums to help prepare children and youth for a sustainable future. We have also considered some of the reasons these institutions might not be doing all they can to further this agenda. Ultimately, we consider out-of-school science education institutions to play an absolutely crucial role in helping humanity transition towards sustainability, and we believe that these institutions should be ready to play much more proactive and transformative roles than has been the case so far. Policymakers across the world have shown us that they are not up to the task of supporting the public's needs. It is up to the museum sector to engage citizens in the profound shift of mindset that characterises transformative science education *as* sustainability.

Acknowledgements

The research presented here was funded by the Novo Nordisk Foundation grant #0052319. An early version of this paper was presented at the 2021 International Conference of Mathematics and Science Education in Indonesia. The present text has been reframed to fit the prospective readership of this book. This reframing includes a revised introduction, new elements in the body of the text, and a completely new discussion. We thank the editors for their insightful comments and guidance in the preparation of this essay.

Notes

1 Caron, R. M., & Serrell, N. (2009). Community ecology and capacity: Keys to progressing the environmental communication of wicked problems. *Applied Environmental Education & Communication, 8*(3–4), 195–203. https://doi.org/10.1080/15330150903269464.

2 Dürr, H.-P., Dahm, D., & zur Lippe, R. P. (2005). *Potsdam Manifesto 2005. "We have to learn to think in a new way"*. Federal Ministry of Education and Research of Germany. http://www.gcn.de/download/manifesto_en.pdf.

3 Block, T., Goeminne, G., & Van Poeck, K. (2018). Balancing the urgency and wickedness of sustainability challenges: Three maxims for post-normal education. *Environmental Education Research, 24*(9), 1424–1439. https://doi.org/10.1080/13504622.2018.1509302.

4 Holfelder, A.-K. (2019). Towards a sustainable future with education? *Sustainability Science, 14*(4), 943–952. https://doi.org/10.1007/s11625-019-00682-z.

5 OECD. (2018). In J. Skovgaard (Ed.), *The future we want. The future of education and skills: Education 2030*. OECD. https://www.oecd.org/education/2030/E2030%20Position%20Paper%20(05.04.2018).pdf.

6 Evans, H. J., & Achiam, M. (2021). Sustainability in out-of-school science education: Identifying the unique potentials. *Environmental Education Research, 27*(8), 1192–1213. https://doi.org/10.1080/13504622.2021.1893662.

7 Spangenberg, J. H. (2011). Sustainability science: A review, an analysis and some empirical lessons. *Environmental Conservation, 38*(3), 275–287. https://doi.org/10.1017/S0376892911000270.

8 Dillon, J. (2017). Wicked problems and the need for civic science. *Spokes, 29*, 1–9. http://www.ecsite.eu/activities-and-services/news-and-publications/digital-spokes/issue-29#.

9 King, H., & Achiam, M. (2017). The case for natural history. *Science & Education, 26*(1), 125–139. https://doi.org/10.1007/s11191-017-9880-8.

10 Anderson, K., & Hadlaw, J. (2018). The Canada Science and Technology Museum. *Technology and Culture, 59*, 781–786. https://doi.org/10.1353/tech.2018.0066.

11 Craps, M. (2019). Transdisciplinarity and sustainable development. In W. Leal Filho (Ed.), *Encyclopedia of sustainability in higher education*. Springer. https://doi.org/https://doi.org/10.1007/978-3-319-63951-2_102-1.

12 Lang, D. J., Wiek, A., Bergmann, M., Stauffacher, M., Martens, P., Moll, P., Swilling, M., & Thomas, C. J. (2012). Transdisciplinary research in sustainability science: Practice, principles, and challenges. *Sustainability Science, 7*(1), 25–43. https://doi.org/10.1007/s11625-011-0149-x.

13 Brandt, P., Ernst, A., Gralla, F., Luederitz, C., Lang, D. J., Newig, J., Reinert, F., Abson, D. J., & von Wehrden, H. (2013). A review of transdisciplinary research in sustainability science. *Ecological Economics, 92*, 1–15. https://doi.org/10.1016/j.ecolecon.2013.04.008.

14 Messerli, P., Kim, E. M., Lutz, W., Moatti, J.-P., Richardson, K., Saidam, M., Smith, D., Eloundou-Enyegue, P., Foli, E., Glassman, A., Licona, G. H., Murniningtyas, E., Staniškis, J. K., van Ypersele, J.-P., & Furman, E. (2019). Expansion of sustainability science needed for the SDGs. *Nature Sustainability, 2*(10), 892–894. https://doi.org/10.1038/s41893-019-0394-z.

15 Funtowicz, S. O., & Ravetz, J. R. (1993). Science for the post-normal age. *Futures, 25*(7), 739–755. https://doi.org/https://doi.org/10.1016/0016-3287(93)90022-L.

16 Aycock, L. M., Hazari, Z., Brewe, E., Clancy, K. B. H., Hodapp, T., & Goertzen, R. M. (2019). Sexual harassment reported by undergraduate female physicists. *Physical Review Physics Education Research, 15*(1), 010121.

17 Gibney, E. (2019). Discrimination drives LGBT+ scientists to think about quitting. *Nature, 571*, 16–17. https://doi.org/10.1038/d41586-019-02013-9.

18 Kaatz, A., Lee, Y.-G., Potvien, A., Magua, W., Filut, A., Bhattacharya, A., Leatherberry, R., Zhu, X., & Carnes, M. (2016). Analysis of national institutes of Health R01 application critiques, impact, and criteria scores: Does the sex of the principal investigator make a difference? *Academic Medicine, 91*(8), 1080–1088. https://doi.org/10.1097/acm.0000000000001272.

19 Global Education Monitoring Report Team. (2020). *Global education monitoring report 2020: Gender report, A new generation: 25 years of efforts for gender equality in education.* UNESCO.

20 Wong, B. (2016). *Science education, career aspirations and minority ethnic students.* Palgrave Macmillan.

21 Dawson, E. (2014). Equity in informal science education: Developing an access and equity framework for science museums and science centres. *Studies in Science Education, 50*(2), 209–247. https://doi.org/10.1080/03057267.2014.957558.

22 Sandholdt, C. T., & Achiam, M. (2018). Engaging or transmitting? Health at the science centre. *Nordisk Museologi – Journal of Nordic Museology, 2018*(2–3), 136–151. https://doi.org/10.5617/nm.6661.

23 Silfver, E. (2018). Gender performance in an out-of-school science context. *Cultural Studies of Science Education, 14*(1), 139–155. https://doi.org/10.1007/s11422-017-9851-z.

24 Rushton, E. A. C., & King, H. (2020). Play as a pedagogical vehicle for supporting gender inclusive engagement in informal STEM education. *International Journal of Science Education, Part B, 10*(4), 376–389. https://doi.org/10.1080/21548455.2020.1853270.

25 Shirk, J. L., Ballard, H. L., Wilderman, C. C., Phillips, T., Wiggins, A., Jordan, R., McCallie, E., Minarchek, M., Lewenstein, B. V., Krasny, M. E., & Bonney, R. (2012). Public participation in scientific research: A framework for deliberate design. *Ecology and Society, 17*(2). http://www.jstor.org/stable/26269051.

26 Sorensen, A. E., Jordan, R. C., LaDeau, S. L., Biehler, D., Wilson, S., Pitas, J.-H., & Leisnham, P. T. (2019). Reflecting on efforts to design an inclusive citizen science project in West Baltimore. *Papers in Natural Resources*, Article 13.

27 Hira, A., Joslyn, C. H., & Hynes, M. M. (2014). Classroom makerspaces: Identifying the opportunities and challenges. *2014 IEEE Frontiers in Education Conference (FIE) Proceedings.*

28 Clark, W. C. (2007). Sustainability Science: A room of its own. *Proceedings of the National Academy of Sciences, 104*(6), 1737–1738. https://doi.org/10.1073/pnas.0611291104.

29 Cheng, K. L., & Wong, S. L. (2014). Nature of science as portrayed in the physics official curricula and textbooks in Hong Kong and on the mainland of the People's Republic of China. In *Topics and trends in current science education: 9th ESERA conference selected contributions* (pp. 519–534). Springer Netherlands. https://doi.org/10.1007/978-94-007-7281-6_32.

30 Estrup, E., & Achiam, M. (2019). The potential of palaeontology for science education. *Nordina – Nordic Studies in Science Education, 15*(1), 97–108. https://doi.org/10.5617/nordina.5253.

31 Irez, S. (2016). Representations of the nature of scientific knowledge in Turkish biology textbooks. *Journal of Education and Training Studies, 4*(7), 206–220.

32 Woodcock, B. A. (2014). "The scientific method" as myth and ideal. *Science & Education, 23*(10), 2069–2093. https://doi.org/10.1007/s11191-014-9704-z.

33 Gorr, C. (2014). Changing climate, changing attitude? *Museums & Social Issues, 9*(2), 94–108. https://doi.org/10.1179/1559689314Z.00000000021.

34 Schläpfer-Miller, J. (2021). Climate Garden 2085: An art-science experiment promoting different ways of knowing about climate change. In M. Achiam, J. Dillon, & M. Glackin (Eds.), *Addressing wicked problems through science education. The role of out-of-school experiences* (pp. 149–165). Springer. https://doi.org/10.1007/978-3-030-74266-9_8.

35 Achiam, M., Glackin, M., & Dillon, J. (2021). Wicked problems and out-of-school science education: Implications for practice and research. In M. Achiam, J. Dillon, & M. Glackin (Eds.), *Addressing wicked problems through science education. The role of out-of-school experiences* (pp. 229–237). Springer. https://doi.org/10.1007/978-3-030-74266-9_12.

36 Rittel, H. W. J., & Webber, M. M. (1973). Dilemmas in a general theory of planning. *Policy Sciences, 4*(2), 155–169. https://doi.org/10.1007/BF01405730.

37 Roberts, N. (2000). Wicked problems and network approaches to resolution. *International Public Management Review, 1*(1), 1–19.

38 Cavender-Bares, J., Polasky, S., King, E., & Balvanera, P. (2015). A sustainability framework for assessing trade-offs in ecosystem services. *Ecology and Society, 20*(1), Article 17. https://doi.org/10.5751/ES-06917-200117.

39 Markley, R. (2012). Time, history, and sustainability. In T. Cohen (Ed.), *Telemorphosis: Theory in the era of climate change* (pp. 43–64). Open Humanities Press.

40 Dansgaard, W., Johnsen, S. J., Clausen, H. B., Dahl-Jensen, D., Gundestrup, N. S., Hammer, C. U., Hvidberg, C. S., Steffensen, J. P., Sveinbjornsdottir, A. E., Jouzel, J., & Bond, G. (1993). Evidence for general instability of past climate from a 250-kyr ice-core record. *Nature, 364*(6434), 218–220. https://doi.org/10.1038/364218a0.

41 Fritts, H. C., Lofgren, G. R., & Gordon, G. A. (1980). Past climate reconstructed from tree rings. *The Journal of Interdisciplinary History, 10*(4), 773–793. https://doi.org/10.2307/203070.

42 Tian, J., Nelson, D. M., & Hu, F. S. (2011). How well do sediment indicators record past climate? An evaluation using annually laminated sediments. *Journal of Paleolimnology, 45*(1), 73–84. https://doi.org/10.1007/s10933-010-9481-x.

43 Janes, R., & Sandell, R. (2019). Posterity has arrived. The necessary emergence of museum activism. In R. Janes, & R. Sandell (Eds.), *Museum activism* (pp. 1–22). Routledge.

44 Salazar, J. F. (2014). Futuring global change in science museums and centers. A role for anticipatory practices and imaginative arts. In F. R. Cameron, & B. Neilson (Eds.), *Climate change and museum futures* (pp. 90–108). Routledge.

45 Dillon, J., Achiam, M., & Glackin, M. (2021). The role of out-of-school science education in addressing wicked problems: An introduction. In M. Achiam, J. Dillon, & M. Glackin (Eds.), *Addressing wicked problems through science education. The role of out-of-school experiences* (pp. 1–8). Springer. https://doi.org/10.1007/978-3-030-74266-9_1.

46 Berg, T. B., Achiam, M., Poulsen, K. M., Sanderhoff, L. B., & Tøttrup, A. P. (2021). The role and value of out-of-school environments in science education for 21st century skills. *Frontiers in Education, 6,* Article 674541. https://doi.org/10.3389/feduc.2021.674541.

47 Xanthoudaki, M. (2015). Museums, innovative pedagogies and the twenty-first century learner: A question of Methodology. *Museum and Society, 13*(2), 247–265. https://doi.org/10.29311/mas.v13i2.329.

48 IPCC. (2021). In V. Masson-Delmotte, P. Zhai, A. Pirani, S. L. Connors, C. Péan, S. Berger, N. Caud, Y. Chen, L. Goldfarb, M. I. Gomis, M. Huang, K. Leitzell, E. Lonnoy, J. B. R. Matthews, T. K. Maycock, T. Waterfield, O. Yelekçi, R. Yu, & B. Zhou (Eds.), *Climate change 2021: The physical science basis. Contribution of working group I to the sixth assessment report of the intergovernmental panel on climate change.* Cambridge University Press. https://www.ipcc.ch/report/ar6/wg1/#FullReport.

49 Purvis, B., Mao, Y., & Robinson, D. (2019). Three pillars of sustainability: In search of conceptual origins. *Sustainability Science, 14*(3), 681–695. https://doi.org/10.1007/s11625-018-0627-5.

50 Brown, K. (2019). Museums and local development: An introduction to museums, sustainability and well-being. *Museum International, 71*(3–4), 1–13. https://doi.org/10.1080/13500775.2019.1702257.

51 Jickling, B., & Wals, A. E. J. (2008). Globalization and environmental education: Looking beyond sustainable development. *Journal of Curriculum Studies, 40*(1), 1–21. https://doi.org/10.1080/00220270701684667.

52 Cameron, F., Hodge, B., & Salazar, J. F. (2013). Representing climate change in museum space and places. *Wiley Interdisciplinary Reviews: Climate Change, 4*(1), 9–21. https://doi.org/10.1002/wcc.200.

53 Hedges, E. (2021). Actions for the future: Determining sustainability efforts in practice in Arizona museums. *Museum Management and Curatorship, 36*(1), 82–103. https://doi.org/10.1080/09647775.2020.1752293.

54 Keogh, L., & Möllers, N. (2015). Pushing boundaries: Curating the Anthropocene at the Deutsches Museum, Munich. In F. R. Cameron, & B. Neilson (Eds.), *Climate change and museum futures* (pp. 78–89). Routledge.

55 Bensted, E., Dominguez, M. C., MNZM, K. F., Gendron, S., Griffith, M., Hughes, A., Lee, H. M., Mann-Lang, J., Werth, J., & Zordan, M. (2020). *Protecting our Planet. Sustainability Strategy 2020–2030*. WAZA. https://www.waza.org/priorities/sustainability/the-waza-sustainability-strategy-2020-2030/.

56 Biermann, F., Kanie, N., & Kim, R. E. (2017). Global governance by goal-setting: The novel approach of the UN Sustainable Development Goals. *Current Opinion in Environmental Sustainability, 26–27*, 26–31. https://doi.org/https://doi.org/10.1016/j.cosust.2017.01.010.

57 Evans, H. J., Nicolaisen, L. B., Tougaard, S., & Achiam, M. (2020). Museums beyond neutrality. *Nordisk Museologi, 29*(2), 19–25. https://doi.org/10.5617/nm.8436.

58 Rieppel, L. (2019). The Smithsonian's new dinosaur hall is a marvel. But its ties to David Koch are a problem. *The Washington Post*. Retrieved May 12, 2020, from https://www.washingtonpost.com/outlook/2019/06/09/smithsonians-new-dinosaur-hall-is-marvel-its-ties-david-koch-are-problem/.

59 Osborne, S. (2021). 'Staggeringly out of step': Science Museum launches climate exhibition sponsored by Shell. *The Independent*. https://www.independent.co.uk/independentpremium/science-museum-shell-climate-exhibition-b1832681.html.

60 Lyons, S., & Bosworth, K. (2019). Museums in the climate emergency. In R. Janes, & R. Sandell (Eds.), *Museum activism* (pp. 174–185). Routledge.

13

INFORMAL LEARNING THROUGH CITIZEN SCIENCE

Authentic, Meaningful, Impactful

Tina Phillips and Heidi Ballard

Citizen science, also commonly referred to as public participation in scientific research, community science, and volunteer monitoring engages the public in authentic scientific research through intentional collaboration with professional scientists.[1] The last few decades have resulted in the creation of thousands of citizen science projects spanning the globe, and engaging millions of people, particularly biodiversity projects.[2] Data collected by citizen scientists have gained widespread acceptance for their ability to address scientifically important issues across large spatial and temporal scales.[3–5] Additionally, citizen science provides multiple pathways for engaging the public in science, across diverse age groups, and in a myriad of informal contexts covering any number of science topics.[6–8]

In addition to facilitation by the Internet, the growth of citizen science has also been spurred on by the informal science education (ISE) field and science education researchers arguing for more hands on, inquiry-based learning experiences to enhance science engagement and science literacy.[9] Citizen science has the potential to engage participants in all aspects of the scientific process, thereby providing an authentic context for public engagement and collaboration with science.[10–12] Thus, citizen science has become a powerful approach to advance the goals of ISE and support voluntary, lifelong, life-wide, and life-deep science learning.[12,13]

We begin with a brief overview of what citizen science is and its entrenched relationship with ISE and highlight the key aspects of the participatory approaches and authentic science that characterize citizen science. For the purpose of bounding this work, we focus largely on environmental-based citizen science projects. We then highlight the evolving trajectory of research that has deepened and expanded over the last decade. We conclude with our views

DOI: 10.4324/9781003145387-15

on the nuanced relationships among engagement in citizen science, science learning outcomes, and the ways in which people identify with science to both reflect on and inform ISE research more broadly.

The evolution of ISE and citizen science

Although the first mention of the term "citizen science" did not occur in print until 1995,[14] efforts to engage people in the scientific process have been documented for several hundred years.[15] Spurred on by the environmental movement of the 1960s and 1970s, the earliest forms of citizen science were largely a disparate set of volunteer monitoring programs, commonly focused on water quality monitoring.[16] Notably large-scale projects such as the Christmas Bird Count, which began in 1900, the US Fish and Wildlife Service's Breeding Bird Survey and the Cornell Lab of Ornithology's Nest Record Card Program, which both began in the 1960s, continue to this day.[17] However, none of these early citizen science projects involved explicit learning goals for participants; they were focused largely on meeting scientific goals and data collection.

In the early 1990s, the National Science Foundation's Informal Science Education Directorate (which had been established for about a decade) awarded the Cornell Lab of Ornithology the first known grant to develop a national scale citizen science project with explicit learning goals, called "Public Participation in Ornithology: An Introduction to Environmental Research." The grant funded three different citizen science projects: Project Pigeon Watch, Project Tanager, and the National Seed Preference Test.[14] The premise of each of these projects was that the people involved would "learn science by doing." In the decades that followed, the NSF continued to support the advancement and evolution of citizen science by awarding more than a thousand grants across its eight directorates. Those early funding sources paved the way not only for the creation of engaging, hands-on science projects across every discipline, they also paved the way for the inclusion of individual learning goals as part of all funded citizen science projects. Since then, other ISE-based organizations such as the Center for the Advancement of Informal Science Education (CAISE), the National Research Council, and the National Academies have helped to advance the field of citizen science by lending theoretical perspectives, disseminating best practices for enhancing informal science learning, and supporting research and evaluation of learning in citizen science, particularly with respect to individual learning outcomes. While it is true that ISE has greatly shaped citizen science in addressing individual science learning outcomes, citizen science has in turn propelled the ISE field to think more broadly about where science learning happens, i.e., in situ, in communities, and online. For example, community-based learning is a common, yet under-studied phenomenon in ISE, and citizen science provides authentic contexts to study how communities learn science together.[18]

In its modern form, citizen science is recognized globally as a collaborative effort between the general public and professional researchers to both answer important scientific questions and promote deep science learning.[1]. Citizen science has evolved to encompass many different forms of engagement, from top-down scientist driven contributory projects to bottom-up, community-driven initiatives.[10,11] While debate about the proper terminology continues, we use the term citizen science here to follow recent discussions[19] and include both scientist-lead and community-lead projects in all their respective forms.

The participatory and experiential nature of citizen science

A recent report by the National Academies of Sciences, Engineering, and Medicine[12] stops short of defining citizen science, but instead characterizes citizen science projects as those that:

> actively engage participants, specifically engage participants with data, use systematic approaches to produce reliable knowledge, meet widely recognized standards of scientific integrity and use practices common in science, engage participants who are (primarily) not project-relevant scientists, seek to use the knowledge gained to contribute to science and/or community priorities, generally confer some benefit to the participant for participating, and involve the communication of results.

With its focus on data sharing and knowledge production, citizen science, in nearly every form, is experiential, collaborative, engaging, and voluntary. Some projects are specifically adapted for particular audiences, such as youth (e.g., LIMPETS or the GLOBE Program), or in particular settings such as museum-lead biodiversity research,[20] while other projects are open to all ages and settings. Participants can engage in projects that last just a few hours, such as a BioBlitz, or be immersed for days or weeks, such as in an Earth Watch expedition. Citizen science projects cover nearly every conceivable scientific discipline from archeology to zoology (see scistarter.org for 1800+ examples). Depending on the project and individual motivation, participants may have opportunities to engage in multiple phases of the science process, sometimes in very easy or very complex ways. Some projects engage participants in a structured, linear fashion, while others allow for non-linear experiences. In many cases projects are designed to tap into intrinsic motivation, curiosity, and exploration,[21] similar to "hobby projects."[22] In such projects, participants may bring certain skills with them, or no skills at all, but the desire and willingness to learn. Other projects, particularly those concerned with pollution or contaminants, are created by and for communities with the explicit goal of improving the livelihoods of community members.[23]

Increasingly, citizen science also provides learning opportunities through digital engagement. Platforms such as Zooniverse enable hundreds of thousands of people to browse a repository of more than 50 active projects to classify data, analyze images, or transcribe historic records. Although this work is still in its infancy, affordances of digital environments often provide more social interaction with other volunteers than traditional, field-based projects.[24,25] Thus, the flexibility of citizen science content, audiences, contexts, and scale, coupled with the focus on data exploration, scientific integrity, and knowledge production, provides for an authentic ISE experience. It should be noted, however, that while citizen science is open for all to engage in all facets of the science process, a recent review of the structure of participation found that the vast majority of data collection happens by a single person or small group, engaging multiple times over a certain period of time, and that the majority of projects are contributory in nature, i.e., most participants are mainly involved in data collection.[26] Further, the demographics of participants tend to be older, white, highly educated individuals that are not representative of the US population.[12]

Key findings from research on the impacts of citizen science on science learning

Studying the learning outcomes from citizen science was started in earnest in 2009 with a landmark report that compared outcomes across a variety of citizen science projects.[10] The report was based on the National Science Foundation's framework of outcome categories for ISE programs in the US designed to collect project-level impacts in a systematic way.[27] The five categories include knowledge, engagement, skills, attitudes, and behaviors. Later, a modified framework was created to be more specific to citizen science learning outcomes and reflective of the focus of science and the environment.[28] These outcome categories include: Interest in science and nature, efficacy for learning and doing science, motivation for environmental action and science learning, knowledge related to content and the nature of science, skills of science inquiry, and environmental stewardship behaviors. These refined categories, stemming directly from ISE documents, have been used in several recent studies[6,7,12,29] to document evidence of learning. We acknowledge, however, that these categories of learning are not exhaustive. For instance, there is nascent work on understanding community learning outcomes[12] as well as outcomes related to trust in and democratization of science through citizen science and other participatory approaches[30] that are beyond the scope of this essay. Below we highlight some of the key findings on learning in citizen science from the last decade aligned to those early ISE learning outcomes.

Interest, self-efficacy, and motivation toward science and the environment. With most hobby-based projects, many people join because of an underlying interest in the topic and thus learning gains related to increased science interest are often small.[28] However, for youth that are often unable to self-select, gains can be more pronounced.[6,31] Changes in self-efficacy for environmental action have been demonstrated in projects focused on air pollution,[32] entomology,[33] and gardening.[34] Motivation for engaging in citizen science has been studied fairly well,[35–37] yet motivation as an outcome, such as motivation for environmental action, has had less attention. In an examination of place attachment and stewardship, statistically significant increases in both internal and external motivations for engaging in stewardship activities were described.[38]

Science content knowledge. Examining the impacts of citizen science participation on participants' science content knowledge gains is the most common area of evaluation and research on science learning impacts, and that's particularly true for environmental and biodiversity citizen science projects. Early work found evidence of increased understanding of specific ecological content knowledge.[39–41] More recently, a systematic review of research on the environmental education outcomes of community and citizen science participation found positive knowledge gains focused on the main topic of the project (insects, birds, biodiversity, water quality, etc.[29,42,43]). Additionally, a recent review of ten different studies of online or digital citizen science projects[24] found that overall, participants had a better understanding of the nature of science, increased science content knowledge, and changed attitudes toward science.

Science inquiry skills. Although critical to assess because of the direct implications for data quality,[44] outcomes related to science inquiry skill gains among participants are seldom reported. This is in part because such assessments are ideally observed (which is difficult, especially in large scale projects) or performance based, which involves a fair bit of investment. As a result, much of what we know about skill gains relies on self-reported data. Nevertheless, there are several examples of increased science inquiry skill gains as a result of participation in citizen science.[28,29,45,46] Interestingly, multiple studies report that the majority of skill gains were reported in activities related to data collection, such as observing/recording species, identifying species, collecting data in a standardized manner, and submitting observations, and that more complex tasks related to scientific reasoning were done far less frequently by participants.[28,29] These findings suggest that citizen science projects need more support to facilitate higher order inquiry skills such as analyzing project data to answer a question, interpreting visual data, conducting statistical analyses, and designing their own study related to project data.[12] Given the lack of studies examining science inquiry skill gains using performance-based assessments, there has been a call for inclusion of embedded assessments as a more accurate way to measure skill acquisition.[47] There is a general consensus that intentional design through scaffolding and facilitation of learning objectives is critical to support participants to gain skills beyond data collection.[7,11,12]

Environmental and stewardship behavior changes. Documenting behavior change is challenging in most settings, and citizen science is no exception.

In a systematic review of 100 studies of environmental education outcomes of community and citizen science, about a third of the studies demonstrated empirical evidence of positive behavioral outcomes related to environmental stewardship. Most of these behaviors involved civic action, management of habitats, and communicating information with others. For example, 95% of participants in butterfly citizen science projects reported increasing their participation in at least one conservation action since they joined a project, with talking to others about butterflies or conservation and involving others in monitoring in conservation being the two most increased actions.[48] Another study reported similar actions related to communicating with others, habitat improvement, and civic engagement.[29] While knowledge alone rarely leads to action, awareness of the decline in turtle populations was deemed important for engaging in turtle-friendly behaviors in a turtle conservation project,[47] highlighting the many inputs required for behavior change.

Identity and agency with science. In citizen science and particularly in community science, people inquire and act on both the physical and social worlds around them, gaining first-hand experience with tools and practices of science. They can come to see themselves as community resources and build linkages between the academic, disciplinary, and everyday communities they inhabit.[49,50] Research on how participants may identify with science, and how they may develop agency with science, is still lacking in citizen science contexts. One study found that young people participating in an out-of-school program studying the water quality of their urban creek displayed agency with environmental science in teaching adults how to properly collect samples and presenting their scientific findings at professional conferences and to their city council.[51] Another study found that adults who'd been trained as naturalists and subsequently volunteered for local citizen science projects increased in their identifying as people who understand and do science in their daily lives and are recognized by others that way as well.[52] Importantly, while we tend to think of all these types of outcomes as separate entities, the reality is they are very much interconnected with one another.

Conclusion

In this essay we expressed the multiple ways that citizen science can provide an ideal informal science learning experience with careful design. Whether one is volunteering as an individual in a scientist-led project or working in a team as part of a community-driven project, citizen science provides rich opportunities for engaging at whatever level of individual agency one finds suitable, with limitless potential for free-choice learning outcomes. The learning outcomes addressed here are by no means exhaustive. Exciting work in and outside of environmentally based citizen science is underway to examine the intersection between engagement and learning, the ways in which communities interact and learn together, the role of social capital in supporting community science literacy, and co-creation and the co-production of knowledge. The potential for citizen science to impact other outcomes related to environmental justice, democratization of science, health and well-being, sense of place, and connection to nature are all ripe for further study. Funding bodies, particularly NSF, have emphasized their support of projects that are community-led, that address issues of power and inequities in science, and expand participation in STEM by historically underrepresented communities. There is also a call for more longitudinal studies of both individual and community learning outcomes as well as inclusion of evidenced-based strategies into the design of new programs.[12]

As a result of these new research trajectories, we argue that research on the learning outcomes and processes that participants engage in through citizen science, while rapidly increasing in scope and depth, is still in its infancy. We further argue that to advance our understanding of these outcomes and processes, educational researchers in informal science learning settings would do well to consider a policy statement addressed to the Board of the National Association Research in Science Teaching (NARST),[53] describing six aspects related to conducting research on, or in, informal science learning settings. First, since learning outside of formal classrooms is often guided by learners' interests or needs, and research in these contexts must include the role that motivation and interest play in the process of learning. Second, the context and physical settings where learning happens must be considered. Third, social and cultural factors that include both individuals and groups as the unit of analysis are important to consider. Fourth, research that utilizes a "learning ecosystems" approach where the experience of learning is considered across time and space are encouraged. Fifth, investigations should emphasize both the process and product of learning. Lastly, research on learning should use a variety of designs, methods, contexts, and analyses, reflective of these dynamic learning experiences. We encourage the ISE field to include these considerations in their work and to continue to explore emerging areas of research in order to unlock the important contributions that well-designed citizen science learning contexts can offer the informal science learning community.

Notes

1 Bonney R., Shirk J., Phillips T. B., Wiggins, A., Ballard, H., Miller-Rushing, A. J., and Parrish, J. K. (2014). Next steps for citizen science. *Science, 343* (6178), 1436–1437. https://doi.org/10.1126/science.1251554.
2 Theobald, E. J., Ettinger, A. K., Burgess, H. K., DeBey, L. B., Schmidt, N. R., Froehlich, H. E., ... Parrish, J. K. (2015). Global change and local solutions: Tapping the unrealized potential of citizen science for biodiversity research. *Biological Conservation, 181*, 236–244.
3 Bonney, R., Cooper, C. B., Dickinson, J., Kelling, S., Phillips, T. B., Rosenberg, K. V., and Shirk, J. (2009a). Citizen science: A developing tool for expanding science knowledge and scientific Literacy. *Bioscience, 59*(11), 977–984.

4 Dickinson, J. L., Shirk, J. L. Bonter D., Bonney R., Crain, R.L., Martin, J., Phillips, T. B., and Purcell, K. (2012). The emergence of web-enabled citizen science: Integrating ecological research and education in the 21st century. *Frontiers in Ecology and the Environment, 10*(6), 291–297. https://doi.org/10.1890/110236.

5 McKinley, D. C., Miller-Rushing, A. J., Ballard, H. L., Bonney, R., Brown, H., Cook-Patton, S. C., ... Soukup, M. A. (2017). Citizen science can improve conservation science, natural resource management, and environmental protection. *Biological Conservation, 208,* 15–28.

6 Bonney, R., Phillips, T. B., Ballard, H. L., and Enck, J. E. (2016). Can citizen science enhance public understanding of science? *Public Understanding of Science, 25*(1), 1–15. https://doi.org/10.1177/0963662515607406.

7 Phillips, T. B., Ballard, H., Lewenstein, B. V., and Bonney, R. (2019). Engagement in science through citizen science: Moving beyond data collection. *Science Education, 103*(3), 665–690.https://doi.org/10.1002/sce.21501.

8 Phillips, T. B., Parker, A., Bowser, A., and Haklay, M. (2022). Publicly generated data: The role of Citizen-Science for knowledge production, action, and public engagement. In Ferreira, C. C. and Klütsch, C (eds) *Closing the Knowledge-implementation Gap in Conservation Science – Evidence Transfer across Spatiotemporal Scales and Different Stakeholders.* Wildlife Research Monographs, Vol. 3. Springer International Publishing, Basel.

9 Bell, P., Lewenstein, B. V., Shouse, A. W., and Feder, M. A., eds. (2009). *Learning Science in Informal Environments.* Washington, DC: National Academies Press.

10 Bonney, R., Ballard, H., Jordan, R., McCallie, E., Phillips, T. B., Shirk, J., and Wilderman, C. C. (2009). Public Participation in Scientific Research: Defining the Field and Assessing Its Potential for Informal Science Education. *A CAISE Inquiry Group Report.* Washington, DC: Center for Advancement of Informal Science Education (CAISE).

11 Shirk, J. L., Ballard, H. L., Wilderman, C. C., Phillips, T. B., Wiggins, A., Jordan, R., McCallie, E., Minarchek, M., Lewenstein, B. V., Krasny, M. E., and Bonney, R. (2012). Public participation in scientific research: A framework for deliberate design. *Ecology and Society, 17*(2), 29. https://doi.org/10.5751/ES-04705-170229.

12 National Academies of Sciences, Engineering, and Medicine. (2018). *Learning through Citizen Science: Enhancing Opportunities by Design.* Washington, DC: The National Academies Press. https://doi.org/10.17226/25183.

13 Falk, J. H., and Dierking, L. D. (2010). The 95 percent solution. *American Scientist, 98*(6), 486–493.

14 Irwin, A (1995). *Citizen Science: A Study of People, Expertise and Sustainable Development.* London: Routledge.

15 Miller-Rushing, A., Primack, R., and Bonney, R. (2012). The history of public participation in ecological research. *Frontiers in Ecology and the Environment, 10*(6), 285–290.

16 Haklay, M. (2017). The three eras of environmental information: the roles of experts and the public. In Loreto, V., Haklay, M., Hotho, A., Servedio, V.C.P., Stumme, G., Theunis, J., and Tria, F. (eds) *Participatory Sensing, Opinions and Collective Awareness.* Springer, Cham, pp. 163–179.

17 Bonney, R. (2008). Citizen Science at the Cornell Lab of Ornithology. In *Exemplary Science in Informal Education Settings.* National Science Teachers Association.

18 Jordan, R. C., Ballard, L. H., and Phillips, T. B. (2012). Key issues and new approaches for evaluating citizen science learning outcomes. *Frontiers in Ecology and the Environment, 10*(6), 307–309. https://doi.org/10.1890/110280.

19 Cooper, C. B., Hawn, C. L., Larson, L. R., Parrish, J. K., Bowser, G., Cavalier, D., Dunn, R. R., et al. (2021). Inclusion in citizen science: The conundrum of rebranding. *Science, 372*(6549), 1386–1388.

20 Ballard, H. L., Robinson, L. D., Young, A. N., Pauly, G. B., Higgins, L. M., Johnson, R. F., and Tweddle, J.C. (2017). Contributions to conservation outcomes of natural history museum-led citizen science: Examining evidence and next steps. *Biological Conservation*, *208*, 87–97.

21 Nov, O., Arazy, O., and Anderson, D. (2011). Dusting for science: Motivation and participation of digital citizen science volunteers. In: *Proceedings of the 2011 iConference*, 68–74. ACM. https://doi.org/10.1145/1940761.1940771.

22 Wiggins, A., and Crowston, K. (2011). From conservation to crowdsourcing: A typology of citizen science. In *44th Hawaii international conference on system sciences* (pp. 1–10). IEEE

23 Ramirez-Andreotta, M. D., Brusseau, M. L., Artiola, J., Maier, R. M., and Gandolfi, A. J. (2015). Building a co-created citizen science program with gardeners neighboring a Superfund site: The Gardenroots case study. *International Public Health Journal*, *7*(1), 13.

24 Aristeidou, M., and Herodotou, C. (2020). Online citizen science: A systematic review of effects on learning and scientific literacy. *Citizen Science: Theory and Practice*, *5*(1), 1–12.

25 Price, C. A., and Lee, H. S. (2013). Changes in participants' scientific attitudes and epistemological beliefs during an astronomical citizen science project. *Journal of Research in Science Teaching*, *50*(7), 773–801.

26 Vasiliades, M. A., Hadjichambis, A. C., Hadjichambi, D., Adamou, A., and Georgiou, Y. (2022). Investigating the participation facets of environmental citizen science initiatives: A systematic literature review of empirical research. *Environmental Sciences Proceedings*, *14*(1), 1.

27 Friedman, A. (Ed.). (March 12, 2008). Framework for Evaluating Impacts of Informal Science Education Projects [On-line]. Available at: http://insci.org/resources/Eval_Framework.pdf.

28 Phillips, T. B., Porticella, N., Constas, M., and Bonney, R. (2018). A Framework for Articulating and Measuring Individual Learning Outcomes from Participation in Citizen Science. *Citizen Science: Theory and Practice*, *3*(2). https://doi.org/10.5334/cstp.126.

29 Peter, M., Diekötter, T., Höffler, T., and Kremer, K. (2021). Biodiversity citizen science: Outcomes for the participating citizens. *People and Nature*, *3*, 294–311.

30 Davis, L. F., and Ramírez-Andreotta, M. D. (2021). Participatory research for environmental justice: A critical interpretive synthesis. *Environmental Health Perspectives*, *129*(2), 026001.

31 Flagg, B. (2016). Contribution of multimedia to girls' experience of citizen science. *Citizen Science: Theory and Practice*, *1*(2), 11. http://doi.org/10.5334/cstp.51.

32 Yen-Chia Hsu, Jennifer Cross, Paul Dille, Michael Tasota, Beatrice Dias, Randy Sargent, Ting-Hao (Kenneth) Huang, and Illah Nourbakhsh. (2019). Smell Pittsburgh: Engaging Community Citizen Science for Air Quality. arXiv preprint arXiv:1912.11936. (Article accepted by ACM Transactions on Interactive Intelligent Systems.) https://arxiv.org/abs/1912.11936.

33 Lynch, L. I., Dauer, J. M., Babchuk, W. A., Heng-Moss, T., and Golick, D. (2018). In their own words: The significance of participant perceptions in assessing entomology citizen science learning outcomes using a mixed methods approach. *Insects*, *9*(1), 16.

34 Sandhaus, S., Kaufmann, D., and Ramirez-Andreotta, M. (2019). Public participation, trust and data sharing: Gardens as hubs for citizen science and environmental health literacy efforts. *International Journal of Science Education, Part B*, *9*(1), 54–71.

35 Hart, A. G., Adcock, D., Barr, M., Church, S., Clegg, T., Copland, S., … Pocock, M. J. (2022). Understanding engagement, marketing, and motivation to benefit recruitment and retention in citizen science. *Citizen Science: Theory and Practice*, *7*(1), 1–9.

36 Rotman, D., Preece, J., Hammock, J., Procita, K., Hansen, D., Parr, C., Lewis, D., and Jacobs, D. (2012). Dynamic changes in motivation in collaborative citizen-science projects. In *Proceedings of the ACM 2012 conference on computer supported cooperative work* (pp. 217–226).Seattle, Washington.

37 West, S. E., Pateman, R. M., and Dyke, A. (2021). Variations in the motivations of environmental citizen scientists. *Citizen Science: Theory and Practice*, 6(1), p.14. DOI: http://doi.org/10.5334/cstp.370

38 Halliwell, P. M. (2019). National Park Citizen Science Participation: Exploring Place Attachment and Stewardship, Prescott College, Ann Arbor. ProQuest.

39 Jordan, R. C., Gray, S. A., Howe, D. V., Brooks, W. R., and Ehrenfeld, J. G. (2011). Knowledge gain and behavioral change in citizen-science programs. *Conservation biology*, 25(6), 1148–1154.

40 Brossard D., Lewenstein B., and Bonney R. (2005). Scientific knowledge and attitude change: The impact of a citizen science project. *International Journal of Science Education*, 27, 1099–1121.

41 Trumbull D. J., Bonney R., and Grudens-Schuck N. (2005). Developing materials to promote inquiry: Lessons learned. *Science Education*, 89, 879–900.

42 Hesley, D., Burdeno, D., Drury, C., Schopmeyer, S., and Lirman, D. (2017). Citizen science benefits coral reef restoration activities. *Journal for Nature Conservation*, 40, 94–99.

43 Haywood, B. K., Parrish, J. K., and Dolliver, J. (2016). Place-based and data-rich citizen science as a precursor for conservation action. *Conservation Biology*, 30(3), 476–486.

44 Stylinski, C.D., Peterman, K., Phillips, T. B., Linhart, J., and Becker-Klein, R. (2020). Assessing science inquiry skills of citizen science volunteers: A snapshot of the field. *International Journal of Science Education, Part B*, https://doi.org/10.1080/21548455.2020.1719288.

45 Lewis, R., and Carson, S. (2021). Measuring science skills development in New Zealand High School Students after participation in citizen science using a DEVISE evaluation scale. *New Zealand Journal of Educational Studies*, 56(1), 101–110.

46 Santori, C., Keith, R. J., Whittington, C. M., Thompson, M. B., Van Dyke, J. U., and Spencer, R. J. (2021). Changes in participant behaviour and attitudes are associated with knowledge and skills gained by using a turtle conservation citizen science app. *People and Nature*, 3(1), 66–76.

47 Peterman, K., Becker-Klein, R., Stylinski, C., and Nelson, A. G. (2017). Exploring embedded assessment to document scientific inquiry skills within citizen science. In *Citizen Inquiry* (pp. 63–82). Routledge.

48 Lewandowski, E. J., and Oberhauser, K. S. (2017). Butterfly citizen scientists in the United States increase their engagement in conservation. *Biological Conservation*, 208, 106–112.

49 Rahm, J., Miller, H. C., Hartley, L., and Moore, J. C. (2003). The value of an emergent notion of authenticity: Examples from two student/teacher–scientist partnership programs. *Journal of Research in Science Teaching*, 40(8), 737–756. https://doi.org/10.1002/tea.10109.

50 Roth, W. M., and Lee, S. (2004). Science education as/for participation in the community. *Science Education*, 88(2), 263–291.

51 Ballard, H. L., Dixon, C. G. H., and Harris, E. M. (2017). Youth-focused citizen science: Examining the role of environmental science learning and agency for conservation. *Biological Conservation*, 208, 65–75.

52 Merenlender, A. M., Crall, A. W., Drill, S., Prysby, M., and Ballard, H. (2016). Evaluating environmental education, citizen science, and stewardship through naturalist programs. *Conservation Biology*, 30(6), 1255–1265.

53 Dierking, L. D., Falk, J. H., Rennie, L., Anderson, D., and Ellenbogen, K. (2003). Policy Statement of the "Informal Science Education" ad hoc committee. *Journal of Research in Science Teaching*, 40(2), 108–111.

14

FROM SCHOOL-BASED CITIZEN SCIENCE TO TRANSITION-DRIVEN ACTIVISM WITH THE COMMUNITY

Tali Tal, Hila Shefet Barkae, and Nirit Lavie Alon

In the past decade, many communities worldwide have taken an active part in science research. These communities are engaged in citizen science or in community science.[1,2] The participants are citizens taking an active part in research as volunteers or even as initiators. They serve as the driving force behind the research in various ways, such as planning the research, collecting and analyzing data, drawing conclusions, and affecting policy of nature conservation, urban planning, and so forth, all while collaborating with researchers.

In this essay, we trace the origins of citizen science, its different models that connect scientists and the public, and its adoption by schools. We then identify various gaps between formal education, informal education, the outdoors and the community and illustrate – through one project – how citizen science can help bridge these gaps.

Apart from scientists and citizen volunteers, citizen science projects include collaboration with organizations, both at the local and national level, such as governmental agencies, industry, academia, community groups and local institutions.[3–5] Many citizen scientists are concerned about particular problems in their environment, like water quality and the conservation of biodiversity.[6,7]

In addition to environmental sciences, citizen science research takes place in many other areas such as astronomy, biology, evolution, health, history, social science, art, physics, and geology.[8,9] Websites like Zooniverse (www.zooniverse.org), for example, offer various citizen science projects in different fields that seek public assistance to promote scientific research.

Collaboration between organizations and communities in citizen science projects is one of the many diverse benefits of citizen science projects that can lead to their success.[10–12] In this way, citizens and stakeholders can partner

DOI: 10.4324/9781003145387-16

in managing natural resources and monitoring the environment;[13] they can also strengthen the relationships or the convergence between science and society and better engage the public with science.[14] A heuristic for citizen science models has been suggested[15] that extends from science-driven to transition-driven models, as we describe in the following sections.

Citizen science models

The research literature on citizen science offers a few typologies or ways to look at collaboration between scientists and the public, looking at the type and the extent of collaboration and at the role of the citizens in the citizen science endeavors.

Contributory projects are designed by scientists. Citizens contribute computing resources or data.

Collaborative projects are designed by scientists, and the citizens contribute data, but they may also assist in refining the project design, analyzing the data, and publishing the findings.

Co-created projects are designed by scientists and citizen collectively. Some of the participating citizens are also involved in most, or in different steps, of the scientific process.[14]

Another typology[17,18] offers four levels of participation and engagement in citizen science (see Table 14.1). This typology argues that extreme citizen science enables citizen science projects to occur with no scientists working in the background; in other words, the entire process is carried out by the citizens.

A special issue of *Conservation Biology*[15] offer heuristics for positioning citizen science that range between various instrumental forms: from being driven by scientists to more emancipatory forms driven by citizens. They distinguish between science-driven citizen science, policy-driven citizen science and transition-driven *civic science* that involves citizens in wicked problems and requires agency.[16]

Figure 14.1 is based on this heuristic.[17] The terms "Simple", "Complex", and "Wicked" refer to the type of problem the research is aiming to address.

In the citizen science described in this chapter, a school-based citizen science project reflects a co-created or even extreme citizen science type, which is transition-driven as well.

TABLE 14.1 Levels of Participation in Citizen Science[11]

Level	Type	Citizen participation
1	Crowdsourcing	Citizens as sensors; volunteered computing
2	Distributed intelligence	Participation in problem definition and data collection
3	Participatory science	Citizens as basic interpreters; volunteered thinking
4	Extreme citizen science	Collaborative science – problem definition, data collection and analysis

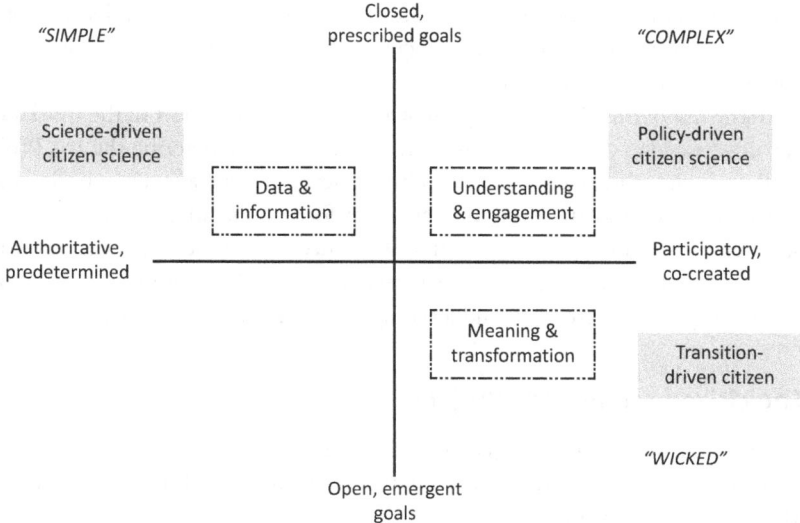

FIGURE 14.1 A heuristic for citizen science.

Citizen science in school: Opportunity for bridging gaps between school, the outdoors and the community

Besides the benefits to scientists and to the participating citizens, which are broadly discussed in the literature, citizen science brings specific benefits for students and teachers. It creates new opportunities for science education that enable students to take part in authentic and hands-on inquiry learning.[17] There is evidence for various learning outcomes of participating students.[18] Citizen science in schools is an emergent field with rather limited understanding of how citizen science supports meaningful learning.[19] Furthermore, there is little evidence on how citizen science bridges the gap between in-school and out-of-school learning and how it connects formal and informal education with schools and with communities. Citizen science in school can, for example, focus on existing scientific problems as well as question and engage the students in simple data collection or analysis, as suggested on many global citizen science platforms (e.g., Zooniverse.org). According to Figure 14.1, such projects can be placed in the first quartile of the scheme.

Alternatively, students can take part in policy-driven citizen science by monitoring air quality with adults from the community. In such a project, held in an air polluted neighborhood, students increased their knowledge and scientific thinking, and adult participants were able to interpret data in daily situations, as well as improved their scientific competencies and public understanding of science.[21,22] Students can help monitor the distribution of small mammals and inform decision makers about ecological corridors. They go outdoors, build "footprint traps", collect footprints, analyze the data and make recommendations to organizations such as the Israeli Nature Parks Authority. According to Figure 14.1, such projects can be placed in the second quartile of policy-driven citizen science.

In the rest of this chapter, we present a citizen science project in the third quartile: co-created and transition-based citizen science. We describe the Iris Project to illustrate how a citizen science can bridge several of the abovementioned gaps: between school and the outdoors, between formal and informal learning, and between the school and its community. We aimed at documenting how a bottom-up co-created citizen science project develops and what were the main citizen science project characteristics and learning outcomes of the participating students

The citizen science Iris Project

The citizen science Iris Project was developed in our *Taking Citizen Science to School* (TCSS) center, funded by the Israel Science Foundation and the Ministry of Education.

The citizen science Iris Project was designed by Hila (one of the co-authors), a science teacher and a graduate student then. Hila has documented the developing characteristics of the project and her students' learning outcomes. She began with an initial design of the citizen science project, but could not anticipate its further development, as this was the result of the development of the scientific investigation and the interaction among students, and between the students and herself, and the other adults who took part in the project during the different stages.

The Iris Project was based on modified steps from the model developed by Cornell Lab of Ornithology[20]: (1) Choose a scientific question. (2) Develop, test and refine protocols, data forms and educational support materials. (3) Recruit participants. (4) Train the participants. (5) Collect the data. (6) Analyze and interpret the data. (7) Disseminate the results.

The students collected data on phenology (the timings of cyclical or seasonal biological events) and on fertilization (fruiting) of *Iris atropurpurea*, which is classified as critically endangered species according to the International Union for Conservation of Nature (IUCN) Red List of Threatened Species.[21] This classification is largely of the product of habitat fragmentation and habitat loss due to massive urban development in the Sharon region of Israel. Referring back to Figure 14.1, conservation in highly populated countries can be considered a "wicked" problem.[6]

The participants were 25 middle-school students in ninth grade, aged 13–14, all enrolled in an honors science class in a school located in the Sharon region that serves middle and upper middle-class communities (Figure 14.2).

The students collected data as part of an inquiry learning unit of the ninth-grade science curriculum. School and outdoor activities took place for two hours per week over a six-month period. Overall, 14 hours were dedicated to data collection in the field. Topics taught in school included nature conservation, biodiversity, habitat loss, and fragmentation, and their connection

FIGURE 14.2 Marking the plots and collecting data.

to extinction, data collection methods and instruments, formulating research questions, and data analysis.

Developing the co-created, extreme, transition-based citizen science project

The idea of citizen science was appealing for Hila, the teacher, but she wanted to engage her students in a place-based fieldwork context. Searching for a topic or an authentic scientific question, we consulted with an ecologist of the Society for Protecting the Nature, the largest Israeli environmental NGO. In the next section, we describe the project phases that reflect the Cornell Model stages[22] but not necessarily in the same order or with the same scope.

1. *Finding a scientific question for the students' inquiry and collaboration with a scientist.* We introduced Hila to the ecologist, who pointed us to an interesting population of the endangered species of iris (*Iris atropurpurea*) in a field right near the schoolyard. The ecologist proposed a worthwhile question to investigate: Why do some plants in the local iris population fertilize and make fruits, while others do not? The question then guided a comprehensive investigation by the students that engaged them, not only in scientific inquiry but also in extensive conservation efforts involving the community and the municipality, due to a planned expansion of an industrial park.

Hila hoped that the project would satisfy three criteria: (1) The project would be applicable to, and feasible for school students, allowing them to collect data and investigate. (2) The data would be useful for ecologists. (3) The project would involve outdoor inquiry[23] within walking distance of the school and would enhance place-based education.[24] After learning about the iris population, and after a short field survey by Hila, she decided that the iris question met her criteria and could be engaging for her students.

2. *Learning about the irises and investigating them.* Hila had learned about irises from different sources. She visited a rare plants refuge garden where irises are grown in order to save them from extinction. She also met two iris researchers and interviewed them to expand and deepen her understanding of the subject. The three authors then thought about what data Hila's students could collect, and how.

3. *Applying for permissions and developing protocols.* Since the irises are protected wildflowers, permission was needed from the Israeli Natural Parks Authority (INPA) to allow the students to investigate the flowers. Collaboration with INPA provided initial data from a survey of rare species done in the area. The survey results recommend a continuous monitoring of the iris population in the site we planned to investigate. Then, consulting with the ecologist, Hila developed the data-collection protocol.

4. *Recruiting the school and developing the educational program.* The authors met with the school principal, the leading administrators, and the science teacher department and they all expressed great enthusiasm for the authentic inquiry and field work of the students and gave their blessing to hold outdoor lessons. Hila then developed the educational program and received the approval of the Ministry of Education for the inquiry topics. Finally, the plan included meetings with people working in nature conversation from different organizations as well as a field trip to an iris nature reserve.

5. *Identifying community volunteers.* Hila met a volunteer from the community who had already marked the irises in order to attempt saving them, and she told him about the project. He informed her about the local municipality's urban and industrial construction plans that might destroy the iris population. Hila then asked him to recruit other volunteers to collaborate with the students' citizen science project. Referring back to the Cornell Model in the Irises project, recruitment occurred in different stages and approached different participants.

6. *Enacting the program.* Hila introduced the program in school and taught the methods for data collection during the flowering season. The teaching unit included the following: indoor and outdoor learning; learning the inquiry process; learning about irises and iris research; biodiversity, and meeting iris researchers. This is when the students collected data in the field, analyzed data in class and presented papers and posters.

7. *Developing students' agency: Saving the irises.* After the students completed the data collection, Hila informed them about the municipality's construction plans. At that stage, the students were strongly committed to the project and to protecting the irises. They acted in various ways to raise awareness of the irises in the community and suggested protest action. Finally, they approached the INPA, requesting to relocate the irises in order to save them from going extinct. Drawing back to Bonney et al.,[25] this can be considered as "making an impact" on different levels: on student learning outcomes, on the community and on the environment.

8. *Collaborating with a researcher.* Up until that point, Hila had led the project, communicated with the community volunteers, collected data with her students, and together they thought about how to protect the iris population. After few months of authentic science learning, monitoring, reporting the data to various forums, and environmental activism of the students and community members, a university plant scientist became interested. Under his guidance and permission, the students learned how to dig the iris tubers out of the ground for relocation. Collaboration with the researcher developed only at this advanced stage of the conservation effort.

9. *Presenting the project to the community.* The students completed a collaborative inquiry paper that concluded the inquiry process. Eventually, they did not get an answer to their initial question due to snails eating all the flowers. In their papers the students presented the data they were able to collect and analyze up to this event. In addition, they presented their conservation efforts. They presented the project to the community, as well, explaining about the importance of conservation. They described the challenges they had dealt with during the project and reported on their actions for protecting the irises and on their overall positive experience.

10. *Relocating the irises.* After the plant scientist's visit to the school, and under his guidance, the students and the volunteers relocated 200 iris tubers and ultimately saved them.

Referring back to the different models of citizen science, we see the Irises project as co-created citizen science or even extreme citizen science with respect to the role of Hila. She initiated the project and planned the activities with some help of the NGO ecologist and with the plant scientists who joined the conservation effort. The project is transition-based[26] because it developed students' agency and inspired them to act.

The characteristics of the citizen science project and learning outcomes

The project aimed at identifying the main features of a place-based extreme citizen science project that involved the community in conservation efforts. Data on student participation and learning were collected during the project by observations of all the activities, from pre/post open-ended surveys, and from students' written reflections. The main characteristics that emerged from the data were collaboration, student hands-on fieldwork, the use of a variety of scientific practices, open-ended inquiry and genuine scientific work, and student-initiated conservation actions that enhanced the students' agency.

1. *Collaboration.* The project enhanced collaboration between different stakeholders during different stages of the project. These stakeholders included SPNI (an environmental NGO), INPA (a governmental organization), a university-affiliated researcher, community volunteers, a science teacher and middle-school students. In the final field activity, the students collaborated with community volunteers in order to replant the irises and ultimately protect them from future building development. The volunteers helped the students by preparing the plots for the tubers, relocating and watering them. The INPA ecologist joined the students in the field as well and explained the importance of their activities for biodiversity.

2. *Hands-on fieldwork.* In order to identify the causes of the lack of the iris's fertility (i.e., flowers making fruits), the students carried out field observations and collected data during the flowering season. They were divided into four groups, each collecting data on specific clusters of flowers, such as measurement their stems, the number of flowering plants, pollination and fertility. The students measured air and ground temperature and relative humidity using scientific instruments. Unfortunately, the flowers were eaten by snails; this event stopped data collection and enabled discussions of the limitations of the investigation and the nature of field studies. In the Post-Project Feedback Questionnaire, two activities were mentioned as being the most meaningful for the students: 34% of the 25 students pointed to data collection and 31% mentioned the field trip to the Iris Reserve.

3. *The use of a variety of scientific practices.* Asking questions, planning and carrying out investigations, analyzing and interpreting data, constructing explanations, and communicating information are all scientific practices that the students used. The educational program was designed to support the development of these practices.

4. *Open-ended inquiry and genuine scientific work.* The project started with a specific research question about possible reasons for the lack of fertilization in the iris population. Unlike many science classes, where students conduct experiments with predictable results and answers, in our project, the students did not know what, if anything, they might discover. By facing the obstacles and trying to mitigate them, the students followed the process of authentic scientific work. This led to frustration, as well as to unexpected or inadequate results, and to the disappointment over the impact of the snails on flowers. Consequently, the focus shifted to how to preserve the population, while learning about the topics of endangered species, nature conservation and activism.

5. *Contribution to nature conservation.* The students initiated different actions in order to prevent the iris population from becoming extinct. All their actions were not planned as part of the initial educational program, but eventually their ideas and actions saved the irises, thus reflecting the values and the responsibility of developing nature conservation.

Students addressed two specific learning outcomes: positive attitudes toward science and scientific work and developing stewardship and activism.

Positive attitudes toward science and scientific work. Meeting people who worked in different scientific disciplines, such as botany and ecology, and working with them in the field, using scientific practices throughout the project and having exposure to scientific research on irises helped the students understand the scientists' work and its importance to our society. Students' written reflections following the interaction with the plant scientist included:

> It was interesting to learn from a researcher and feel like researchers for a day.
>
> *Shani*

> Working in the field taught me what field research is and about scientific research more than any other task in which I would look for data in the Internet.
>
> *Sharon*

> The scientist's work is a very interesting and unusual and full of surprises, it is one of the most interesting jobs in the world.
>
> *Ram*

> In my opinion, the work of the scientist is very important, because he actually conducts a complete research on various subjects, and his discoveries can help in different areas that affect society and the population.
>
> *Or*

> I learned that the scientist's work is highly fascinating and creative. The scientist thinks about various causes for the problem or phenomenon, he has to come up with new ideas and then go to test his hypothesis.
>
> *Mor*

Given the initial lack of interest in plant sciences, and its limited scope in the curriculum, these student statements are outstanding. Furthermore, many responses to an open-ended question referred to curiosity, creativity, persistence and social contribution:

Scientists should be curious, think creatively, be determined, be persistent, be diligent, be clever, and be willing to contribute to humanity.

Shani

Arguments were detailed and addressed the benefits of science to humanity in solving daily problems such as patient care and healing disease, improving the quality of life and extending human life expectancy. Additionally, the students explained that knowing science is important for general knowledge:

Even if we do not use science on a daily basis, we should all have the ability to explain the basic principles of science, upon which our world is based on.

Ofer, questionnaire

Science helps us understanding the world.

Roni

Activism and stewardship. Several variables impact environmental behavior, such as experiences in natural environments, knowledge, attitudes, and values. We have documented behavior and stewardship as actions resulting from engagement in the irises citizen science which were not designed as part of the planned educational activities. The students developed a sense of ownership and responsibility and expressed their enthusiasm about the beautiful flowers:

I love the blooming irises, the blooming is beautiful.

Dan, field observation

Last time we collected data, only one flower bloomed, but today more than 50 bloom!

Ram, field observation

Look. How many flowers do we have here?

Shir, observation

Ownership and responsibility have developed among the students, sometimes expressed as "our irises" or as references to their specific plots. For example,

Dan, wait a minute, how do you walk there without stepping on the flowers? "Omer, you are stepping on our irises!"

Lior, observation

I'm the head of the group that has the greatest number of irises

Yuval, observation

These feelings were also expressed in the students' reflections:

I thought how I can help the irises and avoiding them going extinct.

Dana, reflection

Behavior changes and stewardship, which were not part of the planned program, were evident in various conservation activities and actions the students initiated during the project.

The students demonstrated environmental stewardship throughout the project. They wrote a letter to the mayor explaining the importance of the area and its vegetation, asking him to support its declaration as a nature reserve. They wrote and distributed a petition over social media to recruit the public, they put signs in the field to protect the flowers and to raise public awareness. Finally, they suggested the idea to relocate the irises, and they successfully did this after receiving permission from the INPA, under the guidance of the iris researcher. The reasons students gave for conservation reflected three arguments: preserving biodiversity, the need for sustainable development that secures next generations' rights, and the practical reason that plants absorb carbon dioxide.

Summary

In the introduction, we highlighted bridging the gap between school and outdoor learning, between formal and informal learning environments, and between schools and communities. The Iris Citizen Science Project integrated all these into one complex "multiple road junction" metaphor. The program developed through enactment, and relationships with scientists and community members have been built based on the direction the project took. The action component was not planned in advance, and was led by the students. The outcome of stewardship was not even discussed in the beginning stages. Outdoor learning was implemented in the field investigations, in the field trips to the natural history museum, and in the Iris Reserve. Finally, the students relocated the tubers with the help of community members. Collaboration between stakeholders was only partially planned ahead, especially with the ecologists from the Society for the Protection of Nature in Israel (SPNI) and INPA, although the researcher joined the effort mostly in the project's final stage. This collaboration implied that the Iris Project reflects not only the characteristics of participatory co-created citizen science but also that of extreme citizen science because the two scientists who were involved in different stages of the project acted as facilitators in addition to being experts, because we were involved in choosing the problem to investigate and in designing the data collection and further activities, but mainly because we addressed the different needs of Hila, as a science teacher, her students, community members and the involved scientists.

Finally, referring back to the heuristic in Figure 14.1, the Iris Project can definitely be placed in the third quartile of "transition-driven and civic science". It addressed the wicked problem of conservation vs. development. Many conflicting stakeholder interests emerged, and the students were exposed to the process in real time. Dillon and colleagues[6] argue that in transition-driven citizen science, citizens must have agency, scientific knowledge, local knowledge, and that social learning must occur. All these aspects of deep learning were evident in the Iris Project.

The transition in the Iris Project is reflected in the shift from a pre-planned, pre-structured, school-like inquiry project, to students beginning to investigate an authentic scientific question about fertilization in an endangered species, but ending in learning about and investigating the expansion of an industrial park into an open space where a unique habitat of this plant species' population exists. In addition, Hila invited community members – who were already engaged in protecting the flowers – to collaborate with the students, and eventually, the scientist was impressed with the students' efforts and came aboard to help relocate the flowers. The students lost the battle for protecting the irises in their natural habitat, but they understood that the survival of small patches of isolated populations of the plants was slim and that, at least, the flowers they observed and came to love were saved through careful and professional relocation to a nearby nature reserve.

Acknowledgments

This research was supported the Israel-Science-Foundation grant 2678/17. We would also like to thank our school partners, scientist, and the nature conservation organizations partner.

Notes

1 Carr, A. J. L. (2004). Why do we all need community science? In *Society & natural resources* 17(9), 841–849.
2 Kruger, L. E., & Shannon, M. A. (2010). Society & Natural Resources: Getting to Know Ourselves and Our Places Through Participation in Civic Social Assessment. https://doi.org/10.1080/089419200403866org/10.1080/089419200403866.
3 National Academies of Sciences and Medicine, E. (2018). Learning through citizen science: Enhancing opportunities by design (R. Pandya & K. A. Dibner, Eds.). The National Academies Press. https://doi.org/10.17226/25183.
4 Silvertown, J. (2009). A new dawn for citizen science. *Trends in Ecology and Evolution*, 24(9), 467–471. https://doi.org/10.1016/j.tree.2009.03.017.
5 Whitelaw, G., Vaughan, H., Craig, B., & Atkinson, D. (2003). Establishing the Canadian community monitoring network. *Environmental Monitoring and Assessment*, 88(1–3), 409–418. https://doi.org/10.1023/A:1025545813057.
6 Dillon, J., Stevenson, R. B., & Wals, A. E. J. (2016). Introduction to the special section Moving from Citizen to Civic Science to Address Wicked Conservation Problems. Corrected by erratum 12844. *Conservation Biology*, 30(3), 450–455. https://doi.org/10.1111/cobi.12689.

7 Phillips, T., Porticella, N., Constas, M., & Bonney, R. (2018). A framework for articulating and measuring individual learning outcomes from participation in citizen science. *Citizen Science: Theory and Practice*, 3(2), 3. https://theoryandpractice.citizenscienceassociation.org/article/10.5334/cstp.126/.

8 Bonney, R., Ballard, H., Jordan, R., McCallie, E., Phillips, T., Shirk, J., & Wilderman, C. C. (2009). Public Participation in Scientific Research: Defining the Field and Assessing Its Potential for Informal Science Education. A CAISE Inquiry Group Report. In *A CAISE Inquiry Group Report* (Issue July).

9 Conrad, Cathy. C., & Hilchey, Krista. G. (2011). A review of citizen science and community-based environmental monitoring: Issues and opportunities. *Environmental Monitoring and Assessment*, 176(1–4), 273–291. https://doi.org/10.1007/s10661-010-1582-5.

10 Bonney, R., Ballard, H., Jordan, R., McCallie, E., Phillips, T., Shirk, J., & Wilderman, C. C. (2009). Public Participation in Scientific Research: Defining the Field and Assessing Its Potential for Informal Science Education. A CAISE Inquiry Group Report. In *A CAISE Inquiry Group Report* (Issue July).

11 Haklay, M. (2015). Citizen science and policy: A European perspective. Wilson Center.

12 Sagy, O., Golumbic, Y., Ben-Horin Abramsky, H., Benichou, M., Atias, O., et al. (2019). Citizen science: An opportunity for learning in the networked society. *Learning in a Networked Society*, 17, 97–115. https://doi.org/10.1007/978-3-030-14610-8_6.

13 Keough, H. L., & Blahna, D. J. (2006). Achieving integrative, collaborative ecosystem management. *Conservation Biology*, 20(5), 1373–1382. https://doi.org/10.1111/j.1523-1739.2006.00445.x.

14 Bonney, R., Ballard, H., Jordan, R., McCallie, E., Phillips, T., Shirk, J., & Wilderman, C. C. (2009). Public Participation in Scientific Research: Defining the Field and Assessing Its Potential for Informal Science Education:A CAISE Inquiry Group Report https://files.eric.ed.gov/fulltext/ED519688.pdf

15 Dillon, J., Stevenson, R. B., & Wals, A. E. J. (2016). Moving from citizen to civic science to address wicked conservation problems. *Conservation Biology*, 30(3), 450–455. https://doi.org/10.1111/cobi.12689.

16 Dillon et al. (2016), p. 451.

17 Bonney, R., Shirk, J., & Phillips, T. B. (2015). Citizen science. In *Encyclopedia of Science Education* (pp. 152–154). Springer Netherlands. https://doi.org/10.1007/978-94-007-2150-0_291.

18 Sagy, O., Golumbic, Y., Ben-Horin Abramsky, H., Benichou, M., Atias, O., et al. (2019). Citizen science: An opportunity for learning in the networked society. *Learning in a Networked Society*, 17, 97–115. https://doi.org/10.1007/978-3-030-14610-8_6.

19 Phillips et al. (2018).

20 Bonney, R., Cooper, C. B., Dickinson, J., Kelling, S., Phillips, T., Rosenberg, K. V., & Shirk, J. (2009). Citizen science: A developing tool for expanding science knowledge and scientific literacy. In *BioScience*, 59(11), 977–984. https://doi.org/10.1525/bio.2009.59.11.9.

21 Sapir, Y. (2020). Iris atropurpurea. The IUCN Red List of Threatened Species 2020: e.T13161450A177444302. https://dx.doi.org/10.2305/IUCN.UK.2020-3.RLTS.T13161450A177444302.en. Accessed on 27 February 2023.

22 Bonney, et al. (2009)

23 Tal, T., Aviam, M., Levin-Peled, R., & Lavie Alon, N. (2016). Teachers in the outdoors: Bridging formal and informal practices. In L. Avraamidou & W.-M. Roth (Eds.), Intersections of formal and informal science (pp. 93–109). Philadelphia: Routledge.

24 Smith, G. A. (2007). Place-based education: Breaking through the constraining regularities of public school. *Environmental Education Research*, 13(2), 189–207.
25 Bonney, R., Cooper, C. B., Dickinson, J., Kelling, S., Phillips, T., Rosenberg, K. V., & Shirk, J. (2009). Citizen science: A developing tool for expanding science knowledge and scientific literacy. In *BioScience*, 59(11), 977–984. https://doi.org/10.1525/bio.2009.59.11.9.
26 Dillon et al. (2016).

15

WHERE YOU STAND AND WHY YOU STAND THERE

Cultural Competency as a Lens for Understanding Climate Change in the Minds of Rural Pennsylvania Fairgoers

Taiji Nelson and Mary Ann Steiner

In surveys, science and natural history museums have a large share of public trust because they are seen to be neutral, fact based, and because they present real objects and are research oriented.[1] But rather than resting on that limited notion of what museums can be, there has been interest in the field for a broader social role that acknowledges the fact that science is constantly evolving, not always resolved, and that engagement with it is valuable at all stages.[2] Because of the public trust and popularity of museums, they are well positioned to join the public in facing complex, current socio-scientific issues like climate change.[3]

Some museums have been establishing new roles as convenors for collective action, as collaborators in local identification and resolution of public health questions or as supporters of youth voice in community action on climate change.[4] In this work, the notion of two-way learning, that the institution has as much to learn as the community, is central.[5] Because of this, the passion and responsiveness of staff members as they learn with their audiences and help institutions navigate new roles as public partners in addressing climate change is central to each of these efforts. Taking this step into territory outside institutional expertise and going to locations outside institutional bounds, takes curiosity, vulnerability, and courage. While many museums still feel safer in the neutral stance of edutainment, there are indications that the public believes museums are capable of wider roles than neutral display of non-controversial facts.[6]

Through both urban and rural climate change education work supported by the National Science Foundation's Division on Research on Learning, Carnegie Museum of Natural History has also been taking on a new stance with public audiences around climate change.[7] This chapter presents a first-person account of a museum's efforts to listen deeply and engage in climate change conversations

DOI: 10.4324/9781003145387-17

within a local, rural context. The museum was the source of the interaction design, the context is a Western Pennsylvanian rural community, and the facilitators are a combination of museum and community actors who together broke the ice and started climate conversations. The experience surfaced ideas about how important policy vs science is to the outcome of climate communication; how open the museum can be to a role in a conversation that is curious rather than authoritative; how shifts in engagement with topics like climate change involve emotional connection, respect and interest first; and how museums can move beyond ideas of educating the next generation and call on public audiences of all ages to join them in openly caring about and acting on controversial topics.

Talking climate at the Stoneboro Fair

In early September of 2021, we were part of a small team of educators, scientists, conservation district staff and learning researchers who gathered in front of an exhibit table in the Sportsmen Building at the Stoneboro Fairgrounds in Mercer County, Pennsylvania. Behind us was a simple statement written on a whiteboard – *I am concerned about climate change* – and five clear plastic jars labeled: *Strongly Disagree, Disagree, Neutral, Agree, and Strongly Agree.*

As families and fairgoers wandered through the exhibit space, we invited them to drop a colorful pom-pom in the jar that represented their stance on the statement after assuring them that there was no "right" or "wrong" answer. Once they cast their vote, we asked if they were willing to share what they were thinking or feeling as they made their decision. Over the course of the day (and dozens of conversations) the *Strongly Agree* and *Agree* jars were filled with 65 percent of the votes, with *Neutral* comprising an additional 17 percent.

The growing consensus among participants was clear – a majority of people expressed concern about climate change – and that public response became a starting point for conversations. Among other things, those discussions included feelings of surprise, skepticism, curiosity, exasperation, determination, uncertainty, and hope. The visible evidence that many of their neighbors shared similar concerns and questions seemed to break the silence that people had been keeping. They shared personal experiences of floods and extreme weather, talked about what they'd been reading in the news or hearing from friends and family experiencing wildfires, and asked clarifying questions of educators and scientists from the Carnegie Museum of Natural History's Center for Anthropocene Studies. There was a sense of camaraderie, excitement, and a release of built-up tension among people who engaged in lengthy conversations, sometimes telling their friends that they would catch up with them so that they could stay and keep talking.

The interactive had several goals. The first was to better understand how we might spark and navigate productive, non-judgmental, and heartfelt one-on-one conversations about climate change at a public event. The second was to better understand the different ways that people relate to climate change in rural communities and how those communities grapple with those differences. The third was to engage folks with the right evidence, at the right time, from the right source – evidence that was relevant to, and framed for, them. Finally, we were curious to hear how people in this community described the intersections between climate change and their values, worldviews, identities, behavior, and futures. The design of the activity, the statement, the follow-up questions, and strategies for navigating discussion were all intended to "fit" the rural context, while also surfacing misconceptions, assumptions, and biases that often go unsaid, unseen, or unchallenged.

The Mercer County Conservation District had invited educators and scientists from the Carnegie Museum of Natural History (located 90 minutes south in Pittsburgh, PA) and learning scientists from the University of Pittsburgh Center for Learning in Out-of-School Environments as part of *the Climate in Rural Systems Partnership*, a project funded by the National Science Foundation's Advancing Informal Science Learning program. The Conservation District was a trusted institution within the community that participated in the Fair annually but engaging the public in conversations about climate change in this setting was new. Up to that point, the network of educators, scientists, farmers, artists, sportsmen, conservationists, community organizers, and government officials had been co-developing resources and strategies for climate communication in a safe space with generally like-minded people. The fairgoers represented the broader range of beliefs, attitudes, and sensitivities held by the network members' friends, families, colleagues, and neighbors. Folks were nervous about how the interactions would go, if we had sufficiently prepared for the discussions, and what might change if we shared our stances on climate change publicly in a community that felt deeply divided on the issue.

Knowing where I stand and why I stand there

As a person who was born and raised in rural Pennsylvania a short drive away in the Allegheny National Forest, the vibe of the fair felt reminiscent to me. The fried food being sold as a fundraiser for the local volunteer fire department, the blueribbon animal contests, the music, the rides, the carnival games, the rodeo, and the groups of teenagers spending time away from their parents all conjured memories of what life was like growing up in my hometown. I snapped photos and texted them to friends and family who expressed surprise and wondered what I was doing there. When I told them I was there to talk to people about climate change, there was a mix of concern and admiration.

The crowd was more racially diverse than I remembered rural Pennsylvania being as a kid, or maybe I just wasn't attuned to consciously noticing it in the early 2000s. The music they played during barrel racing crossed cultures and cycled between country, rap, and pop. There were a number of people with open carry fire arms and I noticed myself feeling uncomfortable with the prevalence of the American flag alongside clothing branded with *Blue Lives Matter* and *Make America Great Again*, sometimes spelled out in bullets. It reminded me that rural Pennsylvania communities and culture have been changing with the rest of America and that our relationship with climate change is interconnected with (and confounded by) a wide range of contemporary social, political, ethical, and economic issues. As a queer and mixed-race person, it also reminded me of complicated truths from my childhood related to sense of belonging, pride, and the risk associated with being different and breaking norms in a rural community.

Two years earlier, I left a job coordinating outdoor education programs for teens after nearly a decade because I felt a deep sense of disillusionment that I wasn't doing enough to engage other adults as our country was regressing on issues of social, racial, economic, and environmental justice. I had been designing cross-cultural program spaces that promoted diversity, equity, inclusion, and positive identity formation for young people, but in many ways, it felt that was putting the burden of societal change on the next generation. I felt that our field was dodging tough conversations among adults, instead looking for solutions by projecting our hopes, expectations, and shortcomings onto young people. In many ways our strategy seemed to be that *if we do our job as educators, the next generation will get it right.* In reality we need each generation to get it right, right now, in collaboration and solidarity with younger and older generations.

Furthermore, I recognized a growing sentiment of distrust, division, resentment, aggression, and hate between Americans and felt it permeating the community where I was born and where I live now. In my mind, Western Pennsylvania represents a microcosm of the broader United States. We are situated within Appalachia but are often torn between cultural identities of the Northeast and Midwest. We have a shared heritage of manufacturing, fossil fuel and natural resource extraction, and outdoor recreation. Our legacy also involves current and past colonization (the most recent census reports an American Indian population of more than 12,000, yet there are no federally recognized Indian tribes within Pennsylvania), labor abuses, and environmental injustices that fall unequally on women, the poor and working class, and racial and cultural minorities. As the home of the United States' first capital (Philadelphia), and prominent figures in the environmental movement (Rachel Carson, the author of Silent Spring was a Pittsburgher) and labor movement (Pittsburgh is sometimes called the "cradle of the American labor movement"), there are examples of cooperation, solidarity, and a respect for relationships between people and with land. There's a reason that Pennsylvania is called the Keystone State. Our borders bind communities with many social and political differences together.

Where do we go from here?

This duality and tension represents *a pickle* that makes talking about what we can/should collectively do about climate change intimidating. People are hesitant to talk about climate action because it's a complex and emotional topic. It's a discussion of both morality and mortality. It's existential. Climate communication researchers have already classified some common barriers including overwhelm, isolation, and hopelessness that contribute to a "spiral of silence." Overwhelm is fed by the scope, scale, and reach of the problem, the number of variables, the depth and breadth of expertise our understanding draws upon, and the ever-changing nature of systems. Isolation is fed by social pressure to not be depressing, to not be an alarmist, to not be angry, to not be preachy, to not be hypocritical. Hopelessness is fed by questions and doubts about *what* can I/we do, *will* I/we ever actually do it, *how* do I/we do it, and *does it even matter at this point*. The pickle gets even more sour when the environment to navigate those complexities is uncertain, tense, divided, suspicious, confrontational, and urgent. But as poet and social justice advocate James Baldwin said in his 1962 essay in the *New York Times, As Much Truth as One Can Bare*, "not everything that is faced can be changed, but nothing can be changed until it is faced." If we are going to solve climate change, we must find ways to talk about it.

One path forward involves refining our cultural competency to recognize the ways that rural communities talk (or don't talk) about climate change, and what that reveals about how they relate to and conceptualize the topic. Looking at peoples' relationships with climate change beyond their knowledge of climate science allows people to read between the lines to truly hear what is being said. It's also important to treat these conversations as opportunities for relational and reciprocal learning, rather than opportunities to transmit information, beliefs, or values unidirectionally. We aren't out here to treat people like we have all the answers (though we have some) or to tell them how they need to change (though we all do). If you don't live or work in a rural community, it's important to recognize that your worldview comes with blindspots and biases and that people from the community are the best source of information and insight about what works where they live. In my experience (and with some major caveats), the cultural norms of rural Pennsylvania prioritize harmony, responsibility, individual freedom, humility, self-sufficiency, and hard work.

For outsiders to engage productively, it requires an awareness, intentionality, and cultural competence to support talking across aspects of identity including class, job, political affiliation, religion/spirituality, education, age, culture, nationality, race, gender, sexuality, and ability. Spaces for conversations about climate action require the same level of care in their design and facilitation, because they touch on similar social and emotional trigger points that intersect with aspects of identity. Luckily, there are many examples of resources and practices from programming spaces designed to support diversity, equity, inclusion, and positive identity formation that can be applied to climate communication.

For instance, when responding to the statement "I am concerned about climate change" during the tabling activity at the Stoneboro Fair, many people who responded "agree" or "strongly agree" shared that talking about climate change was seen as controversial, political, and socially risky. They also felt extreme frustration with inaction and a lack of consensus. The discussion touched only briefly on the specifics of climate change and instead focused on what strategies or actions had efficacy and would fit the complex social, political, and economic systems in rural communities. For people who are already concerned, climate communicators can support them by providing information and environments that affirm that others share their hopes and fears, boost self-confidence in their beliefs, connect them with networks to take collective action, and build a sense of agency to match their urgency. These environments may already exist within a community and should always be created by and with the people who live there.

This is tied to the theory of change of the Climate in Rural Systems Partnership. In quarterly facilitated meetings convened by community partners, the network provides dedicated space and time for members to talk about if, how, and why actions to address climate change fit their identities and local context. Museum scientists, educators, and exhibit designers work with community organizations to codesign resources like videos, ARCGIS story maps, educator kits, tabling materials, and data visualizations, which are piloted at the quarterly network meetings. The chance to practice in a safe space ideally makes network members feel more confident and competent (and therefore more likely) to talk about climate change with the public. The discussion and negotiation that happen during that process are equally as important as the products of the discussion, because observing and reflecting with peers help to address overwhelm, isolation, and hopelessness. Convening around a group-worthy problem provides a place to surface sameness and differences and to verbalize your own perspective, expertise, and ideas about the relationships between climate action, your future, and the future of your community. As a network member reminded us, "It may feel like preaching to the choir, but every good choir needs practice."

Fairgoers who voted "neutral" were primarily seeking more information that could help them gain clarity, address skepticism and wariness, and make sense of the complex processes, systems, and issues. Additionally, the design of the statement "I am concerned about climate change" was open-ended so that conversations could spark around feelings, beliefs, values, identities, and experiences and did not position the understanding of ecological concepts or data as necessary prerequisites or starting points. It also allowed participants to engage in conversation around whatever was important to them.

The "neutral" group were curious but cautious, frequently interrogating the sources and credibility of information, and were also mindful of what would resonate or alienate their neighbors. They stressed the importance of respect for the nuances and opportunity costs surrounding climate action that often leave rural communities behind or in a bind. Examples of this were whether or not the production and disposal of solar panels and wind turbines was actually more costly or harmful to the environment than fossil fuels. In these moments, it's important to remember that while many of us absolutely have an agenda to move people toward action on climate, forcing a person to "drink from our well of knowledge" is not effective and often has an alienating effect. Framings to this group should be mindful of rural values including (but not limited to) a strong sense of place identity, connection to nature, self-reliance, self-determination, resourcefulness, moral responsibility, and reservations around governmental regulatory policies.

In discussions with "neutral" fairgoers, listening for relevance and responding with empathy felt especially important. As one educator from the Conservation District advised the group, "no one cares how much you know, until they know how much you care." My advice for engaging this group would be to navigate the discussion through humble inquiry, asking questions to clarify individuals' viewpoints, where they find relevance, where they have uncertainty that they'd like to explore more, and where they find sources of information that they trust. This is also a group where I respectfully shared my own stances and reasoning (where I stand and why I stand there) and, if they were open to it, confronted misconceptions and misinformation with data, stories, and an explanation of why I was confident in my own understanding. I was transparent with my uncertainties, where I lacked personal experience, and with examples of my understanding shifting over time. I was also clear where my stances were resolute on issues of justice and the scientific consensus around human-caused climate change. For people still making sense of conflicting information (or information that conflicts with their values, identities, actions or beliefs), climate communicators should create a safe environment for them to explore the relationship between these things. The goals could be to provide a different perspective that can serve as an opportunity to reflect on cognitive dissonance in a non-judgmental setting, give them next steps for credible information, and have them walk away more curious and empathetic about the issue.

Conversations with fairgoers who voted "disagree" or "strongly disagree" covered a wide range of beliefs, emotions, concepts, and questions about climate change. Our team had spent the majority of our time thinking through how we would engage this group because of anxieties around how aggressive or divisive the conversation might become. Even though our hunch, based on the Yale Climate Opinion Map, was that the majority of fairgoers would share similar concerns about climate change, several members of our group were unsure of their own preparedness to defuse confrontation, address misconceptions or misinformation, answer detailed questions about the underlying science, or maintain a productive conversation space. There was initially hesitation from members of our group about using the words "climate change" in the statement "I am concerned about climate change" for those reasons. These conversations felt socially risky because of the ways that people in small communities interact with one another across blurred lines of personal, professional, and social life where a negative interaction at the fair could impact their broader relationships in the community.

Our engagement with the fairgoers who disagreed with the statement was important to provide real-world experience to check whether our assumptions (and fears) met reality and to practice navigating differences with an acceptable amount of social risk. In these interactions, we applied strategies from deep canvassing which included building rapport and eliciting stories through compassionate curiosity and empathetic listening. When navigating differences, some tips from our partners at PA United, who specialize in deep canvassing, are to listen, ask open-ended questions, and let silence sit. The goal was to get the person to tell a story, to ask questions that surface complexity, to make the person feel heard, to understand what is at the emotional core of their conflict, and to share stories that establish common ground and resolve cognitive dissonance. They also advise that you should try to not interrupt, talk over the other person, talk about facts rather than experiences, or disregard or downplay what the person shares.

In the next paragraph are examples of what we heard from the people who disagreed with the statement "I am concerned about climate change." Some people said that they weren't concerned because the situation seemed hopeless, so they chose not to worry. Others were confident that the problem would eventually be solved by human ingenuity. An extremely small number of people denied that climate change existed. We also noticed examples of common misconceptions and/or misinformation that are frequently shared online through memes. Much of this misinformation relies on wedge issues, biases, and false dichotomies related to economics, education, class, race, religion, nationality, and heritage. In these instances, it was personally important that I state my own values and beliefs as direct opposition to deliberate efforts to spread division, distrust, and violence. My strategy to counter misinformation is to present what I know and how I know it, in ways that don't feel dismissive of their beliefs or that rely on facts alone. Instead, the intent of including evidence is to demonstrate the thinking behind your thinking.

Many in the "disagree" and "strongly disagree" groups questioned the United States' responsibility when compared to emissions from countries like China. Others suggested that it was hubris to think that human activity could interfere with global processes, that the process was largely driven by natural events like volcanic eruptions, or that the scientific consensus wasn't settled. A few said that *if* climate change was happening, it was "God's plan" and that humans couldn't or shouldn't do anything about it. Some folks deflected their concern by turning the focus onto what they saw as hypocrisy from people who advocate for climate but use or benefit from fossil fuels in their own lives. Also, several people we talked to worked in the natural gas industry or leased land for gas wells or had friends or family members who did. These folks see fossil fuel jobs and income as the best or only option for their communities.

Encountering a barrage of doubt, denial, or misinformation can feel confusing and disorienting. I often found myself wondering, "Where do I even begin?" In these situations, colleagues at the Climate Advocacy Lab advise us to pay attention to the physical reaction in our body as we navigate conversations that bring up the urgency we feel for ourselves, for the people and places we love, and for the future generations of life on our planet. Climate conversations touch on both our cognitive and emotional selves, and colleagues at the Climate Advocacy Lab advise that we should be conscious of this when conversations move from difference (which is an important opportunity space for change) to danger (where we lose the trust, respect, and shared understanding).

As a climate communicator, one of the most challenging parts of my job is accepting that some disagreements may never be resolved, that messages of solidarity from outsiders may not be initially trusted, and that sometimes the most productive outcome may be to *agree to disagree*. We have to expect and accept non-closure, but still find the energy, bravery, openness, and conviction to have tough conversations about workable solutions. Peoples' beliefs and understandings can (and do) change by participating in conversations where they listen, share, and feel heard. Because of the energy this takes though, we should be honest with ourselves when disagreements won't be resolved and recognize when it's okay to respectfully exit the conversation and direct our precious time and efforts beyond "*deny and delay*" conversations.

Finally, a thoughtful conclusion to the conversation is as important as the idea or experience that sparked it. I often ended discussions by thanking people for sharing their perspectives and asking if they had any questions for me. The most common one was, "*What can I do?*" I encourage you to think about what you would say to someone who asked you that. Practice explaining it out loud. It may be a hard question to answer, but here's what I have learned from others: There is no "right" or "wrong" way to take action, as long as you're pushing yourself to do your best. We need actions at all scales, from the individual to the systemic. There is little time and we have to start somewhere (though starting *somewhere* doesn't mean starting *anywhere*). Taking action is what will lead us to understanding and give people hope, and our first few steps will teach us what our next steps should be. We cannot wait to take action until we have a complete understanding of the issue. There will always be uncertainty with many opportunities to course correct. Show up with openness, compassion, and a vision for a hopeful future, and be willing to have the tough conversations to get us there.

Find ways to take action that apply your unique talents, interests, and network(s) of influence to something your community needs. Volunteer. Vote. Run for office. Show up to a public hearing. Call your elected representatives. Write a letter to the editor. Show up to a rally. Share a post. Make a donation. Divest your retirement fund. Ask large institutions to divest their endowments. Hold a discussion group at your school or church. Advocate for healthcare and education funding. Share articles with your friends and family. Teach your kids to be kind to nature. Talk to your parents and grandparents about climate change. Eat less meat. Eat local meat. Get to know your farmer. Walk or bike if you can. Take the bus or carpool if you can. Advocate for a sustainable purchasing policy at your workplace. Replace your lawn with native plants. Install solar panels and insulation in your home. When in doubt, try Googling it. Find a group that you vibe with and organize with joy. Advocate for more funding for green jobs, training programs, and sustainable technologies. Stand in solidarity with the homeless, the poor, the displaced, and others who feel climate change first. Work toward economic, racial, and gender justice knowing that it is also working toward climate justice.

None of us can do it all, and none of us can do it alone. Allow yourself moments to suspend disbelief, trust that change is possible, and encourage others to do the same. When we find ways to bring people together in spite of our differences, we can do a lot. And we must.

Acknowledgments

Many individuals and organizations provided inspiration, insight, and connections that influenced these ideas including the Mercer County Conservation District, Powdermill Nature Reserve, the University of Pittsburgh Center for Learning in Out-of-School Environments, the Carnegie Museum of Natural History, CRSP & SCREST network members, CRSP advisors, Pisano Films, the Climate Advocacy Lab, Project Drawdown, PA United, Climate Generation: A Will Steger Legacy, and the Yale Program on Climate Change Communication. Our work is in solidarity with these and other peoples' efforts to work toward a thriving collective future.

Notes

1 Griffiths, J.M. and King, D. (2008). *Interconnections: The IMLS National Study on the use of Libraries, Museums and the Internet, Museum Survey Results*. Institute of Museum and Library Services, Washington, DC. *Interconnections: The IMLS National Study on the Use of Libraries, Museums and the Internet, Museum Survey Results.pdf.* Wilkening Consulting and AAM (2021). *Museums and Trust Spring 2021.* American Alliance of Museums. Museums and Trust 2021.pdf.

2 Chittenden, D. (2011). Commentary: Roles, Opportunities and Challenges; Science museums engaging the public in emerging science and technology. Journal of Nanoparticle Research 12: 1549–1556. https://doi.org/10.1007/s11051-011-0311-5.

Knutson, K. (2019). Science and natural history museums and the challenges of communicating climate change. In Drotner, K., Dziekan, V., Parry, R., and Schroeder, K. (Eds.). *The Routledge Handbook of Museum Media and Communication.* New York: Routledge.

3 Sutton, S. (2020) The evolving responsibility of museum work in the time of climate change. Museum Management and Curatorship 35:6, 618–635. https://doi.org/10.1080/09647775.2020.1837000. Steiner, M.A. and Crowley, K. (2013). The natural history museum: Taking on a learning agenda. Curator The Museum Journal 56:2, 267–271.

4 Hamilton, P. and Ronning, E.C. (2020) Why museums? Museums as conveners on climate change. Journal of Museum Education 45:1, 16–27. https://doi.org/10.1080/10598650.2020.1720375. Hoffman, J.S. (2020) Learn, prepare, act: "Throwing Shade" on climate change. Journal of Museum Education 45:1, 28–41. https://doi.org/10.1080/10598650.2020.1711496. Kretser, J., & Griffin, E. (2020). Taking back our future: Empowering youth through climate summits. In *Teaching Climate Change in the United States* (pp. 143–152). New York: Routledge.

5 Bevan, Bronwyn, (2017). Research and practice: One way, two way, no way or new way? Curator The Museum Journal 60:2, 132–141. https://doi.org/10.1111/cura.12204. McCallie, E., Bell, L., Lohwater, T., Falk, J. H., Lehr, J. L., Lewenstein, B. V., Needham, C., and Wiehe, B. 2009. Many Experts, Many Audiences: Public Engagement with Science and Informal Science Education. A CAISE Inquiry Group Report. Washington, DC.: Center for Advancement of Informal Science Education (CAISE). http://caise.insci.org/uploads/docs/public_engagement_with_science.pdf.

6 Cameron, F. (2018). Stirring up trouble: museums as provocateurs and change agents in polycentric alliances for climate change action. In W. Leal Filho, B. Lackner, & H. McGhie (Eds.), *Addressing the Challenges in Communicating Climate Change Across Various Audiences* (pp. 647–673). Cham: Springer Nature.

7 Snyder, S., Hoffstadt, R. M., Allen, L., Crowley, K., Bader, D., & Horton, R. (2014). City-wide collaborations for urban climate education. In Hamilton, P. (Ed.), *Future Earth: Advancing Civic Understanding of the Anthropocene*, Geophysical Monograph Series, Vol. 197, American Geophysical Union, Washington, DC. National Science Foundation DRL [1906774, 1906368] Carnegie Institute; University of Pittsburgh.

PART III

Places and Spaces for Informal Science Learning

16

LIBRARIES ARE FOR SCIENCE

Judy Diamond

Outside the schoolyard fence along a busy street, teenagers huddle over their phones. They aren't texting friends or watching YouTube music videos or playing Pokémon GO. They are working on math and English school assignments because that is where they can access the internet. The internet, with its vast information resources, is free only as long as one can access Wi-Fi or a direct internet connection. For many Americans, the freedom of the internet is illusory, a selective privilege reserved for the chosen classes. In rural parts of the United States, as many as a fourth of the population have no internet access. Among Native students, 36% have no access to the internet in their homes. Remote schoolwork occurs where free Wi-Fi signals can be had – a parking lot, outside a school yard, or if lucky, a community center. The cost of home internet service exceeds the entirely monthly incomes for many US households. Technology for much of rural America is a tease – an extraordinary resource held just out of reach for many communities.[1]

There are few community resources that are free for all: A neighborhood playground, a park, and a library. Before the 19th century in the United States, there was little expectation that everyday people had a right to free resources from communities. The American Library Association credits the Peterborough Town Libraries in New Hampshire as the first free public library. Funded by the municipality, it opened in 1833 with free access for all members of the community.

Libraries are one of the few institutions that provide tangible resources without expecting payment. Local taxes do help fund libraries, but libraries are free not only just for those who pay taxes, but for everyone. And today, as neighborhoods are stratified by wealth and privilege, libraries make little distinction between users who are prosperous, main street, or homeless. Libraries

DOI: 10.4324/9781003145387-19

invite all walks of life to access their collections. Susan Orlean describes the daily scene at the Los Angeles Central Library: "In the morning, before the library opens, many of the people waiting to enter the building are carrying their worldly goods on their backs."[2] Libraries across the United States are waiving late fees for children, having decided that fees deter parents from letting their kids take books out, and it is more important for children to read books than for the library to collect money when they aren't returned.

Libraries don't loan just books; they provide resources of all kinds. The public library in Burlington, Vermont, loans garden tools and tennis rackets. The Brewerton branch of the Northern Onondaga library loans fishing poles. Public libraries in Ann Arbor and Northampton loan musical instruments, and libraries in Berkeley and Oakland loan power tools. The Arizona Pima County library gives out free garden seeds and, in 2011, partnered with the local public health department to employ a nurse who provided vaccines and first aid. Libraries loan art, they provide meeting spaces and shelter, and significantly, they provide access to computers and internet. Some even loan mobile hotspots so visitors can have broadband internet at home. All for free.

For many users, libraries are the most accessible source for curated resources – and increasingly this includes STEM. Over three-quarter of adult users say that libraries help them find health information and learn new technologies, and this is dramatically more evident for women, older adults, and Hispanic adults. These groups are also more likely to say that libraries help them decide what kinds of information to trust. Maintaining the well-being of self and family requires some understanding of human health, but sources of information about health often contain contradictory and/or misleading information. Libraries help users not only find information, they provide guidance on how to identify trustworthy sources.[3] In this way, libraries are a principal resource for lifelong learning about health, science, and technology.

Librarians increasingly view their institutions, not only as a collection within their buildings, but also as platforms that can unleash STEM expertise within the communities they serve. In this way, librarians help their users obtain access to STEM expertise that exists both within their community and also through the larger community of scientists committed to STEM outreach in underserved areas throughout the country. Libraries across the country have strengthened their role in encouraging engagement not only with technology but also with science and math. They loan science kits, they offer in-house workshops on fossils and plants, and they offer STEM book clubs for teens. Libraries also collaborate with natural history and science museums to offer story, poetry, and activity times.[4]

Over the last several decades, aided by funding from the John D. and Catherine T. MacArthur Foundation, the Institute for Museum and Library Services, and the National Science Foundation (NSF), public libraries have offered creative new approaches to encourage STEM learning.[5] *YOUmedia*, an innovative digital space for teens at the Chicago Public Libraries, has become a model for how specialized learning labs can function in library spaces. These spaces engage middle- and high-school youth in mentor-led learning using digital and traditional media, and they focus on fostering creativity and collaborative skills. Begun in 2009 with initial funding from the MacArthur Foundation, the program continues to expand, adding its most recent site in 2020.

The National Science Foundation has supported a range of innovative programs that strengthen science engagement through libraries. Pushing the Limits: Building Capacity to Enhance Public Understanding and Science Through Rural Libraries (2010–2015) focused on adult learners in libraries. The project saw its role as reinventing rural libraries to become more attuned to their communities by communicating STEM through a hybrid of book-club and scientific café and by enhancing STEM knowledge for librarians. Another NSF-funded program, *The STAR Library Education Network* (2014–2018), used local libraries as a vehicle for bringing museum quality hands-on exhibits on earth, space, and technology to rural communities. *LEAP into Science* (2007–2017), a partnership of The Franklin Institute Science Museum and the free Library of Philadelphia, integrated science content and inquiry engagement into existing afterschool programs through staff training workshops and table-top activities.

Recently NSF funded a paleontology research group to work with Tribal and rural libraries, Tribal schools, and even a Tribally-run senior care center to provide the latest books on climate change, social justice, paleontology, evolution, and biodiversity. A list of available books was assembled by indigenous educators, scientists, and librarians, so that individual partners could select up to $500 of books. This project provided a means for researchers to help strengthen the science resources of underserved communities through local libraries and schools.

Libraries pride themselves on being willing to meet users on their own terms. Elaine Gurian contrasts how library visitors differ from those to museums: "Where a library experience is free and organized for browsing, the expense of museum admissions can lead visitors to try to see as much as possible in the time allowed. Library patrons don't feel obliged to ration their visits, since they often visit repeatedly for multiple reasons. A library stay can be short and focused or leisurely and random."[6]

There are just over 9,000 public libraries in the United States, and the states with the highest per capita availability of public libraries are those with large rural populations: Vermont, Montana, North and South Dakota, Maine, Nebraska, Wyoming, Iowa, Alaska, and Kansas. Iowa has over 500 public libraries, Nebraska and Kansas over 200. In rural states, a librarian can serve many functions, even in relative isolation. The town of Monowi in north central Nebraska contains only one permanent resident, but she maintains a 5,000-volume library along with serving as mayor and bartender. Libraries in rural communities provide access to Wi-Fi and serve as trusted centers for community activities and governance, gaming centers for youth, galleries for local art, and places to socialize. Over 300 million Americans live within a public library service area, and each year half of those utilize library resources and over a hundred million participate in library programs. And libraries constantly change to meet their communities' evolving needs, adding over half a million new programs just between 2015 and 2016, along with a veritable explosion of new electronic and digital resources.

Strictly speaking, the service area of a library is the population within the boundaries of the geographic area the library was established to serve. Although over 90% of people in the United States live within the service area of a public library, there are as many as 17 million people who are left out. Compared to museums that typically charge admission, libraries have a huge leg up when it comes to public access. About half of Americans 16 or older have some interaction with a public library during a typical year – either through an in-person visit, using a library website, or via a mobile app. A total of 23% of all Americans aged 16 and over went to libraries to use computers, the internet, or a public Wi-Fi network. For Black Americans, this percentage was even higher, at 42%. A Pew Research survey of how people use libraries reveals a sense of dynamic change: Between 2015 and 2021, fewer people borrowed books from a library (66% down from 73%) and more used a library to sit and read, study, or watch and listen to media (53% up from 49%). And these more recent library users are more likely to be young, Hispanic, and lower income.[7]

Libraries deserve recognition for their potential to advance STEM learning. But STEM is not independent of the social and equity concerns that pervade in public institutions. Effectiveness in STEM outreach requires equitable distribution of resources, a climate of openness and acceptance, varied role models in significant positions, and a tolerance for diverse learning styles.

Many, if not most, libraries rightly boast that they give equal treatment to all patrons. First approved in 1939 and amended most recently in 2019, the American Library Association approved a "Library Bill of Rights" that includes powerful statements about the rights of all members of the public to access library resources. Despite the strong support for equal treatment by library organizations, in practice, librarians recognize the lack of diverse books for young readers, particularly those by Black, Hispanic or Indigenous authors. And librarians continuously confront attempts by community members to remove books by authors of color and books with themes relating to social justice.

In 1990, the American Library Association adopted a "Library Services for Poor People" policy that promotes equal access to information for all persons and recognizes the urgent need to respond to the increasing number of poor children, adults, and families in America. According to the policy, "These people are affected by a combination of limitations, including illiteracy, illness, social isolation, homelessness, hunger, and discrimination, which hamper the effectiveness of traditional library services. Therefore. it is crucial that libraries recognize their role in enabling poor people to participate fully in a democratic society, by utilizing a wide variety of available resources and strategies." In practice, librarians are caught between their upper middle class patrons who don't want to associate with homeless library users with their "…odor, intoxication, or noise level," and their desire to serve all users with respect and support. Houston's city council, for example, passed regulations for its libraries that ban "offensive bodily hygiene" and "large amounts of personal possessions" policies aimed directly as limiting the use of the libraries by the homeless.[7]

Although they are more common in states with large rural populations, libraries are far from being evenly distributed. Many low income, immigrant, and Native communities lack ready access to local public libraries, and those that are available are sometimes poorly funded. Public libraries are typically allocated and supported by county governments, who have the discretion to deny equal access across their constituencies. Even in large cities, libraries in poor neighborhoods can suffer. Libraries in poorer neighborhoods are more likely to have smaller collections, shorter hours, or to be closed. In rural and poor neighborhoods, libraries have smaller collections, less funding, fewer books per child and books with less recent publication dates.[8]

In June 2020, the Executive Board of the American Library Association urged its members and library institutions everywhere to "join it in condemning the systemic and systematic social injustices endured by Black people and People of Color and in working not only responsively, but also preemptively, to eradicate racism anywhere and everywhere it exists." Despite the show of support for the concerns of the Black Lives Matter movement, relatively few people of color work in libraries. White people are strongly overrepresented relative to their proportion in the US population (for 2017: 81.3% white librarians vs 73% white people) and black librarians are underrepresented (for 2017: 6.48% Black librarians vs 12.3% Black Americans).[9]

Libraries have come a long way since the 1990s when a nurse tried to check out books for HIV-positive readers from a local NY public library branch and library volunteers refused, invoking fears for their safety. In contrast, during the 2020–2021 COVID-19 pandemic, local libraries kept reading rooms open even when schools and restaurants had closed. The demand for remote services that libraries had always provided, such as loaning e-books and online medical reference, skyrocketed. But libraries went much farther than business as usual: Some repurposed book drops as mask drops, they made Wi-Fi available in parking lots during closed hours, and they provided free help filling out unemployment forms.

Libraries already serve as education centers deeply woven into the fabric of communities. As libraries are rapidly expanding their STEM programs and resources, they need to continue to progress in ensuring freedom of access, valuing diversity and tolerance, providing diverse role models, and delivering resources equitably to all communities. Libraries have the potential to become key institutions for informal engagement with science, health, and technology as they expand high quality STEM resources and improve quality of access for all users.

Notes

1 Katz, Vikki S. (2017). What it means to be "under-connected" in lower-income families. *Journal of Children and Media*, 11(2): 241–244. https://doi.org/10.1080/17 482798.2017.1305602. National Indian Education Association. (2019). *Broadband Access and Education Technology in Indian Country.*
2 Orlean, Susan. (2018). *The Library Book.* P. 73. New York: Simon & Schuster.
3 Moore, Elizabeth, Gordon, Andrew, Gordon, Margaret, and Heuertiz, Linda. (2002. *It's Working: People from Low-Income Families Disproportionately Use Library Computers: A Report to the Bill & Melinda Gates Foundation.* US Library Program on a Survey of Library Patrons. Evans School of Public Affairs. University Of Washington, Seattle. Real, B., Bertot, J. C., and Jaeger, P. T. (2014). Rural public libraries and digital inclusion: Issues and challenges. *Information Technology and Libraries, 33*(1), 6.
4 Harrington, E. G., and Beale, H. (2010). Natural Wonders: Implementing Environmental Programming in Libraries. *Children and Libraries.* Spring 2010. Harrington, Eileen. (2019). *Academic Libraries and Public Engagement with Science and Technology.* Chandos Publishing.
5 Philbin, M. M., Parker, C. M., Flaherty, M. G., and Hirsch, J. S. (2019). Public libraries: A community-level resource to advance population health. *Journal of Community Health, 44*(1): 192–199. Neuman, Susan and Moland, Naomi (2019). Book deserts: The consequences of income segregation on children's access to print. *Urban Education, 54*(1), 126–147. Neuman, S.B. and Celano, D. (2001). Access to print in low-income and middle-income communities: An ecological study of four neighborhoods. *Reading Research Quarterly, 36*(1), 8.
6 Gurian, E. H. (2005). Free at last: A case for eliminating admissions charges in museums. *Museum News, 84*(5), 1–11.
7 Horrigan, John B. (2015). Chapter One: Who Uses Libraries and What they do at their Libraries. In *Libraries at the Crossroads Pew Research Center: Internet and Technology.* September 15, 2015. https://www.pewresearch.org/internet/2015/09/15/who-uses-libraries-and-what-they-do-at-their-libraries.
8 Neuman, S.B. and Celano, D. (2001). Access to print in low-income and middle-income communities: An ecological study of four neighborhoods. *Reading Research Quarterly, 36*(1), 8–26.
9 http://www.ala.org/aboutala/offices/extending-our-reach-reducing-homelessness-through-library-engagement-7. Berman, Sanford. (2005). Classism in the Stacks: Libraries and Poverty. Jean E. Coleman Library Outreach Lecture. American Library Association. See also Neuman and Moland 2019. Pelczar, M., Frehill, L. M., and Nielsen, J. U. (2020). *Public Libraries Survey Fiscal Year 2018.* Washington, DC: Institute of Museum and Library Services.

17

NIGHT SKIES AND BUTTERFLIES

Leisure Science Activities and STEM Interests

M. Gail Jones and Megan Ennes

For many people, learning about science is a lifelong endeavor driven by individual interests.[1] There are many ways people can engage in informal science learning including digital spaces, out-of-school programs, science centers, and by engaging in science hobbies.[1] We recently conducted a study with adults who had science, technology, engineering, and mathematics (STEM) hobbies and found that hobby interests often originated from early childhood experiences with friends or family.[2] Unfortunately, not everyone has the opportunity to explore STEM hobbies and many hobby groups struggle with engaging diverse participants. This pattern of participation raises questions about how hobbyist groups, families, and informal organizations can engage youth from diverse communities and better support and nurture budding interests in STEM.

Rationale

It is increasingly clear that there are a number of factors that contribute to the development of STEM interests and career aspirations. As noted above there is also growing evidence that STEM interests begin in childhood.[3] When scientists were asked in a recent study how their initial interests in science developed, they reported that the leisure activities they engaged in as youth were instrumental in shaping their interests and career decisions.[4] These findings raise questions about how interests develop over time and how interests are initially formed. It seems that children whose families engage in birdwatching or fossil hunting are likely to develop leisure interests in these areas, but it is not clear is how children whose families have other leisure interests develop STEM interests. In this chapter we examine research on the development of

DOI: 10.4324/9781003145387-20

lifelong informal and career interests in STEM. Furthermore, we examine the intersection of culture and habitus (how a family engages with science) on STEM interests from several theoretical perspectives.

Background

The number of people who engage in STEM leisure interests (e.g., hobbies) continues to grow. Here we use the term hobby to refer to free-choice learning for pleasure that occurs outside of work or school. STEM hobbies include a wide range of activities such as bird watching, amateur astronomy, fossil collecting, and even home brewing. Hobbies play important roles for individuals that include relaxation, enjoyment,[5–7] learning about STEM,[3,5,6,8,9] opportunities to meet other people and socialize,[3,5–7] as well as other benefits such as improved mental health.[9,10]

Hobby experiences can also influence career decisions. A study of the impact of early hobbies on career choice found a clear relationship between early STEM leisure experiences and a decision to enter an STEM career.[8] Furthermore, those who engaged in an STEM hobby and held an STEM career were more likely to report they excelled in science, mathematics, and technology than those hobbyists who did not have an STEM career.

Who engages in STEM hobbies?

Researchers have increasingly reported significant differences among adults who engage in STEM hobbies. We recently conducted a nationwide study of nearly 3,000 STEM hobbyists engaged in ten different STEM hobbies including astronomy, beekeeping, birding, electronics/robotics, environmental monitoring, falconry, gardening/horticulture, home brewing, model building, and rock/fossil collecting.2 Hobbyists were contacted through hobby groups, informal education organizations, listservs, internet invitations, hobby publications, and science departments at universities. Special efforts were made to recruit hobbyists from minority groups (e.g., contacts to historically black universities). Furthermore, hobbyists were recruited from every state in the US including rural and urban regions.

Our study found that most of the hobbyists were White (93%) and male (74%). When the data were examined by type of hobby, the race/ethnicity demographics were very similar for birding (97% White), falconry (97% White), and home brewing (96% White). These findings have been reported in other studies of leisure activities that also found differences in engagement by race/ethnicity. For example, research in the US suggests that African American and White individuals tend not engage in the same types of leisure activities.[11–13] Variations in leisure interests have also been documented between Hispanic and White populations.[14,15]

In addition to the differential engagement in leisure interests between racial groups, there are also reported gender and geographic differences. In the study reported above, males were more likely than females to report doing model building (99%) and electronics/robotics (91%).2 Environmental monitoring, unlike other STEM hobbies, was reported equally by both males and females (almost 50% male/female). Females were more likely to engage in birding (59%) and gardening (74%) than the other STEM areas. When geographic location was examined, the study showed that most of the participants lived in suburban areas. This was especially true for hobbyists who engaged in rock/fossil collecting, astronomy, home brewing, electronic/robotics, and model building.

What sparks initial interests in STEM leisure and hobbies?

Research has found that families are important in providing that first exposure to science, driving initial interest.[16] Interest in science also occurs earlier when introduced by the family.[17] Parental attitudes toward science also influence youth interest in science, those with more positive attitudes tend to raise children with more positive attitudes toward science.[18] Parents with positive science attitudes also tend to become more involved in their child's science interests and often encourage visits to informal science centers such as museums and libraries.[19] Parental support of children's science interests is vital as parents have the ability to limit or encourage youth interest and participation in science hobbies.[20]

Family support is also vital in maintaining youth interest in science and science hobbies.[20] Research shows that while young people may enjoy science, they develop ideas related to their identity as what is appropriate for "someone like me" as early as elementary school.[19] Children do not develop their science perspectives in isolation but rather form them as part of their family unit. Youth's interest in science and science hobbies is related to their families' science capital and habitus.[19] *Science capital* refers to the science resources (economic, cultural, and social) a family has, while *science habitus* is how a family chooses to engage, or not, with science.[19] Parents who engage with their children in out-of-school science experiences and hobbies are more likely to raise children who have increased levels of science literacy, a greater desire to keep learning about science topics, and an understanding that science is important to them as a family.[21]

Research with families has found that parents can support their children's science interests by emphasizing the importance of the subject, limiting screen time, and taking their children to informal science centers.[22] These are important strategies for supporting youth interest in science and science hobbies that are available to all families. Museums and other informal science centers can help parents support their children's early interest in science through family programs that make "science accessible, meaningful, and relevant for diverse students by connecting their home and community cultures to science."[23] It is important to recognize that not all families have financial resources and access to experiences such as those offered by museums. However, museums are working hard to provide access to all families regardless of financial status.

Challenges to promoting STEM hobby interests

The studies discussed here show that STEM hobbies develop at an early age and are influenced by many factors in the home and the community. Knowing someone who engages in STEM for fun, growing up doing a hobby alongside another adult, and having access to tools and equipment to engage in the activity all contribute to the development of STEM leisure interests.[24]

Many of these factors that support STEM hobby development are related to economic resources such as having funding for tools and equipment (e.g., binoculars or telescopes) as well as having sufficient income to have leisure time to engage in a hobby. Some activities like home brewing or falconry require significant financial resources that many families may not have access to. But others, like gardening or bird watching, can be done with minimal financial resources.

The more difficult challenge to opening up STEM interests to youth is providing opportunities to experience hobbies with an adult or other person who engages in them. Clearly, if you do not know someone who has an interest similar to yours, the opportunities to learn a new hobby are more difficult. Studies by Robinson[25] examined influences on African American birders and found a lack of role models as well as social pressures were impediments to birding. Other studies[3] have found that safety is a significant concern for some STEM hobbyists such as amateur astronomers who must go in isolated areas to view the night sky.

Strategies that work to encourage STEM leisure interests

The Expectancy Value model by Eccles and Wigfield[26] provides a rich framework to examine motivations for engaging in an STEM hobby. This model suggests that previous experiences, socializer's beliefs, perceptions of gender roles, and activity demands are all factors that influence an individual's interpretation of experiences. This interpretation then influences the expectation of success as well as the interest and value they have for an activity. Although the model is typically applied in academic settings, the factors are also relevant for predicting leisure interests and engagement. As we noted above, if you do not perceive others as supporting your engagement in an STEM hobby, do not know anyone who does a hobby like yours, or if you have not had STEM experiences, you are less likely to engage in an STEM hobby.

The encouraging aspect of this model for STEM leisure activities is that many of the variables can be altered to provide youth with support for growing interests. Adult hobbyists want to engage others in learning about their hobby,[2] and we can link families and hobbyists together to explore new interests. Inside of schools, teachers can connect to those in the community who engage in subject-related hobbies (e.g., birdwatchers, robotics hobbyists) to model how to continue interests throughout one's lifespan.[27]

Developing programs to introduce youth to science hobbies within their family and community can have long lasting, positive impacts.[11] Data are emerging that show that parents from groups historically excluded from science often choose family-oriented events so they can be present to make sure their child is safe.[28] However, there are many perceived barriers families face when choosing to engage in science and science hobbies. These include transportation, cost, language barriers, and not being aware of the programs.[28] Addressing these barriers will be important for supporting diverse groups in engaging in science hobbies as a family.

Directions for future research

There are a number of areas where we need to know more about leisure interest development if we are to fully support youth in exploring STEM fields. First, what experiences are most influential in sparking and sustaining an interest? Is it enough to try to build one robotic toy to know if it is interesting or do you need different experiences over time? Is there a differential interest in types of leisure activities by age and development? We know that other individuals (influential socializers) can shape interests, but little is known about how one socializer is more or less influential than another. For example, can a retired White engineer be as influential to a young Hispanic male as a retired Hispanic engineer? Do the characteristics of the influencer make a difference in this interest development process? What about the influence of family members on the development of interests? Does it matter if it is a parent, grandparent, or sibling who engages an individual in a leisure activity? We need to know more about how to sustain interests over time. What types of support (financial, social, or institutional) are needed to move an interest from initial interest to a lifelong interest?

Conclusion

Supporting early science interests through family engagement and informal science centers is an important strategy for encouraging STEM hobbies and careers. By taking a more holistic approach to nurturing and supporting STEM interests, we can better support youth and promote lifelong engagement with STEM. The benefits of STEM engagement are tremendous for the individual as well as for the larger society. The challenge for educators as we move forward is to build educational programs that ensure that all families have access and opportunities to engage in STEM experiences. Additionally, hobbyist groups and organizations such as museums need to intentionally develop programs that seek to address the gender and ethnic biases found in many of these STEM activities. Mentoring or family programs that engage youth with hobbyists from varying backgrounds and gender identities will help youth from all backgrounds see themselves as someone who can engage in and enjoy these STEM hobbies. Not all youth will pursue STEM careers but they can still develop lifelong enjoyment of STEM through hobbies. Increasing participation of women and historically excluded groups in science, technology, engineering, and mathematics (STEM) is vital to increase the creativity, innovation, and diversity of STEM fields.[29]

Notes

1 Sacco, K., Falk, J. H., & Bell, J. (2014). Informal science education: Lifelong, life-wide, life-deep. *PLoS Biology, 12*(11), e1001986. https://doi.org/10.1371/journal.pbio.1001986.

2 Corin, E., Jones, M. G., Andre, T., & Childers, G., (2017). Characteristics of lifelong science learners: An investigation of STEM hobbyists. *International Journal of Science Education, Part B, 8*(1), 53–75.

3 Jones, M. G., Corin, E., Andre, T., Childers, G., & Stevens, V. (2017). Factors contributing to lifelong science learning: Amateur astronomers and birders. *Journal of Research in Science Teaching, 54*(3), 412–433.

4 Jones, M. G., Taylor, A., & Forrester, J. (2011). Developing a scientist: A retrospective look. *International Journal of Science Education, 33*(12), 1653–1673.

5 Cheng, E. H., Patterson, I., Packer, J., & Pegg, S. (2010). Identifying the satisfaction derived from leisure gardening by older adults. *Annals of Leisure Research, 13*(3), 395–419.

6 Murray, D. W., & O'Neill, M. A. (2015). Home brewing and serious leisure: Exploring the motivation to engage and the resultant satisfaction derived through participation. *World Leisure Journal, 57*(4), 284–296.

7 Scott, D., Baker, S. M., & Kim, C. (1999). Motivations and commitments among participants in the Great Texas Birding Classic. *Human Dimensions of Wildlife, 4*(1), 50–67.

8 Jones, M. G., Childers, G., Corin, E., Chesnutt, K., & Andre, T. (2019). Free choice learning and STEM career choice. *International Journal of Science Education, Part B, 9*(1), 29–39.

9 Edwards, P. K. (1981). Race, residence, and leisure style: Some policy implications. *Leisure Sciences, 4*(2), 95–112.

10 Floyd, M., Shinew, K., Mcguire, F., & Noe, F. (1994). Race, class, and leisure activity preferences: Marginality and ethnicity revisited. *Journal of Leisure Research, 26*, 158–173.

11 Robinson, J. C. (2005). Relative prevalence of African Americans among bird watchers. In C. Ralph, & T. Rich (Eds.), *Bird conservation implementation and integration in the Americas: Proceedings of the third international partners in flight conference.* Vol 2. Gen. Tech. Rep. PSW-GTR-191 (pp. 1286–1296). Albany, CA: U.S. Dept. of Agriculture, Forest Service, Pacific Southwest Research Station.

12 Shinew, K. J., Floyd, M. F., & Parry, D. (2004). Understanding the relationship between race and leisure activities and constraints: Exploring an alternative framework. *Leisure Sciences, 26*(2), 181–199.

13 Washburne, R. F. (1978). Black under-participation in wildland recreation: Alternative explanations. *Leisure Sciences, 1*(2), 175–189.

14 Floyd, M. F., Gramann, J. H., & Saenz, R. (1993). Ethnic factors and the use of public outdoor recreation areas: The case of Mexican Americans. *Leisure Sciences, 15*, 83–98.

15 Irwin, P. N., Gartner, W. C., & Phelps, C. C. (1990). Mexican-American/Anglo cultural differences as recreation style determinants. *Leisure Sciences, 12*(4), 335–348.

16 Maltese, A. V., & Tai, R. H. (2010). Eyeballs in the fridge: Sources of early interest in science. *International Journal of Science Education, 32*(December 2011), 669–685. https://doi.org/10.1080/09500690902792385.

17 Dabney, K. P., Tai, R. H., & Scott, M. R. (2015). Informal science: Family education, experiences, and initial interest in science. *International Journal of Science Education, Part B, 8455*(June 2015), 1–20. https://doi.org/10.1080/21548455.2015.1058990.

18 Perera, L. D. H. (2014). Parents' attitudes towards science and their children's science achievement. *International Journal of Science Education, 36*(18), 3021–3041. https://doi.org/10.1080/09500693.2014.949900.

19 Archer, L., DeWitt, J., Osborne, J., Dillon, J., Willis, B., & Wong, B. (2012). Science aspirations, capital, and family habitus: How families shape children's engagement and identification with science. *American Educational Research Journal, 49*(5), 881–908. https://doi.org/10.3102/0002831211433290.

20 Nugent, G., Barker, B., Welch, G., Grandgenett, N., Wu, C., & Nelson, C. (2015). A model of factors contributing to STEM learning and career orientation. *International Journal of Science Education, 37*(7), 1067–1088. https://doi.org/10.1080/09500 693.2015.1017863.

21 Crowley, K., Callanan, M. A., Tenenbaum, H. R., & Allen, E. (2001). Parents explain more often to boys than to girls during shared scientific thinking. *Psychological Science, 12*, 258–261. https://doi.org/10.1111/1467-9280.00347.

22 Lee, O., & Luykx, A. (2006). *Science Education and Student Diversity: Synthesis and Research Agenda.* Cambridge University Press.

23 National Research Council (NRC). (2009). In P. Bell, B. Lewenstein, A. W. Shouse, & M. A. Feder (Eds.), *Learning Science in Informal Environments: People, Places, Pursuits.* Washington, DC: Board on Science Education, Center for Education. Division of Behavioral and Social Sciences and Education. https://doi.org/10.1080/09500690903454217.

24 Jones, M. G., Chesnutt, K., Ennes, M., Mulvey, K.L., & Cayton, E. (2021). Factors predicting future science task value. *Journal of Research in Science Teaching, 58*(7), 937–955.

25 Robinson, J. C. (2005). Relative prevalence of African Americans among bird watchers. In C. Ralph, & T. Rich (Eds.), *Bird conservation implementation and integration in the Americas: Proceedings of the third international partners in flight conference.* Vol 2. Gen. Tech. Rep. PSW-GTR-191 (pp. 1286–1296). Albany, CA: US Dept. of Agriculture, Forest Service, Pacific Southwest Research Station.

26 Eccles, J. S., & Wigfield, A. (2020). From expectancy-value theory to situated expectancy-value theory: A developmental, social cognitive, and sociocultural perspective on motivation. *Contemporary Educational Psychology, 61*, 101859.

27 Ennes, M., & Jones, M. G. (2021). Building a science community. *Science Scope, 44*(4), 29–35.

28 Bruyere, B. L., Billingsley, E. D., & O'Day, L. (2009). A closer examination of barriers to participation in informal science education for Latinos and Caucasians. *Journal of Women and Minorities in Science and Engineering, 15*(1), 1–14. https://doi.org/10.1615/JWomenMinorScienEng.v15.i1.10.

29 Kricorian, K., Seu, M., Lopez, D., Ureta, E., & Equils, O. (2020). Factors influencing participation of underrepresented students in STEM fields: Matched mentors and mindsets. *International Journal of STEM Education, 7*(1), 1–9.

18

CHEMICAL ESCAPE ROOMS

Bridging the Gap between Formal and Informal Science Learning

Malka Yayon, Shelley Rap, and Ron Blonder

Escape Rooms (EsRms) are highly attractive leisure activities that appeal to many audiences. They are games in which a small number of participants (about 4–7 participants) are locked in a room and have to solve a series of puzzles, find clues, and decipher codes, in order to escape from the room in a limited time frame. They require teamwork to solve problems and include enjoyment and cooperative activities, which challenge participants' knowledge and skills.[1] EsRms may be non-virtual, virtual, or hybrid games; the participants have to solve puzzles, think out of the box, work as a team, and solve hands-on puzzles within a certain time frame. Often they must physically escape from a room, hence the name, but in other cases, they need to solve a mystery such as solving a crime, finding a cure, or creating a vaccine.

Recreational EsRms are designed and operated to ensure maximum enjoyment. They create authentic environments by building and decorating the EsRm according to the story and the scenario. The puzzles use technological accessories triggered by the correct solution of a puzzle, such as opening doors or crawling through a laser maze,[2] using virtual reality (VR) to create a sense of fear in a dark and gloomy environment.[3] A person (the master of the game) follows the game from a control room and intervenes only when needed to provide minimum help. All these factors help to immerse and engage participants in the game, so much so that they often report not being aware of the time, having a sense of flow[4] during the activity.

This genre appeared in Japan in 2007 and since then EsRms have spread worldwide like wildfire. Educators discovered their potential and subsequently many different types of Educational EsRms (EdEsRms) have been designed to

DOI: 10.4324/9781003145387-21

spark students' enthusiasm, develop positive attitudes, and teach subject matter content.

At the Chemistry Group in the Department of Science Teaching at the Weizmann Institute of Science, we try to utilize the potential value inherent in EsRms to increase students' enthusiasm and attitudes towards studying chemistry. In this chapter, we describe the research and general principles of EdEsRms and the uniqueness of Chemical EsRms (ChEsRms) that operate in the school context. We have found that ChEsRms function like informal science education activities within formal school settings. Based on these results, we wish to highlight the importance of games such as EsRms in chemistry classes and their implementation by teachers as a possible approach to bridge between formal and informal science education.

Implementation of EsRm format in formal educational settings

Educators and researchers embraced EsRms in formal education systems, even though the time required for planning and designing an EsRm is generally long.[5] Interest in this field is increasing, as reflected by the number of published studies and the high number of members in an open source of EdEsRms such as "Breakout EDU", which includes hundreds of thousands of schools in more than 85 countries.[6]

Many types of EsRms have evolved, varying in their target clientele (age range, type of players, or the number of players), fields (such as medical education,[7] computer science,[8] history,[9] and chemistry[10,11]), and the type of game (physical, fully digital,[8,12] and hybrid, combining the two formats[13]).

In formal education, EsRms are targeted to a certain class, curriculum, and educational purpose. These are the main reasons for the tension between informal characteristics such as playing and having fun, and formal educational characteristics, which tend to be more structured and have less of a playful nature.

Finding ways to impact students' interest and motivation for learning is the holy grail of formal or informal education and is being studied by researchers in the community of science educators[14,15] and educational psychology.[16,17] Inquiry-based experiences and makerspaces that invite the visitors to take an active role were first introduced in informal education settings and later on were adopted by the formal school system. In contrast, EsRms first appeared in the formal education system and only later were introduced to the informal education system.

EsRms in informal education settings

Museums embraced the EsRm format later since EsRms require space used by only a few people at a time. It took time to find ways to cope with this obstacle.[18] EsRms obviously fit the science museums' aims to improve science communication and attract new audiences; moreover, their staffs are capable of building attractive and creative hands-on activities.

Prior to building permanent EsRms, some museums have developed and tested prototypes in their museums. The results were that most participants spent more time[19] and showed more interest[20] while experiencing the EdEsRms compared with students having a free visit to this museum exhibitions. Following these prototypes, EsRms were constructed.

To overcome the drawback of space required by EsRms, which only a few people use for an hour, museums developed several ways in which EsRms are integrated into their floor plan, for example, EsRms that are built into the exhibitions themselves,[21] EsRms that require the visitors to explore a variety of aspects of the different exhibits,[22] or even pop-up EsRms.[23]

Comparing recreational, informal, and formal educational EsRms

These three different types of EsRms share some characteristics but differ in others. In order to better describe the tension between EsRms designed for formal and informal learning settings, we refer to different aspects of EsRm activities: the administrative and the pedagogical aspects.

Administrative aspects. All EsRms are games that require cooperative work and thinking out of the box in order to solve a mystery or assignment. Recreational and informal EdEsRms are usually designed for a specific area, in comparison with formal EdEsRms that usually take place in an existing class or lab at school. Also, they have higher budgets and therefore a higher quality and sophistication of the puzzles and furnishings. All EsRms include hands-on and cognitive puzzles; specifically, chemical EsRms also consist of "wet" puzzles that require chemical lab skills.[10,24] The time of activity, target group, designers, location and accessories and number of participants are summarized in Table 18.1.

Pedagogical aspects. The most obvious difference between commercial and formal EdEsRms is that players choose to participate in commercial EsRms, whereas the EdEsRms are part of the compulsory learning activities, chosen by the teacher. Nevertheless, despite the requirement to participate, students feel a sense of autonomy in the EsRms; the teacher merely provides guidance while the students work independently.[25]

Commercial EsRms are intended to be an enjoyable, social "getaway" activity that requires thinking and teamwork, but pedagogical aspects are not necessarily considered. In EdEsRms there is an additional layer: learning outcomes. These outcomes can be understood as the three categories used in educational research on this topic.

TABLE 18.1 The Administrative Dimension

Categories	Commercial/ recreational EsRm	EdEsRm: informal science education	EdEsRm: formal science education
Time of activity	After school	After school or during field trips	During school hours
Target group	Adults, families, private, company, or team events	Adults, families, private events, and school groups	Students in a specific course
Developed by	EsRm experts	EsRm experts, professional game developers, and/or museum staff	Teachers, educators, and professional game developers
Location	Specially designated spaces	Specially designated spaces (separated or built into exhibitions, inflatable rooms)	In class or in the lab
Accessories	Expensive technological accessories triggered by deciphering puzzles	Usually expensive technological accessories	Less expensive equipment: mobile EsRms, escape boxes, or decoder hardware
Number of participants	Average 3–7 participants	Average of 3–7 participants	Class (24 students), divided into four groups

Educational research on EdEsRms

Four review papers, published in 2019–2021 examined different aspects of EdEsRms.[5,26–28] A consensus emerges from these reviews that these rooms are highly appealing to both students and teachers. The different outcomes can be classified as affective outcomes, general skills, and content-specific knowledge and skills.

> *Affective outcomes.* EsRms generate enjoyment, enthusiasm, and curiosity among students; they promote a sense of flow, and students become intrinsically motivated to learn.
>
> *General skills.* These skills include 21st-century skills, such as critical thinking, collaboration, cooperation, communication, and teamwork.
>
> *Content-specific knowledge and skills.* EdEsRms enable students to apply their understanding of the subject area in a quasi-realistic scene, because in order to solve the puzzles, students must be knowledgeable about the subject.

Design principles of the ChEsRms at the Weizmann Institute of Science: bridging the gap between formal and informal science education

In 2015 at the Department of Science Teaching of the Weizmann Institute of Science, we decided to utilize EsRms to motivate and improve high-school students' attitudes towards studying chemistry. To date, we have developed six ChEsRms for different topics in the high-school chemistry curriculum with three different characteristics: mobile,[29,30] Do-It-Yourself,[10] and virtual ChEsRms.[31] In the design process, we tried to address several foci of the tension to maintain the informal nature of the activity while integrating it into the formal chemistry education system. Below we refer to both the administrative and the pedagogical dimensions of these ChEsRms.

Regarding the administrative dimension, we aimed at reaching as many students as possible. A chemistry lab class in Israel consists of a maximum of 24 students, which is much more than the traditional small groups in commercial EsRms. To overcome this discrepancy, we adjusted the game by dividing the participating class into 4 groups of 4–6 students. This is a good way to adapt to the school system and allow fruitful student teamwork, which is one of the main sources of enjoyment in EsRms. The four teams work independently in the same class, but the last puzzle always requires collaboration from all teams (a meta-puzzle).

In order to reach as many students as possible in the country, the first three ChEsRms were developed *as mobile kits*; teachers can borrow the kits and implement them in class.[29] The process requires a great deal of effort from teachers in order to implement it in school, but still, we could not meet the demand. Teachers first experience the ChEsRm, then borrow the equipment from the Weizmann Institute, arrange the ChEsRm in their lab, and only then implement the activity. A week later they return the ChEsRm. This model, in which teachers experience the ChEsRm themselves, and then watch their students engage for 45–50 minutes in an educational activity while playing, was one of the most important processes that helped teachers, who were used to teach formally, adopt informal activities. The teachers themselves experienced the joy of thinking and cooperating, while engaging in chemistry, and they understood the importance of playful learning (see Figure 18.1).

To meet the growing demand, we developed two Do-It-Yourself (DIY) ChEsRms, one on the Periodic Table of Elements[10] and another on Green Energy. We offered workshops for teachers to build their own kits. The mobile kits have more sophisticated equipment, e.g., suitcases, "bombs" that function with Arduino (an open-source electronic prototyping platform that enables users to create interactive electronic objects) but only two kits can be borrowed weekly. The DIY ChEsRms include less sophisticated equipment, but are available for any teacher who wants to create them, since the instructions are published on the Internet.

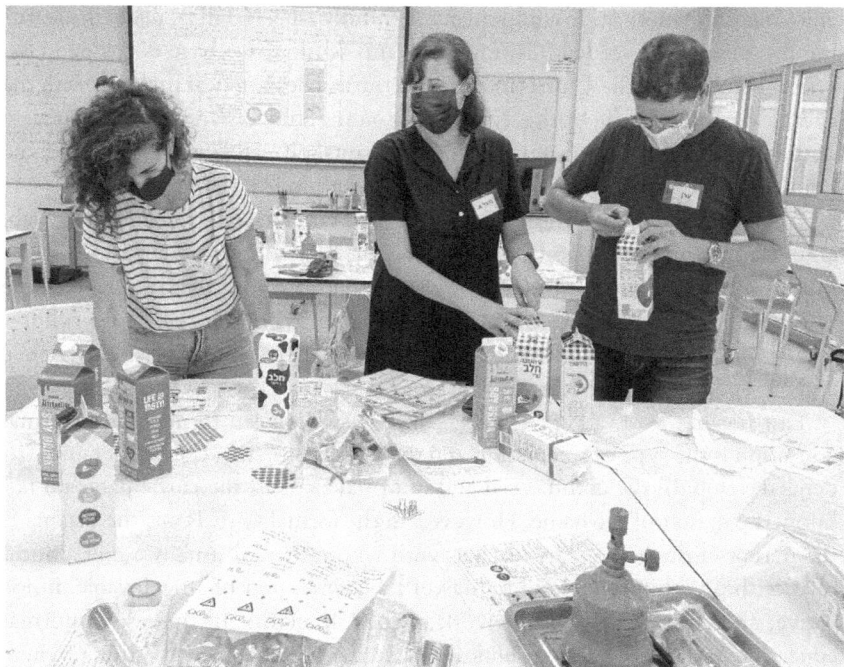

FIGURE 18.1 Teachers prepare a chemistry escape room in a professional development workshop.

One of the major issues regarding the pedagogical aspect that accompanies the design is the tension between the informal nature of the EsRm game (i.e., enjoyment, intrigue) and the formal learning outcomes related to science content and skills. The puzzles are based on the chemistry curriculum to match student knowledge, skills, and their zone of proximal development to challenge and arouse curiosity. This aspect of the ChEsRms is an advantage, because the target population's knowledge is known. The puzzles' mechanisms are often based on aesthetic chemical experiments that students know (e.g., pH color changes that reveal a code, a piston with a key pushed by a reaction that produces gas). The structure of puzzles presents problems for students to solve by thinking "out of the box" and as a team. The puzzles are very different from problems typically given in formal classes. In school, students usually receive feedback as grades, but the ChEsRms, if students solve a puzzle correctly, there is a very positive consequence, such as reveals a code or unlocks a box. This feedback mechanism is more common in games and in informal education than in formal education and is a great example of how these two educational contexts can be connected.

One of the challenges connected to ChEsRms is that their playful nature may lead to neglecting learning goals. On the other hand, emphasizing the learning goals may spoil the game format. We believe in the greater importance of the enjoyment and motivation components; hopefully aspects will facilitate the learning progress of the student. To moderate this tension, we have developed class activities following the ChEsRm in which the teacher can deal in depth with the chemical content of the game. During the game, students are not always aware of the chemistry involved, and not all participants solve all puzzles. However, the follow-up activities insure that all students understand all the puzzles.

All the ChEsRms stories that we designed take place in the chemistry lab, an environment that suits the story, is reliable, and allows students to have an immersion experience. It is very important to create a motivating atmosphere of the story with attractive props and equipment such as locks, cases, rollups, and "bombs" to help students disconnect from their regular class experience and engage in a unique learning environment.

Last but not least, a very essential difference between formal and informal education is the type of interaction and the relationship that the teacher and students develop. In the commercial and informal EsRms, the participants do not know the master of the game. However, in the formal ChEsRms, the teacher is the master of the game. The teacher, with whom there is usually a hierarchical relationship within the school framework, becomes part of an enjoyable, innovative, and popular game framework in the ChEsRms. In this way, informal education reaches the formal education setting. We believe that this teacher-students playful interaction is one of the main reasons for the success of ChEsRms. The effect of this novel relationship is enormous and will be described later.

The impact of the ChEsRm on students and teachers

Hundreds of teachers and thousands of students have experienced our ChEsRms. Based on our observations, as well as on responses and feedback from teachers and students, we can discuss the impact and contribution of the ChEsRms through the three main categories described in the literature: affective outcomes, general skills, and chemistry knowledge and skills.

Affective outcomes. Most of the students reported experiencing enjoyment, no sense of time, and interest; taken together, these outcomes create a sense of a flow experience.[29] In the ChEsRm, students have a sense of autonomy; the teacher is present, but only for security purposes and if the students seek help.

Students' enthusiasm is expressed in statements such as: "This was the most impressive lesson ever, interesting and fascinating. Time went by so quickly. ...." and "This was one of my best experiences of learning chemistry in school". One teacher was surprised to see her 12th-grade students, who are usually quiet and serious, playing like young children and being excited when solving a puzzle together.

In addition, ChEsRms benefited quiet students who usually avoid active participation in regular chemistry lessons. Interestingly, "weak" students expressed immense joy and excitement/enthusiasm when they contributed in solving problems. One teacher said: "The range of tasks allowed everyone to utilize his/ her strengths, and to show that even the 'weaker' students could lead the way and help the group succeed". In this regard, another teacher said: "A student, who usually does not like to work hard and will never stay to ask questions after the class, did not agree to take a break until he finished the puzzle on his own!"

Another very important finding obtained from students' and teachers' feedback was the great appreciation that the students and the school management had for the chemistry teachers as reflected from the following teacher responses: "The ChEsRm revealed the creativity of the chemistry staff to other teachers at school", and "The ChEsRm caused the school administration to greatly appreciate the work of the chemistry staff".

The teachers were perceived as the person responsible for the ChEsRm and were esteemed by their students. This finding has not emerged so far from other studies on EdEsRms, but in our study, it was prominently reflected. The ChEsRm, which introduces informal education characteristics into formal education, enables the students to view their teacher as responsible for this enjoyable, innovative, and popular game framework. In the questionnaire administered to students after implementing the ChEsRm, the item that has received the highest score in every class over the years was "I appreciate my teacher's investment in conducting the activity".[29] Teachers feel this appreciation. Some teachers voluntarily shared with us messages from students and even from their parents who thanked the teachers for conducting the activity.

General skills. The ChEsRm's novel environment empowers students who have skills that normally are not expressed or developed in a regular class (e.g., team work, brainstorming, and thinking "out of the box"). From teachers' feedback, we learned that ChEsRms encourage creative thinking and solving unexpected problems, skills that students do not usually encounter in chemistry classes.

The teachers also indicated that the ChEsRms require cooperation between participants within a group. This is another 21st-century skill that students are required to use in order to "escape" from the rooms; the rooms were designed in such a way that individual's work is not sufficient to solve the puzzles.

In addition, one teacher said: "I discovered that those students whose achievements were not high (i.e., their matriculation scores were mediocre) exhibited original thinking ability and organizational ability". Thus, the novel environment enables students with lower grades to stand out, due to their abilities and skills required to solve the ChEsRm puzzles. Teachers were very happy to see the "dynamics created between the students", which was an important key to solve the puzzles.

Chemistry knowledge and skills. Many students reported that even though recreational EsRms are more sophisticated and have great scenarios, they enjoyed the ChEsRm more because it was based on their chemistry knowledge. While the students performed "cool" experiments and challenging puzzles, they received feedback on their understanding of the chemistry content.

According to the teachers' feedback, they had an opportunity to watch their students think and concentrate on chemistry while playing! For many teachers, one of their satisfying feelings that results from a good and meaningful lesson is when students are engaged in learning. As one teacher wrote: "The students learned. I saw with my own eyes that students can be taught through a game. They concentrated and tried to understand chemistry puzzles for 70 consecutive minutes!" Another teacher added that she will continue to use ChEsRms in the future because she has a "desire to see and show the students' degree of success in solving assignments in a topic we have just finished studying. The EsRm is an opportunity for exploration or alternative assessment".

Concluding remarks

In this chapter we have discussed the process by which EsRms were introduced into the educational system and demonstrated how this novel learning setting enabled the teachers to bring the characteristics of informal learning into a formal education setting. We discussed several necessary conditions that need to be met in order for EsRms to be successfully implemented in a formal education setting. Keeping in mind that the teachers' performance is usually measured by the scores of their students in external examinations, the efforts they invest in borrowing the mobile ChEsRms kit or in constructing the DIY ChEsRms are not to be taken for granted, since most of the resulting outcomes (e.g., affective outcomes and general skills) are not included in these external tests.

Furthermore, during the implementation stage of the ChEsRms, teachers use a gamification pedagogy and focus less on developing the systematic teaching of chemical content; this new pedagogy is usually out of their comfort zone and requires them to play a different role than that of traditional teachers. In other words, teachers act outside of their expertise and invest a lot of effort to operate the ChEsRm, which does not always have a direct impact on the students' formal achievements. The teachers' performances are very much appreciated by people who are close to them, ranging from school management to parents, but their students' appreciation is the most powerful feedback.

We would like to suggest four stages in the ChEsRm implementation process that may contribute to teachers' willingness to invest their time and effort with ChEsRms. First, *teachers must experience the ChEsRm as participants* before they receive the necessary pedagogical and administrative aspects that are necessary for them to manage the ChEsRm successfully in school. From our observations, the teachers participate actively in the ChEsRm activity: they enjoy the game, look for clues, struggle with the puzzles, and think collaboratively and creatively in order to solve them. The teachers deeply experience this informal learning game, as manifested by their enthusiasm and interest in continuing to play the game. Second, in reflective discussions, *teachers are given an opportunity to talk about their feelings* in order to increase their awareness of the experience. Only later do we conduct the third stage: a formal session on the administrative aspects, when *teachers are exposed to the chemical content of the ChEsRm.*

The fourth and final stage, which ultimately motivates teachers to repeatedly implement the ChEsRms in their schools, is when *teachers observe their students participating in the ChEsRms.* During this activity, the teacher wears a shirt with the slogan "I'm not here". The teacher tries not to interfere with the students as they independently endeavor to solve the puzzles. During the hour-long ChEsRm activity, teachers observe their students while they use their chemistry knowledge, enjoy chemistry, and enthusiastically talk about chemistry topics. Although the students' enthusiasm for chemistry is not be measured by an external test, this means a lot for the teachers.

We believe that the teachers' positive experiences as both players in the ChEsRms and as observers of their students playing the game, along with the appreciation they receive for organizing the activity, scaffolds the teachers' efforts to bridge the gap between formal and informal science learning.

Notes

1 Mills, J. & King, E. (2019). Exploration: ESCAPE! learning theories through play. In James, A. & Nerantzi, C. *The Power of Play in Higher Education: Creativity in Tertiary Learning*, pp. 33–41. NewYork: Palgrave Macmillan https://doi.org/10.1007/978-3-319-95780-7_3.

2 Veldkamp, A., Daemen, J., Teekens, S., Koelewijn, S., Knippels, M.-C. P. J., & van Joolingen, W. R. (2020). Escape boxes: Bringing escape room experience into the classroom. *Br. J. Educ. Technol.* 51, 1220–1239.

3 Pendit, C. U., bin Mahzan, M., Basir, M. D.,. Mahadzir, M. & Musa, S. N. (2017). Virtual reality escape room: The last breakout. 2nd International Conference on Information Technology (INCIT). Nakhonpathom, Thailand: IEEE.

4 Csikszentmihalyi, M. (1988) The flow experience and its significance for human psychology. In Csíkszentmihályi, M.& Csikszentmihalyi, I. S. E (eds) *Optimal Experience Psychological Studies of Flow in Consciousness...),* Cambridge: Cambridge University Press, pp. 15–35. https://doi.org/10.1017/CBO9780511621956.002.

5 Taraldsen, L. H., Haara, F. O., Lysne, M. S., Jensen, P. R. & Jenssen, E. S. (2020). A review on use of escape rooms in education – touching the void. *Educ. Inq. 13,* 1–16.https://doi.org/10.1080/20004508.2020.1860284.

6 Breakout EDU. http://www.breakoutedu.com.

7 Hardie, L., Gill, A., Lee, S. & Shively, J. (2021). Escape the womb: A maternal emergency. *Simul. Gaming 52,* 96–103.

8 Lopez-Pernas, S., Gordillo, A., Barra, E. & Quemada, J. (2019). Analyzing learning effectiveness and students' perceptions of an educational escape room in a programming course in higher education. *IEEE Access 7,* 184221–184234.

9 Rouse, W. (2017). Lessons learned while escaping from a zombie: Designing a breakout EDU game. *Soc. Hist. Educ. 50,* 554–564.

10 Yayon, M., Rap, S., Adler, V., Haimovich, I., Levy, H., & Blonder, R.. (2020). Do-It-Yourself: Creating and implementing a periodic table of the elements chemical escape room. *J. Chem. Educ.* 97, 132–136.

11 Ang, J. W. J., Ng, Y. N. A. & Liew, R. S. (2020). Physical and digital educational escape room for teaching chemical bonding. *J. Chem. Educ.* 97, 2849–2856.

12 Makri, A., Vlachopoulos, D. & Martina, R. A. (2021). Digital Escape rooms as innovative pedagogical tools in education: A systematic literature review. *Sustainability* **13**, 4587.

13 Zhang, J. P. (2008). Hybrid Learning and Ubiquitous Learning. In *Hybrid Learning and Education*, Heidelberg: Springer.). pp 250–258. https://doi.org/10.1007/978-3-540-85170-7_22.

14 Vedder-Weiss, D. & Fortus, D. (2011), Adolescents' declining motivation to learn science: Inevitable or not? *J. Res. Sci. Teach.* **48**, 199–216.

15 Fortus, D. (2014). Attending to affect. *J. Res. Sci. Teach.* **51**, 821–835.

16 Hidi, S. & Renninger, K. (2011). The four-phase model of interest development. *Educ. Psychol.* **41**, 111–127 (2006).

17 Renninger, K. A. & Hidi, S. Revisiting the conceptualization, measurement, and generation of interest. *Educ. Psychol.* **46**, 168–184.

18 MacDonald, T. E. *Personal communication, June 2018.*

19 Back, J., Bexell, E., Stanisic, S., Back, S. & Rosqvist, D. (2019). The quest: An escape room inspired interactive museum exhibition. *CHI Play 2019 - Ext. Abstr. Annu. Symp. Comput. Interact. Play* 81–86.https://doi.org/10.1145/3341215.3356987.

20 Peichl, J. (2018). Nutzererfahrungen im Kontext eines Educational Exit Games. Thesis from the University of Freiburg and Offenburg University of Applied Sciences, Germany.

21 Kneale, K., Sear, J., Dangeli, D., Rosqvist, D., Jamous, M., Hunstad, Si., Peleg, R., Yayon, M. & Bakker, L. (2017). Making good museum games. *ECSITE Conference*. Porto, Portugal. https://www.ecsite.eu/activities-and-services/ecsite-events/conferences/sessions/making-good-museum-games-day-1

22 Antoniou, A., Dejonai, M. I. & Lepouras, G. (2019). 'Museum Escape': A game to increase museum visibility. In 342–350.https://doi.org/10.1007/978-3-030-34350-7_33.

23 Agusto, F., Baur, J., Bennett, S., Binford, G., Pillai, S., Holman, B., MacDonald, T., Miller, A., O'Brien-Verhulst, A., Owens, M., Sanchez, E., Thanukos, A., Vega, V., Quimby, C., Gurn, A. & White, L. (2021). STEM Escape: Immersing urban and rural families in a biomedical mystery. https://evolution.berkeley.edu/evolibrary/article/0_0_0/evotrees_zoos_10

24 Avargil, S., Shwartz, G. & Zemel, Y. (2021). Educational escape room: Break Dalton's code and escape! *J. Chem. Educ.* **98**, 2313–2322.

25 Veldkamp, A., Knippels, M. C. P. J. & van Joolingen, W. R. (2021). Beyond the early adopters: Escape rooms in science education. *Front. Educ.* **6**, 1–11.

26 Veldkamp, A., van de Grint, L., Knippels, M. C. P. J. & van Joolingen, W. R. (2020). Escape education: A systematic review on escape rooms in education. *Educ. Res. Rev.* 31, 100364.

27 Fotaris, P. & Mastoras, T. (2019). Escape rooms for learning: A systematic review. *Proc. Eur. Conf. Games-based Learn.* 235–243.

28 Makri, A. & Vlachopoulos, D. (2021). Digital escape rooms as innovative pedagogical tools in education : A sustainability digital escape rooms as innovative pedagogical tools in education : A systematic literature review. https://doi.org/10.3390/su13084587.

29 Peleg, R., Yayon, M., Katchevich, D., Moria-Shipony, M. & Blonder, R. (2019). A lab-based chemical escape room: Educational, mobile, and fun! *J. Chem. Educ.* **96**, 955–960.

30 Yayon, M., Peleg, R. & Katchevich, D. (2017). The great chemical escape. A popular pastime gets education. *Weizmann Wonder Wander.* https://wis-wander.weizmann.ac.il/science-teachingcampus/great-chemical-escape

31 Haimovich, I., Yayon, M., Adler, V., Levy, H., Blonder, R., & Rap, S. (2022). "The Masked Scientist": Designing a virtual chemical escape room. *J. Chem. Educ.* **99**(10), 3502–3509. https://cris.iucc.ac.il/iw/publications/the-masked-scientist-designing-a-virtual-chemical-escape-room

19

STEM APPLIED LEARNING PROGRAM

Infusing the Formal with the Informal

Anne Dhanaraj, Yao Teck Ong, and Tit Meng Lim

STEM education

The movement to promote science, technology, engineering, and mathematics (STEM) as an important interdisciplinary approach, based on a design thinking mindset, has gained momentum worldwide because it is regarded as crucial in creating a knowledge- and innovation-based economy. STEM education is understood as a platform for developing important skills and competences such as research inquiry, problem solving, critical and creative thinking, entrepreneurship, collaboration, teamwork, and communication.[1] This approach is also critical for the personal development of students to enable them to make sense of the world today.[2]

STEM competencies are increasingly required not only within but also outside of specific STEM occupations.[3] Researchers have acknowledged the complexity and inherently multidisciplinary nature of the issues we face in the world today.[4] STEM comprises the content, skills, and ways of thinking of each of its disciplines: STEM. At the same time, the approach also includes an understanding of the interactions and ways each discipline supports and complements the other.[5]

For many years, improvements in education including STEM were targeted at the formal education system, where much has been done, such as developing curriculum and teacher professional development, as well as enhancing school resources and assessment standards.[6] More recently, however, it is recognised that students spend only a fraction of their time in school compared to out-of-school time and learning within organised or extracurricular activities such as camps, clubs, field trips to museums, parks and zoos, time spent online, reading books, watching films and working with parents and other supportive role models.[6] Many out-of-school STEM programmes engage in hands-on

DOI: 10.4324/9781003145387-22

activities, inquiry learning, and connections to real-world applications and phenomena. They also have features – such as lack of assessment, extended engagement over time, and interactions with role models – that help to stimulate and sustain interest in STEM.[7]

Studies support the view that students' interest as well as knowledge and skills in STEM can be influenced by informal learning environments[8] and their likelihood of pursuing an STEM career.[9] Interest and motivation are important factors for sustained student engagement in STEM.[10] STEM informal learning experiences can complement students' learning and engagement in school, by presenting content in a more engaging and interactive manner and making real-world connections[11] and by contributing to more involvement in STEM activities, personal identity, and confidence in science and technology.[12]

STEM education in Singapore: The STEM Applied Learning Progamme (ALP)

In Singapore, the Ministry of Education (MOE) oversees the school curriculum from grades 1 to 12 (7–18 years) and ensures, among other subjects, that students receive a strong foundation in science and mathematics. Singaporean students have good performances in science and mathematics in international benchmarks such as TIMSS[13] and PISA[14] and they display healthy levels of interest in STEM. However, as in many countries, this interest and strong foundational competencies in general education do not translate to a high uptake of STEM options at the tertiary level, with many students opting for careers in finance or business. The other concern is to ensure that STEM education prepares students with competencies to address real-world challenges and to ensure that the country stays economically competitive.

Thus, the strategy and approach taken in Singapore for STEM education is to increase scientific literacy and support lifelong interest and engagement in STEM.[7] This strategy includes harnessing STEM learning in out-of-school settings to complement formal STEM education – hence the genesis of the STEM Applied Learning Programme (ALP) in Singapore schools as a means for students to learn through authentic experiences, learning content, and skills across STEM disciplines.[15] The programme was first conceptualised, developed, and conducted by the Science Center Singapore (SCS) in 2014 and is still running. One of the critical factors for its success was the involvement of relevant industry stakeholders right from the start, to ensure STEM ALP support by an integrated socioeconomic ecosystem in Singapore. Through its extensive links to industry partners, SCS also infused real-world context into its STEM workshops. This central role played by SCS was critical in shaping the character of STEM ALP in Singapore.

The programme offers hands-on and engaging workshops, with sufficient time for students (grades 7–10) to follow their interests and collaborate, with the support of teachers and STEM educators from the Science Centre. The programme started in 73 secondary schools and subsequently expanded to include 65 primary schools. STEM is only one of many applied learning domains in Singapore schools – others including business and entrepreneurship, languages, aesthetics, and humanities. Response from eligible schools to apply to join the STEM ALP programme has been very positive: more than 60% of eligible primary and secondary schools are now in the programme. What started as a three-year pilot programme is now in its ninth year and is likely to remain a feature of STEM education in schools. The universities in recent times have also rolled out integrated curricula and programmes with overt STEM emphasis, as a continuum of students' applied learning journey in Singapore.

The role of the science centre as an informal learning partner

In the first six years of the programme, since starting in 2014, STEM educators from SCS were deployed to each school to assist with the development and delivery of the programme activities; one goal was to ensure that the programme retained its informal learning tone, while providing professional development for the school teachers. The science centre also linked a key industry partner to each school, for example, 3M for materials science, Amgen for the health sciences, Fraser and Neave for food science, and Infineon for semiconductor technology. These school-industry partnerships provide real-world insights and experiences to the students. This core experience is still heavily curated by SCS but now is being carried out by teachers who are supported by the professional development given by SCS.

The programme typically includes weekly workshops, lasting 1–3 hours within curriculum time, over a semester (six months) or a year. The workshop content is flexible, as there is no standard curriculum to be completed. The focus is to develop in students the STEM skills and mindset that focus on nurturing the joy of learning (engaging), aligning learning to real-world context (responsive) and promoting STEM connections (making connections).[16]

When learning is fun and relevant, students are motivated to learn. Hence, apart from being free from examinations, the workshops provide authentic hands-on engaging experiences. The STEM educators ensure that the workshops are enjoyable and that learning is scaffolded according to learner readiness. The workshops focus on the process of learning rather than on the final outcome. They encourage a "dare-to-try" spirit where failure is seen as part of the learning process. This environment is conducive to learning: it encourages students to develop inventive thinking and to innovate in the process of problem solving.

STEM educators from SCS work with schools to develop workshops on a topic that each school has selected. Schools that participate in the STEM ALP programme have the flexibility to select from a range of attractive themes, such as sustainability, health sciences and food, game design and making, future of transportation, and cities and urban landscapes. Alternatively, schools can choose to focus on their own topics of interest. For example, one school chose to focus on fragrances; this focus involves the extraction of essential oils from flowers, spices, and fruits, while using scientific methods and designing perfume bottles using 3D CAD and printing. STEM ALP learning is situated around problem-based learning[17] and set in a real-world context or challenge – a challenge that is usually encountered by the students in their daily lives. The students often use the design thinking process to identify solutions: understanding the problem and the needs of a group of people and brainstorming possible solutions that are based on STEM. Students work in collaborative teams to find solutions that are open-ended.

The programme has two levels: an introductory broad-based level, or Tier1, designed for all students (grades 7 and 8), in order to excite them, create interest and enjoyment, and build competency; and a higher level, Tier 2, designed to cater to students who want to further their interest in STEM through specialised workshops (Figure 19.1).

For example, in Geylang Methodist Secondary School, grade 7 students learn about Urban Heat Islands, built-up city spaces that are much warmer than the suburban areas, resulting from the lack of vegetation and the high density of people, buildings, and vehicles. Students use an infrared camera to view thermal images and measure the surface temperatures of different insulating materials.

Students in grade 8 take up the Urban Agriculture module, where they learn about vertical farming by visiting Sky Greens, an innovative low carbon,

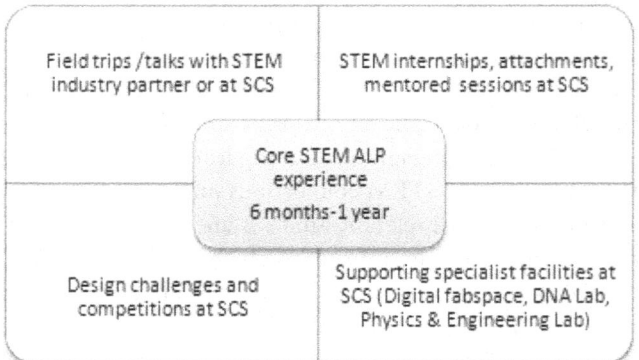

FIGURE 19.1 The STEM ALP programme. This programme is based on a core experience, developed by the SCS, in the participating school. The programme is conducted in a fun, non-graded, hands-on format, allowing students to build skills and interest at their own pace. STEM educators enhance this programme with other out-of-school experiences at SCS, as well as with industry field trips.

FIGURE 19.2 An example of students' visible thinking tools, one of a suite of skills students learn from the Shell NXplorers programme.

hydraulic driven vertical farm. They also get to build their own water wheel, to simulate the system used by Sky Greens.

Grade 9 students, who are interested in going further to address this challenge of urban heat, can opt for a Tier 2 project on roof gardens. At this stage, students have to conduct their own research and put in a lot more time outside of school hours to work on their prototypes. SCS helped link a water environmental expert who provided advice and job opportunities for the students. Students can also use specialised facilities at SCS, in order to develop and refine their projects; the prototypes can then be showcased at competitions organised by SCS, such as Young Sustainability Champion (Figure 19.2).

Another illustration of a productive engagement with an STEM partner is with Shell Singapore and their flagship programme, NXplorers, which teaches students about scenario planning, systems thinking, and the theory of change. Such thinking tools, typically used in corporate settings, were curated by SCS into age-appropriate resources for students. For example, the Ripple Effect tool enables students to think further about implications of a given action, e.g., while promoting cycling might be seen as a positive change for the environment, this activity may have the unintended consequence of raising bus fares if bus ridership drops.

FIGURE 19.3 A student solar race. STEM competitions allow deep engagement and learning, while also promoting the development of collaborative, creative-thinking, critical-thinking, and problem-solving skills.

SCS organises at least ten countrywide STEM competitions annually from preschool to tertiary levels and STEM ALP students can work on prototypes though these competitions. Examples of such competitions are the National Robotics Competition, Singapore Amazing Flying Competition, Drone Odyssey, and SONY Creative Science Award (toy making), to name just a few (Figure 19.3).

Presentation and sharing of project ideas and prototypes by the students is strongly encouraged, in order to build science communication skills. SCS has dedicated exhibition spaces, such as Scientist of a Day and Demo Corners, for students to present their ideas to the public. Community showcases, such as the Youth STEM Empowerment Programme and the Research and Development Mentoring Programme, are organised by SCS to provide other platforms for student teams to ideate, prototype, and present their innovations.

Students from participating schools also engage in enrichment workshops located in specialised labs at SCS, such as the DNA Labs (for those on the health science track) and the Tinkering Studio (for maker programmes). In addition, SCS engages youth in STEM through other interest areas, such as the arts. The Digital Fabrication Space (DFS) × Artist programme invites visual artists to mentor youth in designing sculptures with digital fabrication technologies, such as laser cutting and 3D printing. Students with a deep interest in these topics can join the Singapore Academy of Young Engineers and Scientists (SAYES). This group of passionate youth initiates its own community-related activities, such as teaching youth about flying drones, conducting community star gazing nights, and producing transmedia STEM content.

Research on the STEM ALP programme in Singapore schools

Between 2014 and 2016, more than 10,000 students at the end of the school year (grades 7 and 8) took a 21-item survey (6 point Likert scale) to give their feedback, after experiencing the ALP modules. In these unpublished surveys, students indicated agreement with the following statements: "I enjoyed the STEM ALP" (86.9%), "I learnt a lot in STEM ALP" (86.9%), "The STEM Educator was engaging" (90.1%). Enjoyment came from hands-on activities, working collaboratively with classmates, and working on real-world projects. Close to 80% of the students at the end of grade 7 wanted to continue with the ALP programme. The 14-year-old students also felt they could do well at the tertiary level in subjects similar to those in their STEM ALP (69.9%).

A more recent study[18] sought to check the effectiveness of the STEM ALP programme by measuring five constructs: students' attitudes to STEM lessons, views about STEM, students' self-concept when learning STEM, building of STEM identity, and aspiration to work in an STEM-related career or field. The study sampled students from 13 schools and found that they did not find their lessons too challenging, suggesting they were engaged and well supported in lessons that encouraged the development of their STEM skills. The STEM lessons stimulated their interest and confidence in pursuing a career in STEM. Overall, the students had favourable views towards their STEM lessons, as most expressed a desire for more STEM lessons. This is important in the development of situational interest in STEM, which can lead to the development of individual interest.[19]

Students also appreciated that STEM knowledge could be used and applied in solving many real-life issues and most students also disagreed with the idea that STEM knowledge and skills are only accessible to highly trained individuals. Students' views on such transferability of science are considered a strong predictor of future aspiration and their identity as science students.[20] The students' favourable views on STEM may be a reflection of the success of the STEM ALP programme, since the programme's students were able to gain a better understanding of the importance of STEM.

In terms of STEM identity, students agreed that STEM ALP had positively affected their perception of belonging to the STEM community. In terms of making career decisions in STEM fields, most students recognised the benefits of an STEM career and many expressed interest in such a career. Students also recognised that STEM is a competitive field with good remuneration prospects.

In general, the STEM ALP programme seems to have achieved its goal of positively engaging students in STEM. Students not only enjoyed their STEM lessons but desired more. They better understood the importance of STEM, the characteristics of STEM careers, and how STEM impacts everyday life.

Lessons learned

What we have learned from the surveys, student and teacher interactions, and many iterations of programming is that there are several key factors that play a positive role in STEM ALP:

The informal learning nature of the STEM programme. Incorporating the elements from informal science programming helped the students increase their enjoyment, interest, and understanding of STEM in everyday life. We conclude that it is very important to implement the programme without exams and with plenty of time for the students to play, tinker, discover, and learn. It is also important to provide students with open-ended outcomes based on their own interests and to allow them the opportunity to fail.

The two-tier structure of the core programme. This structure – having a basic level of hands-on workshops and investigations to excite all students and then providing a second level of deeper engagement for developing further STEM skills and competency – worked well.

The support of the Ministry of Education (MOE) and participating schools. A key ingredient for the success of the STEM ALP was the funding of the MOE as well as its openness in working with SCS to roll out the programme. The support of partner schools was also essential; this support insured that all students had access to the programme during curriculum time and also insured that the teachers were on board to support the programme. Since the STEM content area was selected by each school, the whole community could celebrate the achievements of the students through open houses, science days, and parent sessions.

The Professional Development (PD) effort. This effort developed teacher capacity and support at the same time: it insured that teachers had the ability and confidence to help implement the programmes and that the teachers were strong advocates for them.

Moving forward – in STEM ALP and beyond

Although there is general agreement about the importance of including STEM in the curriculum, there have been many challenges to implement this vision in school, due to factors such as lack of time to cover the curriculum, lack of teacher confidence, and lack of available resources.[21] STEM ALP in Singapore offers one way to address these concerns: by infusing formal science learning with informal science learning approaches and resources provided by a science centre.

Building on our experience, we can recommend our approach, based on close collaboration between educators in formal and informal science learning contexts. Such collaboration can build deeper self-efficacy and confidence in STEM students by designing activities that build mastery, while providing appropriate scaffolding for students, as they progress through different levels of difficulty.

Notes

1 English, L. (2016). STEM education K-12: Perspectives on integration. *International Journal of STEM Education, 3*(3), 1–8.
2 Marrero, M., Gunning, A., & Germain-Williams, T. (2014). What is STEM education? *Global Education Review, 1*(4), 1–6.
3 National Science and Technology Council. (2013). A report from the committee on STEM education. Washington, DC: National Science and Technology Council.
4 El Nagdi, M., Leammukda, F., & Roehrig, G. (2018). Developing identities of STEM teachers at emerging STEM schools. *International Journal of STEM Education 5*:1–36.
5 Moore, T., Tank, K, Glancy, A., Kersten, J., Smith, K., & Stohlmann, M. (2014). A framework for implementing engineering standards in k-12. Pre-College Engineering Education Research, 4(1), *1-College Engineering Education Research, 4*(1), 1–13.
6 National Research Council. (2014). *STEM Learning Is Everywhere: Summary of a Convocation on Building Learning Systems.* Washington, DC: The National Academies Press. https://doi.org/10.17226/18818.
7 National Research Council. (2011). *Successful K-12 STEM Education: Identifying Effective Approaches in Science, Technology, Engineering, and Mathematics.* Committee on Highly Successful Science Programs for K-12 Science Education. Board on Science Education and Board on Testing and Assessment, Division of Behavioural and Social Sciences and Education. Washington, DC: The National Academies Press.
8 Denson, C. D., Hailey, C., Stallworth, C. A., & Householder, D. L. (2015). Benefits of informal learning environments: A focused examination of STEM-based programme environments. *Journal of STEM Education: Innovations and Research, 16*(1), 11.
9 Kitchen, J. A., Sonnert, G., & Sadler, P. M. (2018). The impact of college-and university-run high school summer programs on students' end of high school STEM career aspirations. *Science Education, 102*(3), 529–547. https://doi.org/10.1002/sce.21332.
10 Bell, P., Lewenstein, B., Shouse, A. W., & Feder, M. A. (Eds.). (2009). *Learning Science in Informal Environments: People, Places, and Pursuits.* Washington, DC: The National Academies Press. https://doi.org/10.17226/12190.
11 Popovic, G., & Lederman, J. S. (2015). Implications of informal education experiences for mathematics teachers' ability to make connections beyond the formal classroom. *School Science and Mathematics, 115*(3), 129–140. https://doi.org/10.1111/ssm.12114.
12 Falk, J. H., Dierking, L. D., Swanger, L. P., Staus, N., Back, M., Barriault, C., … Verheyden, P. (2016). Correlating science center use with adult science literacy: An international, cross-institutional study. *Science Education, 100*(5), 849–876. https://doi.org/10.1002/sce.21255.

13 TIMSS (Trends in International Mathematics and Science Study), a large-scale international study conducted by the International Association for the Evaluation of Educational Achievement. https://safe.menlosecurity.com/https://timss2019. org/reports/achievement/.

14 PISA (Programme for International Student Assessment), a worldwide study by the Organization for Economic Co-operation and Development (OECD). https:// www.oecd.org/pisa/PISA-results_ENGLISH.png).

15 Lim, T. M. (2019). The Why and How of Teaching STEM. *'STEM Education an Overview' – The Head Foundation Workshop Report* (No. 7), 27–34.

16 National Research Council. (2015). *Identifying and Supporting Productive STEM Programs in Out-of-School Settings*. Committee on Successful Out-of-School STEM Learning. Board on Science Education, Division of Behavioral and Social Sciences and Education. Washington, DC: The National Academies Press. https://doi. org/10.17226/21740.

17 Hmelo-Silver, C. E. (2004). Problem-based learning: What and how do students learn? *Educational Psychology Review, 16*(3), 235–266.

18 Toh, S. Q., Teo, T. W., & Ong, Y. S. (2021). Students' views, attitudes, identity, self-concept, and career decisions. In *STEM Education from Asia: Trends and Perspectives*, Teo, T. W., Tan, A. L., & Teng, P. (Eds.). Routledge pp.144–163. https://doi. org/10.4324/9781003099888-12.

19 Hidi, S., & Renninger, K. A. (2006). The four-phase model of interest development. *Educational Psychologist, 41*(2), 111–127. https://doi.org/10.1207/ s15326985ep4102_4.

20 Jones, M. H., Ennes, M., Weedfall, D., Chesnutt, K., & Cayton, E. (2021). The development and validation of a measure of science capital, habitus, and future science interests. *Research in Science Education, 51*(6), 1549–1565. https://doi.org/ 10.1007/s11165-020-09916-y.

21 Dong, Y., Wang, J., Yang, Y. et al. (2020). Understanding intrinsic challenges to STEM instructional practices for Chinese teachers based on their beliefs and knowledge base. *International Journal of STEM Education*, 7: 47. https://doi. org/10.1186/s40594-020-00245-0.

20

EQUITABLE ACCESS TO MAKING THROUGH PUBLIC LIBRARIES

Bradley Barker

Making is a literacy – a way of reading the world as a collection of resources and materials to be composed, repurposed, and rearranged. Making is "what if?" and "why not?" of positioning oneself as having power – of taking responsibility for challenges and obstacles faced by oneself and one's community and enacting solutions.[1] While the physical manifestation of the maker movement in the form of makerspaces is found in larger communities oftentimes young learners (under 18 years of age) have not been well served by the current iteration of the maker movement.[2] In many instances, these community-based makerspaces are cofounded by adult community members to serve other like-minded supporters to share skills and knowledge as they journey down their own discovery learning paths. Likewise, makerspaces in formal educational environments show minimal support for student-directed learning and display a tension between the school's goals and the maker movement's goals. It is not just young learners who are found on the fringes of the movement, females, members of underrepresented populations and those with physical limitations may not feel welcome to participate in the movement as well. Libraries are in a unique position to offer equity to the movement by embracing and supporting all learners.

As a primer, the modern maker movement is defined as individuals (learners) engaged in the process of creating and sharing products.[3] The movement places great importance on open-source standards whereby the process of design and development of artifacts is shared within and between communities of learners.[4] The maker movement happens in the form of makerspaces in many learning environments from informal like community-based spaces, museums, and libraries to more formalized environments like those found in secondary and higher education spaces including academic libraries. Similarly, approaches to learning vary from highly structured formalized

DOI: 10.4324/9781003145387-23

direct instruction to pure discovery learning based on interest and motivation. While making can be a mix of old (fashion, music, art) and new technologies (digital fabrication), it has been argued that the availability of relatively low-cost digital fabrication equipment (computer network-controlled (CNC) routers, laser cutters, 3-D printers) has democratized the process of personalized product creation and innovation.[3] Makerspaces, FabLabs, Hackspaces, and Tool Libraries are all inclusive of the maker movement as each promotes active participation, collaboration, and open exploration.[4] Unique to makerspaces is the necessity of individuals to come together to innovate at the crossroads of many sciences, technology, engineering, and mathematics (STEM) disciplines and arguable art and design as well. Workshops and laboratories imbued with technology exist but without community, they are not, by definition, makerspaces in the truest form. Likewise, a digital fabrication laboratory without self-directed choice and community is not strictly a makerspace. Learning communities are an important construct in the maker movement, as defined as a network of learning communities that mixes a "can-do attitude" with the willingness to collaborate and inspire others.[5]

Makerspaces are categorized and appear on a spectrum in several different ways.[2] Whether instruction is supervised or organized by members of the maker community, if the space is clean or dirty (i.e., amount of dust or waste), the focus on the space is either on one topic or open to any topic available. Finally, spaces can be established for profit and require a membership fee while other spaces require purchasing or materials while others may be completely free (Figure 20.1).

The maker movement has found purchase in educational environments as educational makerspaces (EM) to improve learning while at the same time breaking down barriers to STEM + arts content knowledge and attainment.[6]

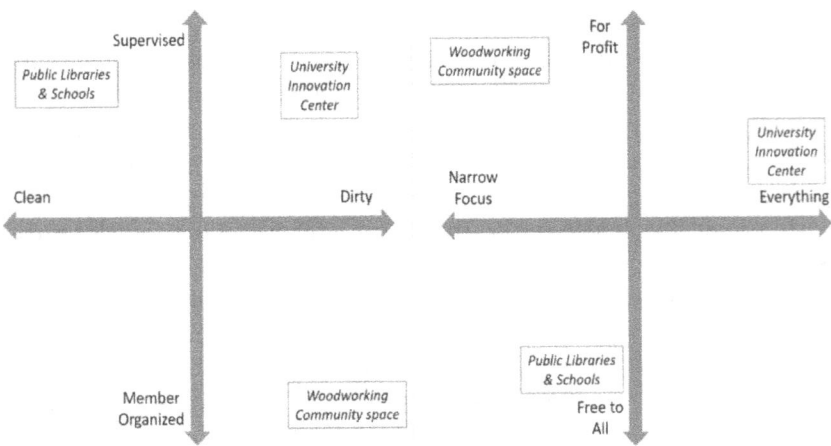

FIGURE 20.1 Typical organization of makerspaces.

These educational environments may be formal as in the case of the Lighthouse Community Public School in Oakland, California where making is found in every classroom, or informal environments like museums, afterschool programs, and libraries.[2] Maker education has been shown to assist youth in the development of positive attitudes toward STEM and creative fields, while also improving skills such as problem-solving, collaboration, and self-expression.[7] The maker movement is built on the constructivism theory of learning whereby learners construct their understanding and knowledge of the world.[8] These theories purport that learners' new knowledge is actively constructed from the world around them through experiential practice and by independent/guided research to integrate new knowledge and to arrive at a deeper understanding.[9] This knowledge is translated into a tangible product, with technologies acting as context, so learners can (re)make knowledge in concrete ways.[10] The product becomes a representation of the learner's thinking and knowledge. In addition, these physical creations can bring experienced and novice learners.[11,12]

The making and the maker movement can also act as a social and cultural edifice that preserves traditional crafts, broadens the participation of makers who face physical limitations, and gives back to the community through job skills and through sustainability practices. For example, the Dundee Contemporary Arts (DCA) collective in Dundee, Scotland is a community-wide makerspace that specializes in the interface of mechanical and digital making processes: in particular, studying how traditional processes can be enhanced with contemporary technology to revitalize and preserve the antique. This mixing of old and new technology is known as hybridity or an unconventional combination of unlikely elements that can be regarded as contradictory and problematic, or novel and original, inducing anxiety or unleashing new possibilities.[2] Hybridity has the added benefit of preserving traditional technologies. In one example, an artist from DCA worked with primary school children who created (drew) their fonts. A CNC router was used to cut out the font blocks which preserved the curves and wobbles in the children's designs. The children went on to use the block to print posters and post around the town. DCA has also explored how new digital processes can help artists with physical limitations. Printmaking is a manual process that relies on dexterity and physical abilities and by using digital fabrication tools the process can overcome some of these limitations. For example, an artist from DCA with a background as a stone mason lost mobility and dexterity in her hands because of the physically demanding work. Using the digital fabrication tools in the space she was able to create digital wood designs and print them using a laser printer she then submerged the cuttings in an Islandic hot spring. These wood cuttings eventually become encrusted in silica and convert into a different sort of artifact than the original laser-cut wood (Figure 20.2).

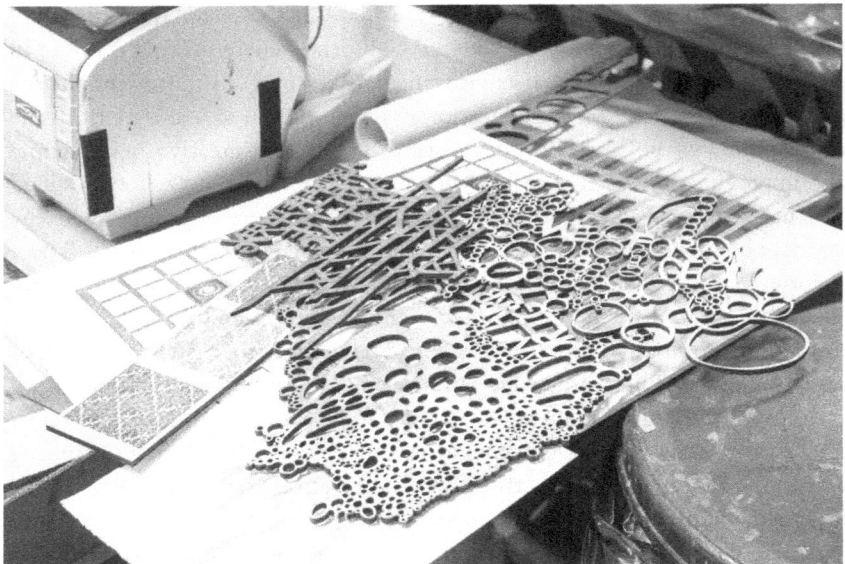

FIGURE 20.2 Images of laser-cut designs based on drawings from the artist.

In another example, the Johnson County Central Resource Library in Johnson County Kansas used 3-D technology to build a prosthetic hand. The hand was designed by a 17-year-old patron for a 9-year-old who was born with a medical condition known as limb difference whereby his right-hand fingers were not well developed. An added benefit of the prosthetic aside from the improved quality of life for the 9-year-old was the low cost of the hand and it could be reprinted as the child grows.[2]

Maker community members also engage in sustainable practices through the renewal and reuse of discarded and broken items. With a focus on sustainability and trade skills, the Fairfield Makerspace at the Maharishi University of Management (MUM) in southeast Iowa is designed to help members learn to create things but also to mend and recycle items. The practices are intended for members to use those skills to make a living.[2] Their "Transformative Tuesdays" workshop brings together members to transform or upcycle items like furniture. In this way, the textile members cross interests with woodworking and metalworking as they bring new life to old objects. As another example, the MADE Makerspace in Barcelona, Spain, supports and encourages members to bring in broken objects to be repaired and renewed. These could be specific items like a pair of headphones or a strip of LED lights that need to be soldered. MADE has a relationship with the "chatarreros" or scrap dealers of Barcelona, the people that go through the rubbish bins and pick out paper and all sorts of recyclable items. Chatarreros have been described as the informal recycling heroes of Barcelona. When Chatarreros find something interesting that can be recycled or reused by the makerspace, members will purchase the item(s). For example, when Chatarreros find non-functioning power tools they can bring them to MADE and make money by selling them to the space, where they can oftentimes be renewed. The Chatarreros also benefit because they receive more money from the space than they would at a flea market. MADE also gives recycled metal coming out of their projects to Chatarreros who can sell it for a profit.

Like so many other educational movements, there are clear inequities when it comes to access to making. Makerspaces lack female representation; most spaces are comprised of affluent males of the dominant culture with ample leisure time to make.[6] The lack of equity is not just a US phenomenon: when European makerspaces were surveyed 10 out of the 26 reported member-ships of over 80% white, and six sites were nearly 100% male.[2] In an over-whelmingly male-dominated space, females may not feel welcomed with a male-dominated attitude that does not condone females as makers. The mak-er's cultural portrayal leads to exclusion of non-dominant cultures as well. Many spaces inherently understand the friction between the maker identity, equity, and inclusion, but few understand how to make meaningful changes. For example, the Station North Tool Library provides low-cost access to tools and information to residents in the Station North Arts and Entertainment District of Baltimore, Maryland. The space offers classes and open times for members to make and create, usually using wood and other construction-type materials (concrete, electronics, metal). Station North has recognized the cul-tural perspective of woodworking as a male-dominated pursuit. While they did consider providing women's classes when they first started the space, it was realized it could then be perceived that women aren't welcome at other times. Staff and volunteers are taught how to build inclusive environments by not singling out females and safeguarding the comfort of all. Ensuring an inclusive environment is a priority in the space if violated members are asked to leave.[2] These efforts have made a positive difference in female participation rates yet there is much more that could be done.

Inequality in the movement has also prevented young learners from expe-riencing self-directed making. For example, unless specifically developed for youth, many makerspaces deny or restrict access to equipment.[2] Conversely, many age-appropriate makerspaces like those found in primary and secondary schools lack the funding and staffing to allow youth to fully engage in making using old and new technologies. Additionally, there are tensions related to the goals of the school and the goals of maker education.[12] Makerspaces in formal learning environments tend to focus on task completion and testing at the cost of student-driven learning. These spaces will also emphasize end prod-ucts rather than the process of design, iteration, and innovation. Researchers have termed the focus on the end product as the *keychain syndrome*, where a short activity (i.e. printing 3-D keychains) based on one piece of digital technology is used in short introductory workshops and learners never build additional skills.[13] School-based makerspaces also test the fortitude and time boundaries of teachers as they give up control and allow students to drive the learning process.[14] Teachers may also be tested when space is narrowly focused on STEM concepts, reducing the willingness of unprepared teachers to utilize making in the classroom. Geographic location is another important access consideration, as rural libraries are less likely to have a dedicated makerspace than urban libraries. In a survey of libraries in 12 US states with large rural populations, it was found that maker equipment was only available to 11% of rural libraries.[15] Only one library reported hosting an active maker group while other libraries worked with community members to bring in a 3-D printer and audio/video equipment for patrons. Additionally, rural libraries offer fewer informal educational programs than their urban counterparts with roughly 25% offering after-school programs and less than 20% offering makerspaces. The lack of accessibility to advanced-technology businesses and higher education institutions is another challenge for 12 rural library makerspaces.[16] While not an easy proposition, there are examples of successful makerspaces in rural library spaces. In one collaborative project, the University of Nebraska-Lincoln (UNL) 4-H Extension and the Nebraska Innovation Studio (NIS) makerspace along with the community of Sidney, a rural town in Western Nebraska, established a rural makerspace. To support the community and learners, the project utilized a virtual reality (VR) platform to cooperatively create and explore 3D objects between experts at UNL and rural learners. Once complete, a process termed "physical manifestation" enabled the artifacts to be printed using digital fabrication equipment.[17] Two telepresence robotics, one housed at UNL and the other at the Sidney Public Library some 400 miles distant, were used to lead workshops and training at the Sidney makerspace with staff from UNL. The results of the project were encouraging, and the youth described their "imagination on the loose" and enjoyed being able to rapidly sculpt and "grab things" using the Oculus system.[17] Participants and instructors in Sidney were interviewed to determine their perceptions of using telepresence robots. In terms of teaching, the robots were used to teach part of the class, while the staff was in the physical space to answer questions and do some of the hands-on components. The robots were also used to support the Sidney makerspace with technical questions on the digital fabrication equipment. The library reported that the robot was capable of traversing the entire library and interacting with library patrons. They noted: "Anytime we have the opportunity to show the robot in a real-life setting with people we try to do that. The kids get very excited when the robot comes out too. Then the novelty wears off, and they learn to interact with it, and ask questions as they would with other adults."[17]

While there are examples of successful makerspaces in schools,[2] informal learning environments like libraries are uniquely positioned to provide youth access to the maker movement. Libraries have traditionally been the catalyst for connecting the public to information and resources. They are well positioned to add the sharing of resources and, in this case, old and new technology tools to increase knowledge and encourage community making. The makerspace shifts the role of the public library from a place of information consumption to a place of knowledge production, though "libraries have always had maker programs even if they weren't called that."[18] Norman[1] notes that the makerspace joins an extensive list of public library programs and services for which the provision of space and equipment for participation and learning is the focus, such as story-times, craft sessions, computer training for seniors, homework help, and author talks.[19] Moreover, libraries have a physical space, a place for community members, including youth, to experience making. Multimillion-dollar technology budgets are not required; a library makerspace can consist of small allocations of old and new technology. What is important is to have a space that invites patrons to explore, create, and share their learning and achievements.

Returning to the maker educational movement, libraries are also well equipped to overcome many of the tensions found in school-based makerspaces. When examining the issue of inequity and inclusivity in makerspaces, libraries have an opportunity to broaden the participation of females, non-dominant cultures, rural audiences, those with physical limitations, and youth by providing an open and welcoming space for all learners. By adopting an equity-oriented stance, ensuring that all cultures are represented in the process and outcomes of making, libraries can ensure broad participation. One exemplary program used storytelling and technology for an indigenous equity-oriented approach to making.[20] One bit of caution from the authors is that while many spaces and activities are motivated by economic growth through the development of a product, not all cultures subscribe to this motivation.

Public libraries are well suited to makerspaces and the application of the maker mindset for creativity and learning, and there are exemplars found in large urban libraries like the Bubbler in Madison, Wisconsin, and The Johnson County Resource Center Makerspace just to name a few. Why then do we not see makerspaces in all libraries? By all outward appearances, the lack of resources to obtain the necessary equipment and supplies is a major barrier to the proliferation of makerspaces. However, as Markgraf and Hillis[21] detailed in their "stone soup" article, a well-developed and thought-out approach can be the key to establishing a makerspace. In the article, they noted even in times of dwindling budgets makerspaces are possible. The authors point out numerous steps to establishing a space, such as getting community input, developing a vision with a task for communicating with internal and external audiences including donors, and building upon existing resources, sustainability plans, and incremental growth.

When makerspaces provide equitable access and follow the tenets of the maker movement there can be a tremendously positive effect on youth and their interest in STEM, arts, and innovation, as well as impacts on the community. Libraries are uniquely situated to offer makerspaces that embrace community-based equitable learning by providing space, equipment, community, and an equitable learning environment that celebrates individual quests for understanding. While understandably not an easy task for public libraries, especially those in rural areas, there are methods and models to encourage and support the formation of makerspaces. It was shown how the Made makerspace and the Fairfield makerspace provided the space for members to connect through the upcycling and reusing of materials. Likewise, hybridity can lead to the preservation of old technology and lead to new and unexpected outcomes. Making can also be used as a way to broaden participation from those who have been excluded due to physical limitations. Finally, youth can also be supported by allowing full and unfettered access to equipment and communities of learners. Youth may also be well served by inviting public experts into the space to share their knowledge and understanding of making based on their experiences and culture. Libraries can also support communities of practice in the makerspace whereby learners can share and learn and identify themselves as part of a community. In the process of defining themselves as learning institutions, public library makerspaces need to capitalize on their strength of flexibility and the tenet of providing resources to their communities.

Notes

1 Holbert, N. (2016). Leveraging cultural values and "ways of knowing" to increase diversity in maker activities. *International Journal of Child-Computer Interaction, 9–10*, 33–39. https://doi.org/10.1016/j.ijcci.2016.10.002.
2 Shane Farritor, founder of the University of Nebraska Innovation Studio. Barker, B. (2019). *American Perspectives on Learning Communities and Opportunities in the Maker Movement*. IGI Global. https://doi.org/10.4018/978-1-5225-8310-3.
3 Halverson, E. R., & Sheridan, K. (2014). The Maker movement in education. *Harvard Educational Review, 84*(4), 495–504. https://doi.org/10.17763/haer.84.4.34j1g68140382063.
4 Rosa, Paulo, Guimarães Pereira, Ângela, & Ferretti, Federico. (2018). *Futures of Work: Perspectives from the Maker Movement*. https://doi.org/10.2760/96812.
5 Peppler, K., Halverson, E., & Kafai, Y. (2016). Introduction to this volume. In K. Peppler, E. Halverson, & Y. Kafai (Eds.), *Makeology: Makerspaces as Learning Environments* (Volume 1, pp. 64–69). New York, NY: Routledge.
6 Calabrese Barton, A., & Tan, E. (2018). *STEM-Rich Maker Learning Designing for Equity with Youth of Color*. Teachers College Press Columbia University.
7 Kalil, T. (2013). Have fun—learn something, do something, make something. In *Design, Make, Play: Growing the Next Generation of STEM Innovators*. https://doi.org/10.4324/9780203108352.

8 Piaget, J. (1972). Intellectual evolution from adolescence to adulthood. *Human Development*, *15*(1), 1–12. https://doi.org/10.1159/000271225.

9 Barker, B., Melander, J., Grandgenett, N. & Nugent, G. (2015). Utilizing wearable technologies as a pathway to STEM. In D. Rutledge & D. Slykhuis (Eds.), *Proceedings of Society for Information Technology & Teacher Education International Conference 2015* (pp. 1770–1776). Chesapeake, VA: Association for the Advancement of Computing in Education (AACE).

10 Papert, S. (1980). *Mindstorms: Computers, Children, and Powerful Ideas*. New York: Basic Books. Retrieved from http://dl.acm.org/citation.cfm?id=1095592.

11 Lave, J., & Wenger, E. (1991). Situated learning: Legitimate peripheral participation. *Learning in Doing*, *95*, 138. https://doi.org/10.2307/2804509.

12 Vygotsky, L. S., Embong, A. R., & Muslim, N. (1978). *Mind in Society: The Development of Higher Psychological Processes*. Cambridge: Harvard University Press.

13 Rouse, R., & Rouse, A.G. (2022). Taking the maker movement to school: A systematic review of PreK-12 school-based makerspace research, *Educational Research Review*, *35*, 100413, ISSN 1747-938X, https://doi.org/10.1016/j.edurev.2021.100413.

14 Blikstein, P., & Worsley, M. (2016). Children are not hackers: Building a culture of powerful ideas, deep learning, and equity in the maker movement. *Makeology*, *1*, 1–14.

15 Hughes, C., & Boss, S. (2021). How rural public libraries support local economic development in the mountain plains. *Public Library Quarterly*, *40*(3), 258–281. https://doi.org/10.1080/01616846.2020.1776554.

16 Petrin, R. A., Schafft, K. A., & Meece, J. L. (2014). Educational sorting and residential aspirations among rural high school students: What are the contributions of schools and educators to rural brain drain? *American Educational Research Journal*, *51*(2), 294–326. https://doi.org/10.3102/0002831214527493.

17 Barker, B., Valentine, D., Grandgenett, N., Keshwani, J., & Burnett, A. (2018, October). Using virtual reality and telepresence robotics in making. In *E-Learn World Conference on E-Learning 978-1-939797-35-3*, pp. 566–568

18 Landgraf, G. (2015). Making room for informal learning. *American Libraries*, *46*(3), 32–34.

19 Gahagan, P. M., & Calvert, P. J. (2020). Evaluating a public library makerspace. *Public Library Quarterly*, *39*(4), 320–345.

20 Barajas-López, F., & Bang, M. (2018). Indigenous making and sharing: Claywork in an Indigenous STEAM program. *Equity & Excellence in Education*, *51*(1), 7–20.

21 Markgraf, J., & Hillis, D. (2021). The 'stone soup' approach to creating a library makerspace. *College & Undergraduate Libraries*, *27*(2–4), 305–325.

21

THE STREET CODE PROJECT

Computational Literacy and the Performing Arts

Izaiah Wallace and Michael Horn

The field of Computer Science (CS) struggles with exclusionary culture and image problems that begin at the K-12 level and persist into college and beyond.[1] The result is a chronic lack of diversity along racial, ethnic, and gender lines that has remained largely unchanged for more than forty years. This matters because computing and technology mediate so many aspects of our human experience (from health and wellness, to media and entertainment, to economics and finance) that it's hard to find a part of everyday life that isn't in some way impacted by an app, platform, or algorithm. Computing has become a *language of power* in our society, but it's also a literacy that's far from equitably distributed.

Among the many approaches that researchers and educators have explored to address these problems, the concept of STEAM – or the integration of the arts into science, technology, engineering, and mathematics – has been especially influential. Since its origins in the early 2000s, STEAM has become thoroughly entrenched in the popular imagination as a way to think about technology-infused learning spaces. The term justifiably calls out the overemphasis of STEM at the expense of the arts and other creative domains in education. It also reminds us that it's important to provide humanizing learning experiences for young people by drawing on a broader range of social, cognitive, and creative skills and then directing those skills toward socially relevant topics. In this way, the "A" in STEAM can help educators make more intentional decisions about how and what to teach in both formal and informal learning spaces.

However, as the idea of STEAM has grown in popularity, there have also been concerns about how the acronym gets taken up in practice. It's not only that "the arts" adds a dizzying array of creative traditions (each with

DOI: 10.4324/9781003145387-24

its own established history and culture) to the jumble of disciplines already encompassed in the STEM acronym. It's also that the arts are mythologized as being able to help students identify with what would otherwise be considered technical, intimidating, and alienating STEM fields. This framing starts to feel problematic for its characterization of STEM in relation to the arts: Aren't math, science, and engineering already creative domains in their own right? Can't STEM disciplines also be avenues for personally meaningful self-expression? Can't the arts be just as technical and intimidating as STEM fields[2]? And, can't the arts also be taught in ways that are as dehumanizing as we imagine many STEM fields to be? The caricature of the *creative* arts as helping students identify with *alienating* STEM fields feels especially problematic when applied in stereotyped ways for learners who have been historically and systematically marginalized from those very same fields.

In this essay, we hope to add to this conversation by providing a concrete example of the integration of STEM and the arts. Our broader goal is to create humanizing learning experiences for young people that work toward a more inclusive computational future. To do this we'll take a quick detour into the world of spoken word poetry and then work our way back to STEAM to describe the work we've been doing at the intersection of music and computer programming. We'll finish up with the perspective of Rosa, a middle schooler who participated in one of our after-school programs during the pandemic. Rosa's reflections on her experience in the program will help keep us honest about what really matters to learning at the intersection of technology and the arts.

Slam poetry and street culture

It was almost 40 years ago that a construction worker named Marc Jacobs upended the world of poetry by having everyday people act as judges for spoken word competitions hosted in a bar in Chicago. This deceptively simple innovation helped create an energetic new art form that came to be known as *slam poetry*.[3] In the decades since, poetry slams have spread throughout the world and energized new generations of poets drawn to the movement's identity-affirming spaces and cultures. As the art form expanded, it also had an impact on the world of education. Spoken word programs have been able to carve out "spaces of refuge" within (but apart from) formal K-12 structures.[4] Adult facilitators take on the role of mentors and coaches for young people learning to express themselves critically through new forms of literacy. Students are motivated to excel not for a letter grade or test score, but for the authentic experience of performing personally and socially meaningful work for a live audience of peers.

Slam poetry's origin story was part of a much larger cultural movement playing out across the country in the 1980s and 1990s. As Jacobs was hosting his first poetry competitions, artists such as Basquiat, Keith Haring, and

Blade were bringing new vitality to *street* art; skateboarders and breakdancers were creating new forms of *street* athletics; and fashion designers such as James Jebbia, Shawn Stussy, and Dapper Dan were breaking with the establishment to elevate *street* wear. On a grander scale, Hip Hop was redefining music, art, poetry, dance, and fashion against a backdrop of racial and economic oppression. In each of these examples, we see the emergence of a global *street* culture led by racially and economically marginalized young people. Common themes include a *subversion* of established literary and artistic conventions, a break from the mainstream culture, and to varying degrees, an appropriation and democratization of technology for new creative purposes. We note that the term "street" is often used in derogatory and racialized ways that erases the vitality and importance of street artists in shaping mainstream culture and giving voice to marginalized communities. Our use of the term is intentional and is meant to acknowledge and celebrate the legacy of these artists and activists.

Street Code

In our own work, we ask if we can do for CS education what slam did for poetry. Can we help foster alternative cultures centered in the performance arts that help young people construct *computing* identities while engaging in meaningful learning experiences? In thinking about this question, we imagine a process of "computational enculturation" in which people are immersed in practices, values, languages, and traditions of computing cultures outside of school. Much in the way that street culture involves a subversion of dominant literary forms and conventions, our vision is that children might develop computational identities over time through a web of influences including family, peer groups, and mentors. And, when they enter mainstream academic CS as students, they will be empowered by their established sense of belonging in an authentic computing culture.

With this goal in mind, we have been pilot-testing out-of-school youth programs in which teams of students work with coaches to develop art pieces consisting of computer code, music, video, poetry, and dance that they perform at *Street Code* festivals. The best way to understand these festivals is to imagine poetry slams – celebrations of culture, language, and identity – mixed with robotics competitions – celebrations of engineering and CS. The result is an experience in which we try to elevate coding to the level of performance art as teams of students showcase their compositions for a live audience. In the long term, we hope to be part of a larger movement in which coding becomes a culturally resonant *language* of empowerment for young people whose voices have been systematically marginalized in STEM fields.

Our use of the word "language" is deliberate. We adopt definitions of *computational literacy* offered by scholars such as Andy diSessa[5] and Annette Vee,[6] who argue that literacy implies a broad-based social-material intelligence that fundamentally shapes how we think and understand the world. Becoming a literate person is a complex social, technical, and cultural phenomenon that goes well beyond taking a course or two in school. Literacies also have profound implications for societies across a wide range of human endeavors. Because literacies can be applied in many different ways by many different people, diSessa has criticized current CS education movements as being too narrowly tied to the specific field of CS. Or, as Vee writes, "because programming is so infrastructural to everything we say and do now, leaving it to computer science is like leaving writing to English or other language departments."[7]

Because CS is largely a domain of science and engineering, certain concepts are given prominence over others. When we teach coding, we are often preoccupied with writing elegant, reusable, efficient, and error free code. We are taught to be neat, to write thorough comments, and to favor top-down abstractions over bottom-up familiarity with computational materials.[8] In short, we are taught to follow rigid literary forms and conventions. It's not that these are bad things, especially when we write code for the *purpose* of science and engineering. But, by emphasizing science and engineering, we miss out on the fact that coding is as much an art as it is a science, as much a craft as it is an engineering discipline, and as much a personal passion as it is a career choice. People who are good at coding tend to *love* coding for its own sake. We might find it more productive from the point of broadening participation to think of coding as more like the humanities (literature, art, music, or poetry) than the sciences. It's something we do because the craft of coding enriches the human experience and provides a personally meaningful form of self-expression.

In thinking about computational literacy, it's important to remember that literacies, and the educational systems that support them, have historically been used as tools to maintain inequitable social, class, and economic power structures.[9] There are many examples in which those in power have dictated what counts as literacy while repressing or marginalizing indigenous languages and cultural knowledge systems. Given this history, it's not surprising that social and political movements of empowerment have often been marked by a subversion of dominant literary forms and conventions. For example, Hip Hop and Jazz can be thought of as not just musical genres but also transformations of poetry, dance, and art that emerged from a rejection of mainstream culture. For such movements, creating new forms of literacy can become an act of reclaiming power and identity.

Street Code Jam Festival

In spring and summer of 2021, we piloted the first Street Code Jam Festival with 45 middle school youth from five public schools and two community organizations. We recruited and hired a team of undergraduate coaches, all with strong backgrounds in CS and music. After training on our curriculum, the coaches worked with middle school students in after-school clubs with support of local teachers and staff. Due to the pandemic, the after-school clubs were conducted over video conference. The involvement of local teachers was crucial in that they supported recruiting, helped with parent buy-in, and ensured students were able to log in from home. Over four months, the coaches worked with their teams to learn the basics of music composition such as beat structure, melody, and harmony, and to develop foundational skills in Python computer programming.[10] With these tools in place, young people created TikTok-style music videos to showcase their final projects. Throughout this period, we iterated on curriculum design and worked with coaches in weekly reflection sessions to improve their pedagogy and practice.

In June 2021, with support of our community partners, we hosted an in-person event at Chicago's Studio 2112. The festival gave students and parents a behind-the-scenes tour of music production studios, meetings with professional producers and artists, and a live-streamed broadcast of 21 student TikTok videos. We organized the videos into three segments: a focus on Computation; a focus on Culture; and a focus on Creativity. Between each segment, the first author conducted live streamed interviews with participants about their creative process and experience in the clubs. The coaches also prepared their own videos to cover popular songs.

Figure 21.1 shows segments of one team's video created for the festival. For this project, a group of eighth-grade students who had spent the last year of middle school attending classes from home, composed a poem that they recorded into TunePad and mixed with their own beat coded in Python. The video featured segments of the girls hanging out together at school interspersed with screenshots of their project and code.

FIGURE 21.1 Segments of one team's video featuring a poem recorded into TunePad, segments of the students' own video, and screenshots of their Python code.

Their poem was simple and powerful – a message for all of us coming out of months of COVID isolation:

You have your entire life.
Things take time.
Relax.
You will graduate.
You will be successful.
You will become an adult.
You will find someone who loves you.
You have your entire life.
Things take time.

Reflection

Our pilot festivals were a time of learning for us as much as for the students. Working online with middle schoolers during the pandemic was often demoralizing for our coaches. Without being able to fluidly interact with students in the room, it was difficult to know what they were getting out of the clubs. Our curriculum was in a constant state of flux as we had to adjust our understanding of what online learning could look like in the after-school context. Seeing each other in person for the first time at the festival was an eye-opening experience for everyone because of the stark contrast between video and real life. It was more than the "I didn't realize you were so tall" comments. There was a sense of being able to read emotions and feel more fully human connections. In the weeks after the festival, the first author conducted phone and Zoom interviews with as many of the participants as he could. Analyzing transcripts of the interviews, which often lasted over an hour, has helped bring several themes into focus.

Theme 1: Deep integration

Our first theme aligns with what the original proponents of STEAM had in mind when they started advocating for the inclusion of the arts. Deep work in any domain is creative and can reveal integrated relationships across math, technology, engineering, and science. In our case music is full of mathematical and algorithmic structure. The emotional quality of chords and note intervals follows from the mathematical ratios of note frequencies. Compelling rhythms are formed by subdividing beats and measures. Musical pieces are composed by layers of repeated patterns, motifs, and modular structures that can be abstracted and parameterized. In this way, music is an excellent domain for thinking about *algorithms*, step-by-step sets of instructions that can generate or describe complex patterns. Some of the deeply integrated nature of math, technology, and music seemed to come through in students' learning. One of the students, Rosa, described her experience this way:

> Yeah like since … we started learning how to code. Just the basics. And like you can also do this kind of code for example, add more than one note in one measure in one line and it was like, oh I'm going to hear how this sounds sound like plays together at the same time and that was like ok I really like that and I feel like that was way more creative than just have one note being played at a time in like one measure which really started like getting more creative and then I think it was also like the loop.
>
> The loop was really kind of hard for me to learn but uhm I didn't end up using it but it was like more creative in a way because it would be like I can put these notes like in repeat while also having these other notes that aren't in repeat to see how it sounded. And also it was it was like this time there was one of my classmates in the group that did kind of like a piano style and it was really amazing so and I also got inspired by that so I decided to do rather than just kick and like open hat it would have been like use all of them at the same time with like drums and bass and it was really, really cool.

What Rosa talks about here integrates computational concepts like loops and lists (how she's able to play more than one note simultaneously) with musical terminology (like "kicks" and "open hat") while also describing a rudimentary creative process that includes finding sources of inspiration. In her case, technical competence (knowing how to code music using lists and loops) led her to "feel like that was way more creative."

Theme 2: The power of community

A second theme was the power of community in shaping the experience for young people. Because we were working outside the boundaries of a normal school day, we had the space and freedom to "waste" time getting to know each other through activities like games, ice breakers, and talking about shared interests. One club of students even volunteered to join a session on a Friday morning when they had a school holiday, just to hang out with their coaches. Rosa put it this way:

> […] and then we were start every meet with like, song recommendations or like, or just songs in general. And we would ask about each other's days, which was like, really fun, because that is how we started to bond with each other and how we started to know each other more.

The importance of building community and relationships with young people is well understood in the out-of-school learning spaces, but the hyper-focus on disciplinary content encompassed in the STEAM acronym can often make it easy to forget the power (and necessity) of this relationship work. In a sense, learning the basics of how to code in Python was one of the least important outcomes of the camp for students. Having a welcoming space to get to know the coaches and peers was foundational. The other aspect that we see in Rosa's description of the club is a sense of growing self-esteem and even surprise by what she was able to accomplish in the club. The visibility of her peers in the space clearly became a motivating factor for her.

> Yeah, what really surprised me was, there was this video about how some-body did, I forgot the song, but a code to one of like a popular song. It was like, how could they do like, the codes in the beat sounds so similar to that song. And then we there was this moment in our class that we took, like a piece of a song and mixed it with our own code. And that was really cool because it was like, just a piece of this music with like, a piece of our code mixed together, which ended up really amazing. And it was like a moment where you would feel like, disbelieved by yourself, because you're like, I didn't know I could do that. Or like, Wow, I did that. Yeah. [...] Or like surprised, by like, under, like, under we underestimated our-selves, but at the same time, we just didn't know we could do that.

In our other interviews, we also see the relationship between community con-nections and a growing sense of self-esteem or being "disbelieved by yourself."

Theme 3: Catalyzing alternative coding cultures through community

A final theme goes back to the heart of STEAM learning, but also reveals what we see as a corrosive mischaracterization of STEAM education in prac-tice. The theme is that the arts can provide a culturally meaningful domain of inquiry and learning for young people. In our case, music is something that youth can relate to and something that they can feel inspired to explore. Rosa describes it this way:

> Um well um music is like most people do like music and music is basically like a part of our daily lives like if you turn on the radio obviously like music with podcasts and also like music in general and then um when I um when I found out that I really could do a little bit of music with the code I wanted to um get inspired by like the songs I heard they didn't have to be like my favorite artists they just had to be like songs I heard and the type of tunes that I heard and like mix them together to see if I could get like inspired and then um I started just following in with like what our coaches was teaching us and it was like I did not want to miss a day in it

The corrosive mischaracterization is that by using a relatable art form like music as "the hook", we can somehow get kids interested in learning intimidating STEM content like CS and coding. On the contrary, what we've found with our programs is that music can be just as intimidating (if not more so) than code. Unfortunately, music, as it's often taught in schools, can be a dehumanizing experience of rote memorization, mindless hours of practice, and strained recitals – an experience drained of joy and meaning. Of course, this is not true of all music education, and, of course many people fall in love with music on their own terms. What we've seen, though, is that code can be a compelling language of music creation that helps make the mathematical and algorithmic concepts much more alive and accessible.

More important for us, however, is that the goal of the arts should not be to make STEM culturally relevant. When we think about the integration of computational literacy with the performing arts, we hope to create spaces where new, alternative computing cultures, grounded in community, creativity, and culturally resonant expressions of identity, can begin to flourish. In other words, the goal is to find ways to normalize coding within youth cultures as opposed to trying to force code to be culturally relevant by applying it to music. The distinction is subtle but important. To catalyze these cultures, you first need to build a community. Within communities, it's important to foster emotional and social connections that affirm identity, creativity, and inspiration. With our clubs and festivals, we hope to accomplish these goals while blending elements of street culture (autonomy, freedom, control, individuality, originality) into CS learning communities.

Acknowledgments

Our funders and community partners included the Verizon Foundation, My Chi. My Future, the National Science Foundation, the Center for Creative Entrepreneurship, Studio 2112, the James R. Jordan Foundation, the Digital Divas Program, and Chicago Public Schools.

Notes

1 According to the Computing Research Association's 2020 survey of Computer Science Departments, Black students accounted for only 3.1% of bachelor's degrees awarded in North America, while Hispanic students accounted for 8.5%. For advanced degrees, Black students represented 1.1% of Master's and 1.2% of PhDs earned, while Hispanic students accounted for 2.8% of Masters and 1.6% of PhDs (Zweben & Bizot, 2021); Zweben, S., & Bizot, B. (2022). CRA 2021 Taulbee survey: CS enrollment grows at all degree levels, with increased gender diversity. *Computing Research News, 34*(5), 2–82.
2 Music, for example, can be extremely technical, with Western music theory on its own drawing on centuries of confusing terminology and unintuitive representations derived from at least four different languages.
3 Woods, S. (2008). Poetry slams: The ultimate democracy of art. *World Literature Today, 82*(1), 18. Hoffman, T. (2001). Treacherous laughter: The poetry slam, slam poetry, and the politics of resistance. *Studies in American Humor,* (8), 49–64.

4 Muhammad, G. G., & Gonzalez, L. (2016). Slam poetry: An artistic resistance toward identity, agency, and activism. *Equity & Excellence in Education, 49*(4), 440–453. Gregory, H. (2008). The quiet revolution of poetry slam: The sustainability of cultural capital in the light of changing artistic conventions. *Ethnography and Education, 3*(1), 63–80.

5 DiSessa, A. A. (2018). Computational literacy and "the big picture" concerning computers in mathematics education. *Mathematical Thinking and Learning, 20*(1), 3–31.

6 Vee, Annette. (2017). *Coding literacy: How computer programming is changing writing.* MIT Press.

7 From Vee (2017), p. 7.

8 Turkle, S., & Papert, S. (1990). Epistemological pluralism: Styles and voices within the computer culture. *Signs: Journal of Women in Culture and Society, 16*(1), 128–157.

9 See Vee (2017).

10 We used TunePad, a music + Python coding website designed by our team at Northwestern University (https://tunepad.com). However, as we develop the concept of street coding, we will incorporate a variety of other creative coding platforms for art, robotics, and computational fashion.

22

BRIDGING INFORMAL SCIENCE LEARNING WITH SCHOOLS

The Open Schooling Model

Sofoklis Sotiriou and Franz Bogner

The EU Educational Policy "Science Education for Responsible Citizenship"[1] introduced the concept of Open Schooling as a promising approach to transform schools into innovation hubs within their local communities, highlighting the importance of cooperation with the informal learning sector. In an open school environment, external ideas need to challenge traditional internal views and, in turn, to benefit its students as well as the community it serves. Such environments may foster learner independence – and interdependence – through collaboration and the provision of opportunities for learners to understand and examine their place in the world.

Open Schooling encourages science students to see how real science is applied in practice *outside* of school. In this way, students better understand how science is applied in real life.[2] The overwhelming majority of these practical activities occur with the active cooperation and co-design of out-of-school informal and non-formal science learning settings, research institutions, civic society institutions, and industry. This approach enables open schools to become agents of community well-being.[3]

Characteristics of the open schools

In general, Open Schools concept is building upon an innovation agenda that aims to create a science learning continuum between formal and informal education.[3] More specifically, it:

1. Promotes the collaboration with non-formal and informal education providers to ensure relevant and meaningful engagement of all societal actors with science and increase the uptake of science studies and science-based careers.[4,5]

DOI: 10.4324/9781003145387-25

2. Promotes partnerships that foster expertise, networking, sharing, and applying science and technology research findings and bring real-life projects to the classroom. Open Schools aim to develop and promote innovative educational applications, share and apply frontier research findings, support competencies through creative problem-solving, discovery learning by doing, experiential learning, critical thinking, and creativity, including projects and activities that simulate real scientific work.[6]

Design features of innovative cooperation schemes

The Open Schooling approach builds on the essential features of creative learning, including exploration, dynamics of discovery, student-led activity, engagement in scientifically oriented questions, priority to evidence in responding to questions, formulations of evidence-based explanations, the connection of explanations to scientific knowledge, and communication and justification of explanations.[3,7] Open Schooling presents an integrated framework for the proposed innovation partnerships and the emerging activities,[3] which have the following features:[7]

> *They are placed.* The activities are located, either physically or virtually, in a world the student recognises and seeks to understand.
> *They are purposeful.* The activities feel authentic, and they absorb the student in actions of practical and intellectual value and foster a sense of agency.
> *They are passion-led.* The activities enlist the outside passions of both students and teachers, enhancing engagement by encouraging students to choose areas of interest that matter to them.
> *They are pervasive.* The activities enable the student to continue learning outside the classroom, drawing on family members, peers, local experts, and online references as sources of research and critique.

These four criteria describe a school as an organisation offering more engagement: a place-based curriculum, purposeful projects, passion-led teaching and learning, and pervasive opportunities for research involvement and constructive challenge. This environment enhances the creation of new partnerships with the informal learning sector to foster improved science education and contribute to a learning continuum for all. It seeks to promote partnerships of museums and science centres with educators, scientists, researchers, and other stakeholders in science-related fields to work together on real-life challenges and innovations within local communities with a view to engage them in teaching and learning processes and to promote science education as part of local community development.[8,9]

To illustrate the Open Science Schooling approach, students may analyse their home river water quality and the air quality at different locations and cooperate for deeper analyses with scientists and experts from natural history museums. In another scenario, students may meet with scientists to learn about butterflies and moths in order to prepare themselves for a popular science event called "Nights for Biologists".[10]

Assessing methods

Assessing the effectiveness of the Open Schooling approach needs an appropriate tool that is sensitive to the key characteristics of these environments. By focusing on three identified areas of school innovation – school management, school process and teachers' professional development – a specific instrument (the Self-Reflection Tool, or SRT) offers the opportunity to school community stakeholders to describe in detail the current school situation related to openness, while at the same time to translate the findings into specific recommendations for future actions and development. By applying the SRT to the school principals, the organisational change during the implementation of the Open Schooling approach is observable.[7] The instrument allows for the assessment of the innovation level of each school for one full academic year by following a pre-/post-test schedule (one before their involvement and one after their engagement in their transformational journey).

Science Motivation (SM) is an instrument that monitors adolescent motivation and attitudes to science on a Likert scale based on five subscales (Self-Determination, Intrinsic Motivation, Career Motivation, Self-Efficacy, Grade Motivation).[11,12] The SM can detect an improvement of student perceptions of science and of their motivation towards science, as the result of their involvement in the proposed approach. For monitoring changes, the questionnaire was completed twice within a pre-post-test design. As the tool is regarded as gender-sensitive, potential gender differences regarding science motivation, for instance, towards self-perception in science become visible.[12,13] Thus, the SM-subscale "Self-Efficacy" is of interest, as it may measure gender differences in the status quo pre-tests that may be bridged in the context of implementation.

School Motivation (SchM)[12–14] as an instrument was developed to measure more general motivation and attitudes to learn. It is a 15-item Likert-type test with four subscales (Self-Determination, Intrinsic Motivation, Self-Efficacy, Grade Motivation)[15] that can measure an improvement of students' willingness to perform at school in general. The subscales are similar to the ones of SM. "Career Motivation" was omitted, since the students of this age group had not yet developed their concepts about their future careers.[16]

Emotions such as anxiety or interest can be distinguished as trait (biographically generated) or state (caused by situational context) characteristics.[16] Positive situational emotions seem to influence learning processes positively, whereas negative situational emotions, such as boredom, do the contrary. Emotions are also a part of motivation and cognition processes, which is why they need consideration in educational instructions.[17]

Results

The data from 316 schools demonstrate that the introduction of Open Schooling produced a significant impact on the innovation levels of the participating schools. The results showed an extremely successful journey in the diffusion of innovation in school settings. The data demonstrate significant growth in innovation (almost 12% on average), while the growth is much higher for less advanced schools (goes up to 40%) in a one-year intervention.[7] The graph plots the post- versus the pre-values for the 316 schools. To provide a prediction of the expected annual development for school leaders who have adopted the Open Schooling approach and to provide suggestions on how school leaders and practitioners can employ the assessment tool in their administration, we are modelling the expected value as a function of the initial school score. Clearly, there are many factors (e.g., size of the school) that could significantly affect the final school score, but our intention in this study is to demonstrate the net contribution of the School Innovation Model to the development of schools, independently of the local conditions and other factors that can affect the final result (Figure 22.1).

We observed a significant increase of 11.79% of the mean innovation value between the pre- and post-measurement, after one year of implementing the

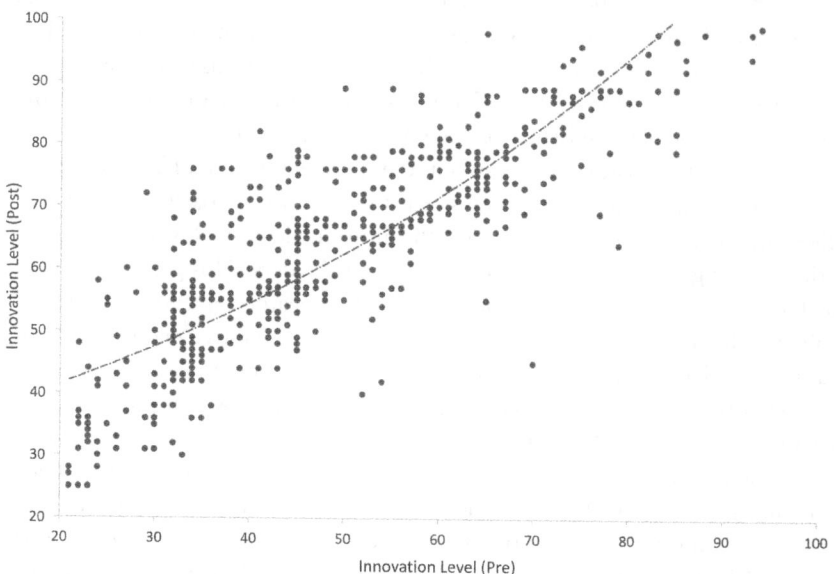

FIGURE 22.1 The graph presents the pre- and post-innovation levels of the 316 schools that participated in the pilots, after one academic year of implementation of the Open Schooling activities. The fit line represents the average growth of the participating schools according to their pre-measurement.

Open Schooling approach in the school settings. The specific increase in the innovation level indicates that schools integrated many of the aspects of the proposed approach in their day-to-day activities as well as into the Development Plans.

In the SM, for all subscales, there was a significant increase in scores (Wilcoxon, $p < .001$). In Grade Motivation (GM), hardly an increase is seen because the median remains at the high score of 4. The increase in motivation is reflected in the reduction of the standard deviation towards the low scores, while the median remains stable, although shrinking downward the standard deviation.[15]

Discussion

All these results are in line with the basic principles of an Open Schooling environment: it is an environment to facilitate the process for envisioning, managing, and monitoring change in school settings by providing a simple and flexible structure to follow, so school leaders and teachers can innovate appropriate for school needs. Our results are also in line with the expectations of the EU Policy document "Science Education for Responsible Citizenship"[1] where collaboration between formal and informal educational is generally earmarked to enhance meaningful engagement of all societal actors with science and to increase uptake of science studies and science-based careers to improve employability and competitiveness. Open Schooling demonstrates how formal schooling and informal learning experiences can complement and strengthen each other. Our aim is to demonstrate that such cooperations can support people in acquiring a scientific way of thinking, so that they can understand and correctly use scientific information to which they are exposed.

Analysing the relationship between the different scales offers innovative insights and brings into light a transformational process, based on the Open Schooling approach, which combines the implementation of student projects with active engagement of informal learning stakeholders. This process enables a quick testing of the self-reflection process, which is regarded as a powerful way to facilitate school development. Thus, the process seems sensitive enough to communicate evidence of the school development progress and to provide insights to both schools and the informal learning stakeholders. More specifically, the participating schools (n = 316) demonstrate a significant increase on their innovation levels which goes up to 35–40% for schools with lower scores at initial assessment. Nonetheless, schools that have higher scores in the first measurement demonstrate increased final scores. During one academic year, the participating schools involved 1,500 teachers and 22,000 students in more than 320 projects promoting cooperation with more than 100 science centres and museums.

Students' motivation was significantly higher after they participated in the project activities, which were not limited to STEM subjects, as might be expected in such initiatives. Students reported generally higher school motivation after one year. The project's motivational impact is in line with earlier expectations that motivation and improvement of student learning would be promoted primarily by working on the openness and innovation level of schools.[18]

Appropriate professional development initiatives therefore are promising triggers for having a positive effect on students.[19,20] Flexible teaching practice is needed to make educators more readily adapt to the ecological, social, and economic challenges that humanity currently faces.[21] Through the SRT-feedback, the need for own professional development and school development becomes intrinsically encouraged. Their effect on student motivation is obvious. A special feature is the reaction of Grade Motivation, which allows a more differentiated view. All motivation factors scored significantly higher in post-test schedules except grade motivation. We judge this pattern as a reorientation away from external assessment towards the students' responsibility for their individual learning by enacting creativity or curiosity.[21,22] A drop in grade motivation and the simultaneous increase in Self-Efficacy and Self-Determination indicate that our approach effectively promoted students' self-responsibility.[23]

Furthermore, the correlation of the SRT results demonstrates that when a school increases its openness level (meaning that it proceeds with organisational changes and that its cooperation with external stakeholders becomes the norm) the student's motivation will increase as well. To investigate the extent to which the motivation effect is related to school development, the correlation of the development of schools and their students was studied. Looking at the correlation of SRT with motivation development, the direct effect of school innovation modelling on learner motivation becomes even clearer. The enormous impact of involving schools in out-of-school activities in STEM subjects is particularly evident in the general school motivation pattern. The results of SRT correlate even more strongly with school motivation than with science motivation.[12,15] The strong correlation of both motivation tests needs emphasising as they have a common theoretical ground (although they do not measure the same concept). Thus, the science-focused interventions were also able to increase overall motivation to learn, which is a promising result. Science interventions through the implementation of the Open Schooling model can increase general motivation to learn and achieve, so that other subjects also benefit. General school motivation is an important indicator for the effectiveness of STEM projects, which indirectly support other subjects as well.

Applying the Open Schooling model has made it clear that schools have much to gain by fostering connections between formal and informal learning, between existing providers of education and new entrants. Such a process can best support schools in their attempt to evolve, transform, and reinvent their structures towards a more open, innovative, localised and socially responsible learning environment, following the example of informal learning settings. This process highlights that one of the core elements of the schools' success in raising achievement is a robust focus on the tracking and monitoring of development progress and the use of assessment data for progress tracking, target setting and support for schools slipping behind their targeted interventions. Such models can be used effectively by educational policy stakeholders to answer questions about current standards, trends over time, progress made by the schools and to set high expectations in case study schools.[22] The Open Schooling model offers the opportunity to revisit the effective and in-depth engagement of the informal learning stakeholders to formal curriculum by (a) highlighting their significant contribution to students' achievement and (b) developing ways to certify their contribution.

The Open Schooling Model designed, developed, and tested in the framework of the OSOS initiative (www.openschools.eu) has managed to demonstrate effective ways to open up schools, specifically to get the management staff on board and to provide quality Professional Development for teachers. By providing numerous examples of school-informal science learning institutions that go beyond the traditional school visits to these institutions and their collections, we have offered a rich repertoire of working models and approaches that can be used in different settings and cultures.

For example, one of the most popular activities that was reproduced in multiple ways was the concise introduction to ground-breaking scientific advances in astronomy, from Galileo's first telescopic discoveries in 1600 (the first scientific revolution) to the recent Nobel prize-winning direct observations of gravitational waves by the VIRGO experiment. In this activity, informal STEM learning was considered the outcome of the learner's engagement with activities in different learning contexts (school, Museo Galileo, VIRGO Research Facility) where the formal and informal approaches were in continuous interaction by forming a science learning continuum. The roles of the instructor changed many times: between the teacher, the museum guide, and the scientist – all of whom shared the responsibility for promoting student learning. The different science learning contexts were characterised by their diversity, redundancy, and local adaptations, while the overall experience contained a wide variety of activities, across a range of institutions and places, allowing learners many different ways to engage with science.

Such initiatives could act as incubators and accelerators on school innovation. Practitioners and school heads could advance the expected learning outcomes in the framework of such activities as they are expected to increase motivation and interest. Schools, in close cooperation with informal stakeholders, facilitate open, more effective and efficient co-design, co-creation, and use of educational content (both from formal and informal science education providers), as well as tools and services for personalised science learning and teaching that form the basic ingredients for innovative student projects.[24] This work differs from the work in engagement and affect in terms of time-scale. Whereas engagement and affect often manifests in brief time periods – as short as few seconds – motivation and interest are more long-term stable aspects of students' experience with the expectation to connect students learning experience with their values, leading to greater degrees of completion of school projects and related tasks.[25,26] Within this context, projects related to students' personal interests are likely to influence students to work faster, become less often disengaged, and learn more.[26-28] Thus, the overall self-reflection process could help education leaders and curriculum designers to reflect on the ways they facilitate the proposed innovation framework, both through their individual actions as well as through the structures they work to create and maintain. The work of the OSOS Initiative is documented in practical terms for school leaders and innovative teachers.[3] In addition, detailed strategies for the involved stakeholders are provided to help implement the framework of effective cooperation schemes between schools and informal learning institutions.[29]

Notes

1 European Commission. Directorate General for Research and Innovation. (2015) Science education for responsible citizenship: Report to the European Commission of the expert group on science education. Publications Office, Brussels.

2 Montero, S. M., Baranowski, A., & Gejel, J. (2019). Open science schooling: Rethinking science learning. *EDULEARN19 Proceedings*, pp. 9159–9164. https://doi.org/10.21125/edulearn.2019.2263.

3 Sotiriou, S., Cherouvis, S., Zygouritsas, N., Giannakopoulou, A., Milopoulos G., Bogner, F.X., Verboon, F., & de Kroon, S. (2017). *Open Schooling Roadmap: A Guide for School Leaders and Innovative Teachers*. Retrieved November 5, 2022, from https://rri-tools. eu/-/open-schooling-roadmap-a-guide-for-school-leaders-and-innovative-teachers.

4 Goddard, R., Goddard, Y., Sook, K. E., & Miller, R, (2015) A theoretical and empirical analysis of the roles of instructional leadership, teacher collaboration, and collective efficacy beliefs in support of student learning. *American Journal of Education*, 121, 501–530. https://doi.org/10.1086/681925.

5 Wenner, J.A., & Campbell, T. (2017). The theoretical and empirical basis of teacher leadership. *Review of Educational Research*, 87, 134–171. https://doi.org/10.3102/0034654316653478.

6 Sotiriou, S. A., Bybee, R. W., & Bogner, F. X. (2017). Pathways: A case of large-scale implementation of evidence-based practice in scientific inquiry-based science education. *International Journal of Higher Education*, 6(2), 8–19.

7 Sotiriou, M., Sotiriou, S., & Bogner, F.X. (2021). Developing a self-reflection tool to assess schools' openness. *Frontiers in Education*. https://doi.org/10.3389/feduc.2021.714227.

8 Salmi, H., Sotiriou, S., & Bogner, F. (2010). Visualising the Invisible in Science Centres and Science Museums. In Karacapilidis, N. (Ed.), *Web-Based Learning Solutions for Communities of Practice*. IGI Global, pp. 185–208. https://doi.org/10.4018/978-1-60566-711-9.

9 Sotiriou, S., Riviou, K., Cherouvis, S., Chelioti, E., & Bogner, F.X. (2016). Introducing large-scale innovation in schools. *Journal of Science Education and Technology*, 25(4), 541–549. https://www.learntechlib.org/p/176163/.

10 Grau, D., Torram, I., & Mancho, F. (2020). *Open Science Schooling: Guide for Secondary Schools*. https://openscienceschooling.eu/intellectual-outputs/io1/. Accessed 25 July 2022.

11 Schumm, M., & Bogner, F.X. (2016). Measuring adolescent science motivation. *International Journal of Science Education*, 38, 434–449. https://doi.org/10.1080/09500693.2016.1147659.

12 Schumm, M., & Bogner, F.X. (2016). The impact of science motivation on cognitive achievement within a 3-lesson unit about renewable energies. *Studies in Educational Evaluation*, 50, 14–21. https://doi.org/10.1016/j.stueduc.2016.06.002.

13 Conradty, C., & Bogner, F.X. (2016). Hypertext or textbook: Effects on motivation and gain in knowledge. *Education Sciences*, 6, 29. https://doi.org/10.3390/educsci6030029.

14 Ferdous, A.A., & Plake, B.S. (2008). Item response theory–based approaches for computing minimum passing scores from an Angoff-based standard-setting study. *Educational and Psychological Measurement*, 68, 778–796. https://doi.org/10.1177/0013164407312605.

15 Conradty, C., & Bogner, F.X. (2022). Measuring students' school motivation. *Education Sciences*, 12, 378. https://doi.org/10.3390/educsci12060378.

16 Ainley, M. (2006). Connecting with learning: Motivation, affect and cognition in interest processes. *Educational Psychology Review*, 18, 391–405. https://doi.org/10.1007/s10648-006-9033-0.

17 Randler, C., Hummel, E., Glaser-Zikuda, M., Vollmer, C., Bogner, F.X., & Mayring, P. (2011). Reliability and validation of a short scale to measure situational emotions in science education. *International Journal of Environmental and Science Education*, 6, 359–370.

18 Guskey, T.R. (2003). What makes professional development effective? *Phi Delta Kappan*, 84, 748–750. https://doi.org/10.1177/003172170308401007.

19 Hattie, J., & Hattie, J.A.C. (2009). *Visible Learning: A Synthesis of Over 800 Meta-analyses Relating to Achievement*. Routledge, London.

20 Conradty, C., & Bogner F.X. (2019). From STEM to STEAM: Cracking the code? How creativity & motivation interacts with inquiry-based learning. *Creativity Research Journal*, 31(3), 284–295. https://doi.org/10.1080/10400419.2019.1641678.

21 Gnambs, T., & Hanfstingl, B. (2016) The decline of academic motivation during adolescence: An accelerated longitudinal cohort analysis on the effect of psychological need satisfaction. *Educational Psychology*, 36, 1691–1705. https://doi.org/10.1080/01443410.2015.1113236.

22 Black, A.E., & Deci, E.L. (2000). The effects of instructors' autonomy support and students' autonomous motivation on learning organic chemistry: A self-determination theory perspective. *Science Education*, 84, 740–756. https://doi.org/10.1002/1098-237X(200011)84:6<740:AID-SCE4>3.0.CO;2-3.

23 Bandura, A. (2002). Social cognitive theory in cultural context. *Applied Psychology*, 51, 269–290. https://doi.org/10.1111/1464-0597.00092.

24 Earley, P., & Greany T (Eds) (2017). *School Leadership and Education System Reform*. Bloomsbury Academic, London, New York.

25 Randler, C., & Bogner F.X. (2009). Efficacy of two different instructional methods involving complex ecological content. *International Journal of Science and Mathematics Education*, 7(2), 315–337.

26 Conradty, C., Sotiriou, S., & Bogner, F.X. (2020). How creativity in STEAM modules intervenes with self-efficacy and motivation. *Education Sciences*, 10(3), 70. https://doi.org/10.3390/educsci10030070.

27 Sotiriou, S., & Bogner, F.X. (2011). Inspiring science learning: Designing the science classroom of the future. *Advanced Science Letters*, 4(11–12), 3304–3309. https://doi.org/10.1166/asl.2011.2039.

28 Salmi, H., Sotiriou, S., & Bogner, F. (2010). Visualising the Invisible in Science Centres and Science Museums: Augmented Reality (AR) Technology Application and Science Teaching. In N. Karacapilidis (Ed.), *Web-Based Learning Solutions for Communities of Practice: Developing Virtual Environments for Social and Pedagogical Advancement* (pp. 185–208). IGI Global. https://doi.org/10.4018/978-1-60566-711-9.ch014.

29 The OSOS Open Schooling Model. https://www.openschools.eu/open-school-model/.

23

BENEFITS OF ENGINEERING BEYOND THE SCHOOL DAY

Insights from Research

Cary I. Sneider and Mihir K. Ravel

As conceived in *A Framework for K-12 Science Education*[1] and the *Next Generation Science Standards*,[2] engineering is a life skill having wide-ranging value for dealing with issues in the workplace, for effective citizenship in a participatory democracy, and for solving the problems of daily life in a structured way. Inspired by the potentially high impact on youth, the last several years of our professional work have focused on supporting educators by introducing engineering into both formal and informal learning environments. Our activities have included developing curriculum materials and formative assessments, conducting workshops for teachers, and a wide range of webinars for various audiences.

In doing this work, we have had the privilege to collaborate with a growing cadre of educators who are deeply committed to supporting this vision. Over the course of these collaborations the following questions frequently arose:

What are the benefits of engineering education as a complement to science learning?

What are effective methods for engineering education in informal learning environments?

What are long-term effects of informal engineering education?

What are the challenges of introducing engineering into after-school and summer programs?

To answer these questions, we conducted regular searches of the literature, and as our body of research grew, we decided to summarize our findings for broader audiences in the form of a literature review. The result was a landscape

DOI: 10.4324/9781003145387-26

study of 263 research studies,[3] including 70 that took place in informal learning environments. In this essay we offer a summary of our findings with a few illustrative examples.

Findings

Both classrooms and informal learning environments are effective venues for introducing engineering, but their focus is somewhat different. The development of knowledge and skills is the primary goal of most teachers in schools, while instilling an interest and thirst for learning is more compatible with the goals of most educators in after-school and summer programs. To illustrate the value of engineering for informal learning and provide examples of the research studies we reviewed, we begin with a pair of studies – one quantitative and one qualitative – of a single program that has grown and flourished over more than 20 years.

Techbridge is one of many programs for engaging youth – and girls in particular – in engineering activities during after-school and summer programs. Starting in 2000 at a small science center in Oakland, California, Techbridge has since spread through Girl Scout Councils nationwide, reaching hundreds of thousands of girls. Techbridge "Activities in a Box" aim to motivate girls with topics such as Thrill Builders (designing a playground), Power It Up (electronics), and Engineers to the Rescue! (solving problems on a camping trip.) Like 4-H and other youth programs, Techbridge depends on volunteers as frontline staff. Below is an online invitation to potential volunteers:

> Girl Scouts on a BLISS Journey define their dreams, set goals to achieve them, and find ways to help others do the same. Engineers work in similar ways by defining problems and coming up with solutions. Adding in some fun, hands-on engineering activities gives girls a chance to create, work as teams, and problem solve. You will inspire girls to discover their talents and dreams, feel confident about pursuing engineering careers, and be excited about how they, as engineers, can make the world a better place. Girls can explore techniques and strategies to use on this Journey and throughout their lives, whether or not they pursue a career in science, technology, engineering, and math (STEM).[4]

A longitudinal quantitative study of Techbridge collected data from a self-report survey of 362 girls from predominantly underrepresented groups in Oakland, California, who had participated in Techbridge in prior years. The study[5] found that 67.7% plan to take or have already taken advanced or honors mathematics and/or science classes in high school and 97.5% either plan to attend college or are already attending college. This extraordinarily high proportion of girls who say they plan to go to college is remarkable given that the average *high school* graduation rate in Oakland at the time of the study was only 68.3%. Furthermore, the more years that girls participated in Techbridge, the more likely they were to report planning STEM-related jobs and the less likely they were to choose non-STEM jobs.

In a complementary qualitative study, the same researchers interviewed 13 girls several years after they had participated in Techbridge with questions about what they considered the most valuable aspects of the program to be. A consistent finding was that participating in hands-on projects was by far the most important. Also important were field trips and role models.[6]

The Techbridge studies were just two of the 70 studies that took place in informal environments included in our research review. In the remainder of this essay, we provide a broad-strokes overview of the findings that are relevant to informal STEM education.

Benefits of engineering as a complement to science learning

While each study in our review had unique measures of outcomes and tested widely varying instructional methods, we were able to draw the following broad conclusions about the value of integrating engineering as a part of science.

Benefits to learning science concepts and skills. Engineering is highly interactive. If the design challenge is sufficiently compelling and the goals are clear, initial engagement rapidly deepens into commitment to meeting the challenge. Students learn the science and mathematics built into the lesson not because it is required, but because they are motivated to learn.

Benefits of design skills. Engineering in the new standards is defined by a set of core ideas and practices, which include the ability to define a problem, generate, and test solutions, then continue to iterate until a solution is found. This set of capabilities, commonly referred to as an engineering design process, is a valuable set of skills, useful in the workplace as in everyday life.

Benefits to learning 21st-century skills. Engineering is a team effort. Students naturally develop team-building skills such as collaboration and communication as they work toward a common goal. It is also a highly creative activity since design challenges present novel problems requiring novel solutions. Figure 23.1 shows one such design challenge.

Benefits of expanded career opportunities. By solving authentic problems, students gain a sense of agency, and as they learn about the wide diversity of engineering fields, they encounter opportunities to develop an intense interest that has personal meaning and offers a potential career goal.

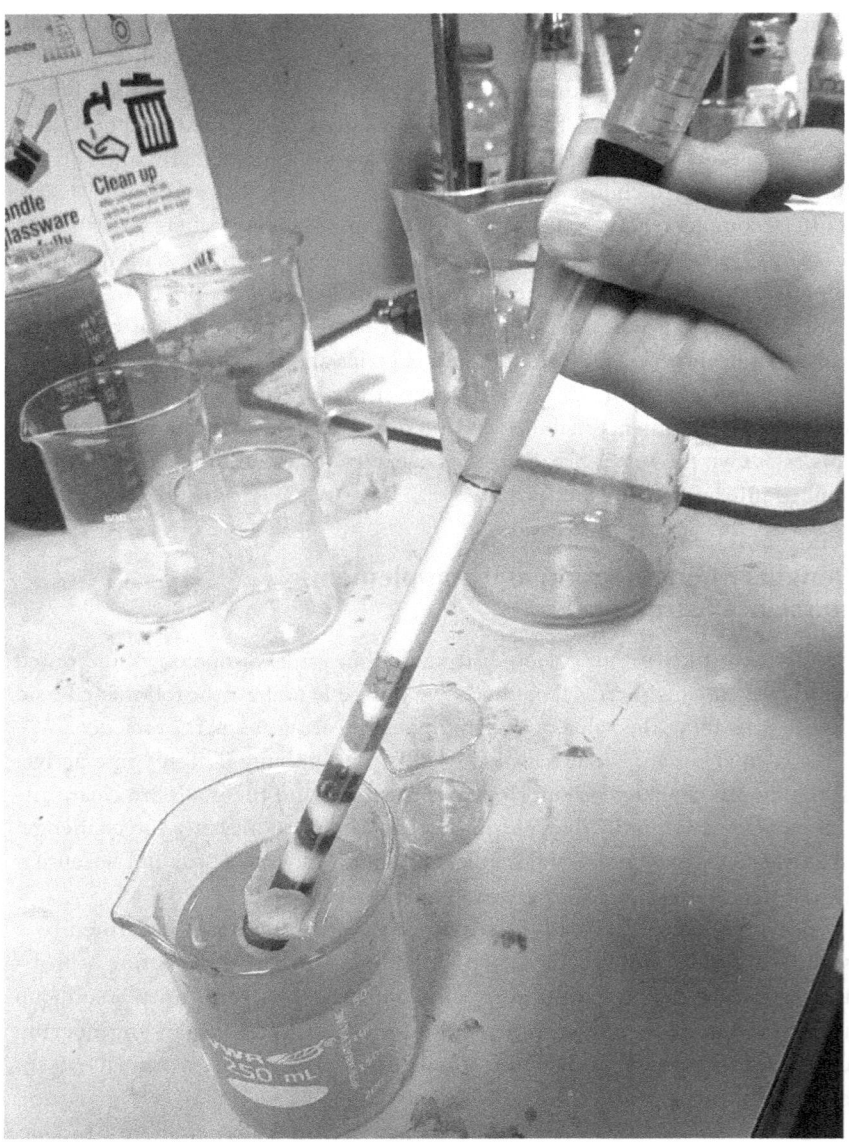

FIGURE 23.1 In this design challenge, students must create an emergency water purifier that filters water from a contaminated source. To do so, they measure the filtering properties of various materials and then use these data to engineer a layered filter structure that can be tested and improved.

Effective methods for engineering in informal learning environments

Role models and mentoring can be very effective. An important aspect of the Techbridge program was a major effort to introduce girls to women of various ethnic backgrounds who have succeeded in careers in engineering, science, or related fields. In many of the studies, stories about role models were built into the curriculum. Additionally, volunteers were encouraged to invite female professionals or engineering students from their communities to meet with the youth.

Mentors can also serve as role models, especially when they match the youth in terms of sex and ethnic background, but their relationship is much deeper. An example is an outreach program in which volunteer undergraduates from Rice University went to local high schools at least once per week to mentor students after school over 5–7 weeks. The youth worked with their mentors on engineering design projects culminating in a competition on the Rice University campus. More than 90% of the mentees were Latinx and 3% were African American. A total of four cohorts had participated at the time of the study, with 23–34 youth per cohort. A large proportion of the volunteer mentors were themselves members of underrepresented minority groups, enrolled in engineering. The study found that the mentorship program was very successful in improving students' understanding of physics, knowledge of what engineers do, and knowledge of what it takes to prepare for and survive in college.

Another excellent example is MESA, an abbreviation for Mathematics, Engineering, Science Achievement, which was started in California during the 1970s to provide pathways to STEM careers for students from populations underrepresented in STEM fields. MESA activities take place primarily after school and during summers. An essential component of MESA is involvement of mentors, commonly college students enrolled in engineering programs. Many MESA participants begin their involvement in middle school and continue to participate through high school and into college. A research study of the nearly 50-year-old program was conducted in 2017. Data were collected from focus group interviews of 30 former MESA participants and surveys from 484 former MESA youth. Results showed a positive influence of MESA activities on students' self-efficacy, interests, and perceptions related to engineering ($p < .001$). The most influential factors were hands-on activities, followed by meeting with professionals (i.e., role models), student advisors (mentors), and field trips. An important finding was that participants talked more about their roles as mentors with youth rather than the mentoring that they themselves received from MESA teachers and advisors.[7]

For educators in informal programs who want to have a significant impact, the extra effort to find role models and mentors with whom the children and youth can identify is well worth the time and effort.

Summer Camps allow more time for STEM, which in turn leads to deeper impact. Although many studies demonstrated the value of summer engineering camps by using various types of tests and surveys, the strongest evidence was provided by two research studies that used comparison groups.

In a study of the long-term impact of a two-week middle school engineering summer camp for girls, researchers conducted telephone interviews of 88 students who had attended the camp as sixth graders 4–7 years previously and 41 students who had applied for the camp but did not attend. The camp was taught by female role models/mentors and focused on hands-on learning, building self-efficacy, collaboration, and teamwork and emphasized how engineers make the world a better place. Camp participants were found to be significantly more likely to enroll in elective computer science classes (p < .02) and other science/engineering classes (p < .05) in high school than the control group.[8]

Another team of researchers compared the impact of a weeklong robotics/geospatial technologies summer camp for 147 middle school youth with 141 middle school youth who participated in a three-hour robotics intervention. Both groups were given pre-post tests on computer programming, mathematics, geospatial technologies, engineering, and attitudes. The weeklong summer camp led to significant pre-post-test learning gains for males and females (p < .001), although males scored higher than females on both tests. The short-term intervention did not result in cognitive learning gains but did positively impact the students' attitudes toward the value of science, mathematics, robotics, and geospatial technologies.[9]

These studies provide compelling evidence that summer programs can not only provide an excellent STEM learning experience but can also make a big difference in attitudes that can have significant long-term impact.

Robotics can be appealing to both boys and girls. Our review of the engineering education literature, in both school and non-school settings, showed that the most common type of program to be reviewed was robotics. In fact, there were more robotics programs studied than all other types of engineering programs combined. A 2019 study summarized findings of 147 such studies.[10] Given the high visibility of robotics programs, an important question is whether focusing on robotics might be one of the reasons that more boys are attracted to engineering than girls. Some evidence suggesting that is the case was a study that collected self-report surveys from 502 elementary, middle, and high school students who had participated in various robotics competitions, to determine how the experience shaped their interests in programming. In the youngest groups (entry level competitions) girls were heavily involved in programming, but in more advanced competitions fewer girls were involved, even after controlling for prior programming experience.[11]

Another study demonstrated that with some modifications, robotics can be just as engaging for girls as for boys. The research study, which involved seven ninth-grade teachers and their students (N = 361), compared students' responses to engineering robotics using traditional rigid robots versus soft robots made from materials such as silicone and fabrics. Twenty-two classes were randomly assigned to one of the two conditions. Boys and girls had positive gains (p < .001) for interest, general self-efficacy, experimental self-efficacy, tinkering self-efficacy, and design self-efficacy. The only significant gender difference was on tinkering self-efficacy, which is interpreted as comfort with the manual aspects of engineering, such as assembly, disassembly, manipulating devices, and similar tasks. Boys performed significantly higher than girls (p < .05) in the rigid robotics condition, but not in the soft robotics condition. The study concluded that "engagement with the soft robotics experience led to an increase in tinkering self-efficacy for girls, which relieved gender differences in that element of engineering perception."[12]

Competitions allow for individuals or teams to participate, with or without guidance. There are a great many engineering competitions that provide opportunities for students to tackle engineering design challenges, either on their own or as part of a team. Many of these allow for individuals to participate at no cost, while others require adults to help organize and fund opportunities for participation. One website[13] lists 65 engineering competitions.

Perhaps the best-known engineering competition is FIRST Robotics. Although elementary and middle school versions exist, the high school competition, which requires several thousand dollars to fund a single team and requires an adult with significant engineering skills to supervise, is the most widely researched. A five-year longitudinal study of 289 FIRST participants found that in comparison with 162 controls, the program had a statistically significant positive impact on STEM-related attitudes and interests, including life and workplace skills, STEM career interests, and postsecondary aspirations.[14]

As an alternative to the FIRST model which requires significant materials and cost, ExploraVision is a very successful approach that requires no equipment and no cost to participate. Students write essays describing inventions that they think would solve current problems. A study of a representative sample from more than 15,000 submissions over 18 years found that students most frequently recommended changes in communications and medical technologies, and a great many students were concerned with negative impacts of changes in technology, such as the loss of jobs.[15]

Competition can be a double-edged sword. For individuals with a competitive bent it can be a great motivator; but for others it can create undue pressure. Team (rather than individual) competition can go a long way to accentuate the positive side of competition. It's helpful to keep in mind that there is a wide variety of types of competitions suitable for many different personalities and interests.

What are long-term effects of informal engineering education?

Although most of the studies we examined provided data on the immediate effects of programs, several examined deeper long-term impacts.

Engineering can help reduce the gender gap. Technovation is an after-school program designed to reduce the gender gap by providing young women with an accessible, entry-level coding experience as well as entrepreneurship and business leadership with the assistance of female mentors. The program features an annual challenge to design an app that addresses a local community problem. Researchers conducted a pilot study (N = 117) and main study (N = 653) of the program in which participants completed self-report surveys at least four months and up to five years after the end of the program. Participants were primarily high school youth. Technovation experience increased girls' interest in computer science, entrepreneurship, and business leadership. Among prior participants who were now in college, 26% were majoring in computer science, a rate 65 times higher than the national average. Thirty-three percent who were not in computer science were in some other STEM major with engineering most common.[16]

Early OST experiences can lead to later STEM college enrollments. While many studies examined youths' intentions to pursue STEM careers, one group of researchers examined the impact of out-of-school STEM experiences by analyzing data from 15,847 college students who participated in the Science, Technology, Engineering, and Mathematics Talent Expansion Program (STEP) and took a survey about their career intentions and prior experiences. The findings indicated that out-of-school experiences increased the odds of students choosing engineering disciplines. Experiences traditionally stereotyped as masculine and more often reported by men, such as tinkering, increased the odds of choosing engineering disciplines with a higher representation of men. However, some experiences equally reported by men and women, such as mixing chemicals or engaging with chemistry in the kitchen or talking with friends or family about science, predicted higher odds of choosing engineering disciplines (i.e., chemical, biomedical, and environmental engineering) with a higher representation of women.[17]

What are the challenges of introducing engineering into after-school and summer programs?

It has been nearly a decade since many states adopted new educational standards that include engineering at the same level as the traditional sciences. Nonetheless, it has been slow to catch on.

Few teachers include engineering. In a national sample of about 10,000 teachers, the percentage who reported emphasizing engineering was 8% of elementary teachers, 10% of middle school science teachers, and 5% of high school science teachers.[18] Although there is no similar dataset for informal education, the percentages are unlikely to be higher.

English and math dominate the curriculum. For most of the 20th-century science education was an essential aspect of the curriculum. Engineering was considered "just" applied science and was rarely taught. In recent decades, even science is struggling for a place in the curriculum, as it is nudged aside by English and mathematics. And those two subjects also tend to be emphasized in after-school and summer programs that aim to help kids "catch up."

Adults have common misconceptions about engineering. One of the greatest challenges to the growth of engineering in after-school and summer programs is a set of misconceptions among adults generally. Many people think of engineering as a highly technical field that only the most brilliant people can aspire to, while others think of it as a boring occupation for number crunchers, and still another misconception is that it's only for those who like manual hands-on tinkering work. As an activity for kids, it's common to envision engineering as limited to robots and computers. While it's possible to overcome these misconceptions through engaging workshops, webinars, YouTube videos and articles, getting the leaders of informal educational programs to pay attention to these resources is one of the biggest barriers to overcome.

More instructional materials and professional development programs for teachers are needed

Once informal educators become aware of the value of engineering, the next step is for them to find accessible, inclusive, and affordable sources of curriculum materials and professional development. Introducing youth to engineering requires more than presenting a design challenge. It's also important for learners to define solvable problems, to generate alternative solutions before fixating on one idea, to test their chosen solution, and then redesign based on the results of the tests. Just as learning the process of scientific inquiry requires guidance from a capable and knowledge teacher, so does learning the process of engineering design.

Fortunately there are many opportunities for professional learning, especially though such organizations as the National Science Teaching Association (NSTA) and the International Technology and Engineering Education Association (ITEEA). There are also online sources for curated engineering activities[19] for educators who want to connect with others and share resources. Making teachers aware of these resources is critical if the vision of integrating engineering into science education is to be fully realized.

Despite the challenges, we believe that as more informal educators recognize the value of engineering as a learning opportunity for youth, uncover the available resources, and have their first engaging experiences, seeing the positive experiences for learners will provide an intrinsic motivation for continuing this effort. The most effective advocates for engineering education are the young people who find that engineering activities are not only exciting and fun but are purposeful and powerful agents for bettering the world around them.

Notes

1 National Research Council (2012). *A framework for K-12 science education: practices, crosscutting concepts, and core ideas.* Committee on a Conceptual framework for New K-12 Science Education Standards. Board on Science Education. Washington, DC: National Academy Press.

2 NGSS Lead States (2013*) Next generation science standards: For states by states.* Washington, DC: National Academy Press.

3 Sneider, C. I., & Ravel, M. K. (2021). Insights from two decades of P-12 engineering education research. *Journal of Pre-College Engineering Education Research (J-PEER)*, 11(2), 1–38, Article 5. https://docs.lib.purdue.edu/cgi/viewcontent.cgi?article=1277&context=jpeer

4 From https://www.girlscoutsoc.org/content/dam/girlscoutsoc/documents/Program/Engineer%20Your%20Bliss%20planner.pdf.

5 Ancheta, R. (2008). Report on quantitative longitudinal evaluation of Techbridge. San Francisco: Report to the Gordon and Betty Moore Foundation.

6 Ancheta, R. (2008).

7 Denson, C.D. (2017). The MESA study. *Journal of Technology Education*, 29(1), 66–94.

8 Hubelbank, J., Demetry, C., Nicholson, S.E., Blaisdell, S., Quinn, P., Rosenthal, E., & Sontgerath, S. (2007). Long term effects of a middle school engineering outreach program for girls: A controlled study. In *Proceedings of the American Society for Engineering Education Annual Conference & Exhibition*, Honolulu, Hawaii.

9 Nugent, G., & Barker, B. (2010). Impact of robotics and geospatial technology interventions on youth STEM learning and attitudes. *Journal of Research on Technology in Education*, 42(4), 391–408.

10 Anwar, S., Bascou, N.A., Menekse, M., & Kardgar, A. (2019). A systematic review of studies on educational robotics. *Journal of Pre-College Engineering Education Research* (J-PEER), 9(2), Article 2.

11 Witherspoon, E.B., Schunn, C., Higashi, R.M., & Baehr, E. (2016). Gender, interest, and prior experience shape opportunities to learn programming in robotics competitions. *International Journal of STEM Education*, 3(1), 1–12.

12 Jackson, A., Mentzer, N., & Kramer-Bottiglio, R. (2021). Increasing gender diversity in engineering using soft robotics. *Journal of Engineering Education*, 110(1), 143–160, p. 153.

13 University of Maryland. Student competitions. http://competitions.umd.edu/directory.

14 Burack, C., Melchior, A., & Hoover, M. (2019). Do after-school robotics programs expand the pipeline into STEM majors in college? *Journal of Pre-College Engineering Education Research* (J-PEER), 9(2), Article 7.

15 Eisenkraft, A. (2011). Student views of technology. *Technology and Engineering Teacher*, 71(2), 26–30.

16 Vega, V. (2006). Technovation lookback report: A retrospective survey of five Technovation cohorts (2010–2014). Rockman et al.

17 Godwin, A., Sonnert, G., & Sadler, P.M. (2016). Disciplinary differences in out-of-school high school science experiences and influence on students' engineering choices. *Journal of Pre-College Engineering Education Research* (J-PEER), 6(2), Article 2.

18 Banilower, E.R., Smith, P.S., Malzahn, K.A., Plumley, CL., Gordon, E.M., & Hayes, M.L. (2018). Report of the 201718 national survey of science and mathematics education (NSSME+). Chapel Hill, NC: Horizon Research, Inc., p. 109. Retrieved from http://horizon-research.com/NSSME/wp-content/uploads/2019/06/Report_of_the_2018_NSSME.pdf.

19 See <teachengineering.org> and an online community <linkengineering.org>.

24

DIGITALLY MEDIATED LEARNING MODALITIES FOR COMPUTATIONAL THINKING

Promises and Challenges for Informal STEAM Learning

Ted M. Kahn

Stimulated by the former USSR's surprise launch of Sputnik in 1957, the US began a massive effort to revise K–12 science and math education. Since the early 1960s, digital computers, multimedia, communications networks, and social media have played an increasingly important role in the world of STEM – and more recently, STEAM – education. We can call this type of learning Digitally Mediated Learning (DML) and can identify four modalities as they relate to learners:

DML 1: Learners as STEAM content consumers
DML 2: Learners as programmers, digital media designers, and application creators
DML 3: Learners as video and computer game players and creators
DML 4: Learners as networked virtual community participants and organizers

The purpose of this essay is to briefly trace the development of these four DML modalities, especially as they relate to STEAM disciplines in general, and informal STEAM learning in particular, in order to identify both their promises and their challenges. In this brief historical overview, we will see that these modalities are not distinct from one another, since learners have been increasingly engaged in many or all of them.

As these DML modalities have developed and evolved, they have promoted Computational Thinking[1] and related digital literacies. For our purposes, Computational Thinking may be defined as: "…an interrelated set of skills and practices for solving complex problems, a way to learn topics in many disciplines, and a necessity for fully participating in a computational world."

DOI: 10.4324/9781003145387-27

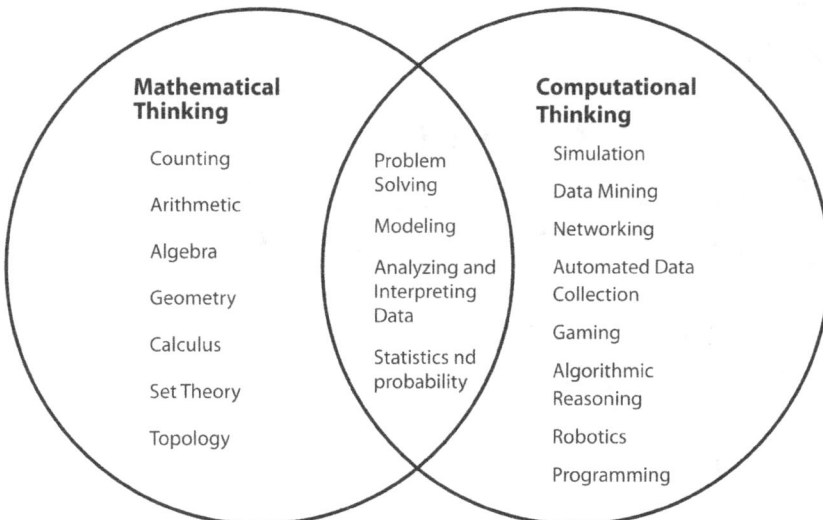

FIGURE 24.1 Venn diagram of mathematical and computational thinking. Used with permission.[3]

Computational Thinking is now considered a key set of skills (which should also encompass computational learning),[2] as a major focus of STEM educational strategies for the 21st century. Figure 24.1 shows how Computational Thinking relates to mathematical thinking, as one key set of STEM skills.

We will now take a brief journey through the history of each of the above four modalities to highlight the richness and power of each to engage and sustain the STEAM interests of learners and the development of Computational Thinking. We will see how increasing crossovers between these modalities supports "convergence" education that is in line with an emerging National Science Foundation (NSF) to advance convergence R&D, in order to prepare learners to address complex global problems. After tracing the development of each modality, we will review its key connections with Computational Thinking. Finally, we will discuss some of the promises of these modalities as well as its challenges, in order to effectively engage in and sustain informal STEAM learning, now and in the future.

DML 1: Learners as STEAM content consumers

Only about 20% of K-12 students' waking hours are spent in school classrooms and most learning throughout our lifespan is informal and out-of-school. But there was very little early funding support for R&D of learning resources informal STEM learning. The earliest major federally funded DML projects in STEM focused on providing digital content to formal K-12 classrooms and teacher professional development opportunities. The first NSF-funded effort in the area of computer-assisted learning was the PLATO (Programmed Logic for Automatic Teaching Operations) Project, from the University of Illinois at Urbana-Champaign, starting in the early 1960s.[4]

The PLATO system included use of interactive terminals connected to large mainframe computers for asynchronously delivering interactive multimedia lessons in STEM for both K-12 and postsecondary students. By the early 1980s, PLATO supported thousands of student terminals worldwide. In addition, many modern concepts in multi-user computing were developed for or matured under PLATO, including forums, message boards, online testing, email, chat rooms, instant messaging, remote screen sharing, multimedia, and multiplayer video games, many of which are summarized here via other DML modalities.

Other early R&D efforts relating to the development of digital learning content included Computer-Assisted Instruction (CAI) that focused on teaching elementary math and reading skills using early mathematical models of learning and teletype terminals in schools connected via time-sharing computers. Many of these early CAI efforts were based on a programmed instructional pedagogy, based on behaviorist psychologist B. F. Skinner's "teaching machines."[5]

Catalyzed by the birth and rapid growth of personal and home computers in the late 1970s and the 1980s, a diverse set of interactive learning programs began to emerge from commercial educational software companies, such as The Learning Company (early childhood education), Sunburst, and Wings for Learning, followed by more traditional educational publishers like Scholastic and National Geographic. The earliest museum-based interactive DML programs included interactive games for the Chevron-sponsored *Creativity: The Human Resource* traveling museum exhibition (1979–1981), the Atari Institute sponsored "Communications" Exhibit at the Capital (now National) Children's Museum in Washington, DC, and even "personal computer kiosks" for young children in settings like Sesame Place.

Open education resources (OER). The birth and growth of the Internet and Web in the 1990s led to major international efforts to provide open access to a rapidly growing collection of STEM learning resources in both formal and informal learning settings. These learning resources – freely available, open-licensed, customizable, "remixable," and globally accessible – became known as OER.[6]

Beginning with the OpenCourseWare teaching materials of the Massachusetts Institute of Techology (MIT)[7] scores of OER resources and programs emerged at universities and colleges. In parallel, the NSF funded the creation of the National Science Digital Library project, which collected and provided free access to thousands of high-quality STEM-related learning resources.[8] As of November 2022, the user co-curated OER Commons' global learning resource provided more than 80,000 freely web-accessible interactive lessons, activities, and projects available for K-16 students, teachers, and parents.[9]

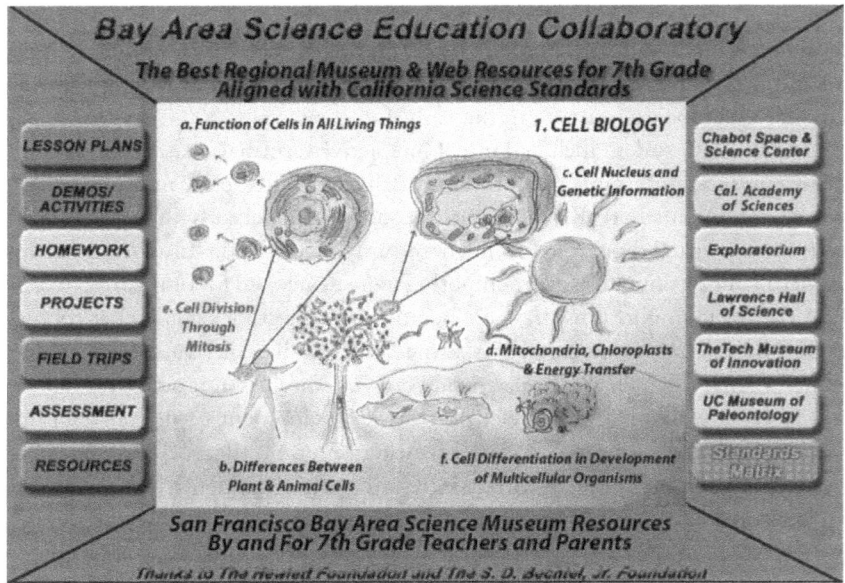

FIGURE 24.2 Bay Area Science Education Collaboratory. Used with permission, DesignWorlds for Learning.

In 2002, my colleagues and I began to design and create one of the first OER aggregations of DML informal science learning resources, co-curated and federated by San Francisco Bay Area teachers and museum science educators. The goal of the Bay Area Science Museum and Learning Collaboratory (2002–2008)[10] was to bring the richness and diversity of multiple museums' digital assets to middle school (grades 4–8) science teachers, parents, and science educators.[11] This project was based largely on an emerging need to "bridge the gap" between informal and formal science education (Figure 24.2).[12]

Public media and museums have also been sources for similar STEM resources. The Public Broadcasting System and National Public Radio, TED and TEDx talks, and many YouTube channels, such as MinutePhysics, have thousands of hours of content-related STEM resources. In addition, STEAM summer camps have increasingly become popular ways to augment formal STEM learning content in classrooms. Furthermore, many interactive exhibits at science-technology, children's museums, and art museums use computers as a delivery platform. These organizations also produce their own podcasts and share videos of exhibits, lectures, and experiments that can be viewed at home or in classrooms.

Reflection: Connection of DML 1 to Computational Thinking. The main connection of this learning modality to Computational Thinking lies in providing free-choice and free access to these shared learning resources in STEAM – by students, teachers and learners of all ages – outside formal classrooms. There are also over 600 OER Commons resources related to computer programming alone, and many thousands of others related to all STEAM topics, including simulations and mathematical modeling. Using these resources to supplement formal classroom STEM learning, as well as to support individual and collaborative Project-Based Learning (PBL), has major growth potential, provided these resources can be maintained, updated, and effectively validated by their respective learner communities.

DML 2: Learners as programmers and digital media designers and creators

The earliest computer programming languages, such as FORTRAN, were used primarily as tools for solving complex math and physics problems. The first time programming introduced to students in public schools anywhere in the world is believed to have been in the early 1960s when an innovative young high school math teacher, the late Dr. Irwin Hoffman, and a young entrepreneurial computer systems analyst, Bob Albrecht, collaborated to teach programming to a small group of bright high school math students at George Washington High School in Denver.[13]

Shortly thereafter, mathematician and early AI pioneer, Seymour Papert, and his colleagues at MIT created the Logo computer programming language for students as a "conceptual framework for teaching mathematics."[14] But Logo had a very different pedagogical focus from most math programming. As explained in his 1980 book, *Mindstorms*,[15] Logo was designed to help even young children understand what Papert called "powerful ideas," even seemingly abstract Computational Thinking concepts, such as recursion.

The Logo project created both a new pedagogical philosophy and methodology for DML known as "constructionism," as a pioneering introduction to computational learning.[16] Later collaborations between the Media Lab and companies like Atari and Lego resulted in commercial versions of Logo for personal computers, and Lego Mindstorms software and hardware as a way of introducing robotics to elementary and middle school students.

A second major innovative effort of DML learners as programmers was the development of the Smalltalk, the world's first object-oriented programming language and a key part of the development of world's first personal computer system. Smalltalk was a revolutionary concept in computer programming, based on new metaphor for defining classes of programmable objects which could communicate with one another by passing messages. Inspired by his pioneering vision of a personal programmable "Dynabook,"[17] Alan Kay and his colleagues created the Learning Research Group (LRG) at Xerox Palo Alto Research Center (PARC) during the 1970s. LRG envisioned emerging personal computers as "personal dynamic media," rather than just computational devices, for "children of all ages."[18] As a graduate student doing my doctoral research with LRG, I personally saw just how much elementary, middle, and high school students were active research associates and co-developers in this adventure of "inventing the future."[19]

A key pedagogical focus of LRG on students learning to program using Smalltalk was simulation, with programming being a means of creating, testing, and revising mental models of how phenomena in the real world work through their programs. These are now some of the key skills of contemporary Computational Thinking.[1,2,3]

These early innovative efforts continued to evolve under Alan Kay and Dan Ingalls leadership at companies like Atari, Apple, Disney, and lastly, Viewpoints Research Institute. MIT Media Lab's Mitch Resnick and many Media Lab, as well as Seymour Papert's graduate students, helped develop personal computer versions of Logo,[14] and Smalltalk-descendant visual programming languages and environments targeted for children and pre-teens, such as Apple's Cocoa (later Stagecast), Squeak, Snap, Scratch, and E-Toys.[20]

These early innovative programming languages were paralleled by the explosive growth of the BASIC programming language, first developed at Dartmouth in 1964. BASIC became the first programming language to be widely adopted by almost all early personal computers during the late 1970s and early 1980s.

With colleagues at the University of California, Berkeley's Lawrence Hall of Science, I co-founded the Math and Computer Education project in 1970–1971, the first such public access and Computational Thinking learning program in the world, based at a science museum, where we introduced programming and recreational computer learning experiences to thousands of students, teachers, and parents. This project rapidly became one of the Lawrence Hall's most popular education programs, especially after the introduction of personal and home computers in the late 1970s and 1980s.

Reflection: Connection of DML 2 to Computational Thinking. A conventional approach for this DML modality is that all students should learn coding and programming as a basic skill. However, this focus alone often misses the benefits of informal and recreational use of programming and digital media design, such as creative problem-solving, model building, and expression through digital multimedia, plus the social sharing of creative ideas through the building of and participation in virtual communities. These important aspects of scientific and engineering practices also have strong connections with Computational Thinking.

The corrective and complementary activity to this conventional approach to programming has been to encourage students to create their own multimedia. For example, Dave Master, an animation teacher at the low-wealth Rowland High School east of Los Angeles, successfully used and quickly integrated personal computers into one of the most successful animated film communities for students. Students created thousands of animated movies, winning more awards than any other similar program, and with many of these former students moving directly into working for major Hollywood studios. When my DesignWorlds colleagues and I designed the education content for Apple's web site at the launch of its iMovie software, we based much of our work on Dave Master and his students' experiences in animated movies as "good stories, well told."[21]

Since the earliest days of BASIC, programming, multimedia content design, and coding have largely been accomplished through informal sharing with others. Books that published copies of computer code that anyone could enter into their personal or home computers were extremely popular in the 1980s.[22] Modifying existing code and sharing with others is an early example of the core values of the open source software movement, which matured in the late 1990s and early 21st century. The excitement and energy of small groups of students working together to develop new software, simulations, games and tools is still present today in current "hackathons" and engineering design team competitions, such as FIRST Robotics and the Tech Interactive's annual "Tech Challenge" competitions. These are all essential key social and collaboration practices of Computational Thinking.[1,10,16]

DML 3: Learners as computer and video game players and creators

Games and play are an integral part of cultural folklore as well as early childhood and recreational lifelong learning. They can also be an engaging way to simulate and explore powerful ideas in STEAM.

The first interactive computer game was allegedly SpaceWar, which emerged from the MIT computer community in 1962. Following the 1969 NASA Apollo 11 mission, a set of related interactive computer games, called "Lunar Lander," became popular on timeshared computers. Shortly thereafter, the first commercial coin-operated video game, Pong, launched both Atari, and later, an entire industry. Video and computer games have since become one of the largest commercial entertainment industries in the world.

Gaming has been one of the most popular DML modalities of informal and recreational learning for decades. Because computer games are often played repeatedly, early research on these games identified three intrinsic motivational variables: challenge, fantasy, and curiosity.[23] More recently, research supports the notion that making these games engaging and effective for teaching requires paying attention to a core set of design principles. For example, math activities should not be separate from the game itself and should not encourage students to answer quickly, without reflection.[24] Building on this research, there is now increasing interest in the "gamification of science."[25]

More recently, museums, such as the Computer History Museum (CHM), have collaborated with game design platforms to launch informal STEM simulation/exploration outreach programs. In just two months since its launch, CHM and Roblox's *TechQuest*, which includes activities like exploring a tropical rain forest jungle river while searching for elusive pink dolphins via drones, has already gained more than 1.8 million player visits worldwide.[26]

Many learners who enjoy gaming have now also become designer-programmers of their own games, and several NSF and other funded research projects have explored how this activity helps students develop both STEM and CT skills.[27] One of the best examples of a game-like environment and community that promotes students as programmers and content creators is Minecraft, a constructivist game backed by a huge virtual world and community. Educational researchers have begun to explore how to best understand and develop STEM and Computational Thinking within the Minecraft virtual world environment.[28]

Reflection: Connection of DML 3 to Computational Thinking. In general, coding games are widely used to teach Computational Thinking. Educational research is increasingly providing evidence that there is a causal connection between the two.[29]

DML 4: Learners as networked virtual community participants and organizers

Virtual learning, design and collaboration communities grew with the evolution of the Internet and Web in the mid-late 1990s. The exponential growth of the "open source" movement in software development and the emergence of resources like Wikipedia are examples of informal learning and knowledge sharing in practice. "Open innovation" challenge projects and networks, like InnoCentive.com and OpenIdeo and more than 3,000 Citizen Science projects,[30] have opened innovations in STEM R&D to the general public, including individuals and small groups responding to research and design challenges sponsored by companies, universities, or non-profit research organizations.

"Virtual Worlds,"[31] such as Second Life, and other MUVEs (Multi-User Virtual Environments) emerged in the early 2000s. When my colleague, Kat Galas, a teacher at the Corrine Seeds University Elementary School at UCLA, introduced her class to the Whyville MUVE to supplement lessons on epidemiology and spread of infectious diseases, her students joined a global online community of 550,000 "Whyvillians" who daily explored, learned, created – and had fun – while gathering and sharing knowledge of physics, biology, and world history.[32] Other early MUVEs that focused on developing STEM higher order thinking skills included Harvard's "River City" and Indiana University's "Quest Atlantis."[31]

Two major technology shifts between 2000 and 2020 have provided support for virtual learning, design and collaboration communities: the growth of mobile "smart" phones and the growth of digital video. Smart phones enable text-messaging, GPS-based geolocation, digital voice and high quality sound capture, high-resolution photo and video communication, augmented reality (AR) and games, and thousands of other applications. Digital video – through YouTube, Vimeo, and digital video archives of millions of Zoom video conferences – also creates design and collaboration communities. As an example, our OER Mindseum[33] project has collected and visually "tagged" more than 350 freely available digital video and related digital media assets from museums, YouTube, and TED.com videos, as a cross-disciplinary way to explore the inter-connections between mind-brain-technology and the arts. Sharing these resources among students who are interested in these topics has created a community of users in the project.

Reflection: DML 4 and Computational Thinking. The evolution of virtual worlds and associated networked virtual communities has provided one of the most creative opportunities to see the convergence of all four DML modalities in support of both Computational Thinking and collaborative, creative computational learning. Each of the other DML modalities highlight how important "communities of practice" are to learning, collaborative problem-solving and creative activity. The other modalities also highlight new ways to share knowledge and Computational Thinking practices, incorporate the visual and media arts, attract participants, and sustaining participation in these communities.

Promises and challenges

In our brief historical journey and exploration of the four modalities of DML, we have seen several general trends relating to the coevolution of technologies, learning, and social practices. The first trend is from the development of the individual learner to learners engaged in sharing and participating with others in common creative enterprises. The second trend is from course-centric *pedagogy*, with a focus on traditional formal modes of teachers or experts transmitting knowledge to learners, to *andragogy*, an approach based on the self-directed and project-based agency of learners in different communities.

In addition, the current major educational focus on convergence education is also directly connected with a movement toward transdisciplinary research and learning (Figure 24.3). Both research and learning highlight why Computational Thinking needs to be connected to helping learners of all ages engage in solving complex and sometimes "wicked" problems. For example, in our DesignWorlds for College & Careers work, we have encouraged our high school and college student clients to engage actively with what are known as the Millennium Project's 15 Global Challenges for humanity in the 21st century.[34]

These challenges are interconnected, requiring a greater focus on thinking about future scenarios, systems thinking, and learning across traditional

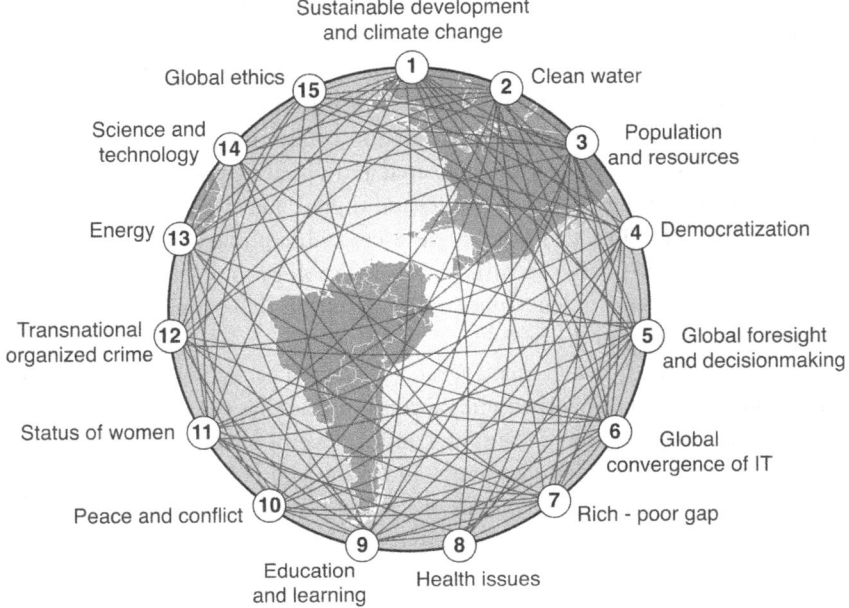

FIGURE 24.3 The Millennium Project's 15 global challenges. Used with permission.

subject matter boundaries, college degrees, and separated knowledge communities. The interconnectivity of humanity's global challenges is our greatest opportunity to connect Computational Learning with other kinds of learning, both formal and informal.

A set of four learning and design principles, taken from the Clubhouse Network, may help us in this quest.

> *Learn by design.* People learn best when they are actively engaged in designing, creating, and inventing, not just passively receiving information.
>
> *Follow your interests.* When people care about what they are working on, they are willing to work longer and harder, and they learn more in the process.
>
> *Build a community.* When people collaborate with others of diverse ages, cultures, genders, and backgrounds, they gain new perspectives for understanding the world and themselves.
>
> *Foster respect and trust.* In places where everyone's ideas and opinions are respected, people are more likely to take risks and experiment and thus more likely to learn and innovate.[35]

Along with the above promises, we also have many challenges, including a variety of technology-related dilemmas – from violations of data privacy to widespread false information and even fake identities. As AI and other advanced technologies, such as CRISPR (genetic editing), become more prevalent in our lives, we need to be conscious of not only what technologies *can* do – but also what they *should or should not* do. These dilemmas fall into the area of ethics, as well as public policy.

In the movie, "Field of Dreams," the protagonist, Ray Kinsella (played by Kevin Costner) and his family followed their dream to build a baseball diamond in their cornfield. One of their guiding visions is "If we build it, they will come…" And in the movie, after building the baseball field, the ghosts of former famous baseball players appear, to relive their days of glory and to play together. Perhaps we need to rephrase the movie's guiding vision in slightly different words:

> If we build it together, can we all use and learn effectively, creatively and ethically from it—with one another—to help solve our global challenges?

Acknowledgements

This chapter is dedicated to Frona, Yoni & Alison, and Aaron, and especially, to our grandson, Micah, who may benefit from these "DesignWorlds for Learning" – and to the memory of my teachers and mentors, Dean Brown and Moshe D. Caspi.

Notes

1 Digital Promise (2022). What is computational thinking? https://digitalpromise. org/initiative/computational-thinking/about/.
2 Barba, L. (2014). Computational thinking is computational learning. Keynote address at SciPy 2014 Conference, https://lorenabarba.com/gallery/prof-barba-gave-keynote-at-scipy-2014/ and slides at: https://figshare.com/articles/presentation/ If_there_s_Computational_Thinking_there_s_Computational_Learning/ 1096344/3.
3 Sneider, C., Stephenson, C., Schafer, B., & Flick, L. (2014). Exploring the Science Framework and NGSS: Computational Thinking in High School Science Classrooms. *The Science Teacher*, 81(5), 54. Copyright © 2014 Cary Sneider, Ph.D.
4 Alpert, D., & Bitzer, D. L. (1970). Advances in Computer-based Education: The PLATO Program Will Provide a Major Test of the Educational and Economic Feasibility of This Medium. *Science*, 167(3925), 1582–1590.
5 Skinner, B. F. (1961). Teaching Machines. *Scientific American*, 205(5), 90–106.
6 "Open Education." The William and Flora Hewlett Foundation, https://hewlett. org/strategy/open-education/#overview
7 d'Oliveira, C., Carson, S., James, K., & Lazarus, J. (2010). MIT Open Course Ware: Unlocking Knowledge, Empowering Minds. *Science*, 329(5991), 525–526.
8 National Science Digital Library (NSDL), https://www.nibib.nih.gov/content/ national-science-digital-library-nsdl and now at https://nsdl.oercommons.org/ (last accessed 10/24/22).

9 OER Commons. https://www.oercommons.org/ (last accessed 10/24/22).

10 Bay Area Science Museum Learning Collaboratory (2002–2008). www.design worlds.com/collab/ and http://www.designworlds.com/collab/collab-7th/7th-Cell.html.

11 Kahn, T. M. (2007). Science Museum Learning Collaboratories: Helping to Bridge the Gap Between Museums' Informal Learning Resources and Science Education in K-12 Schools. Bearman, D., & Trant, J. (Editors) (2007). *Museums and the Web. Proceedings.* Toronto, Ontario, Canada): *Archives and Museum Informatics*, published March 31, 2007 at http://www.archimuse.com/mw2007/papers/kahn/kahn.html (last accessed 10/9/22).

12 Hofstein, A., & Rosenfeld, S. (1996). Bridging the Gap between Formal and Informal Science Learning. *Studies in Science Education, 28*, 87–112.

13 Kahn, T. M. (2008). Journeys in Inventing the Future: The Central Role of "Kids of All Ages" in the Design, Development and Growth of Interactive Computing and DigitaMedia. Stanford University, Media X Forum presentation, May 12, 2008.

14 Solomon, C., Harvey, B., Kahn, K., Lieberman, H., Miller, M. L., Minsky, M., Papert, A., & Silverman, B. (2020): History of Logo. *Proceedings of the ACM on Programming Languages*, 4, No. HOPL, https://dl.acm.org/doi/abs/10.1145/3386329.

15 Papert, S. (1980). *Mindstorms: Children, Computers and Powerful Ideas.* New York: Basic Books.

16 Barba, L. (2016). Computational Thinking: I do not think it means what you think it means. Lorena Barba Group blog post: https://lorenabarba.com/blog/computational-thinking-i-do-not-think-it-means-what-you-think-it-means/. Retrieved October 30, 2022.

17 Kay, A. C. (1972). A personal computer for children of all ages. In *Proceedings of the ACM National Conference*, Boston, August 1972.

18 Kay, A. C., & Goldberg, A. (1977). Personal Dynamic Media. *IEEE Computer* 10(3). Goldberg, A. & Kay, A.C. (1975) *Personal Dynamic Media.* Xerox Palo Alto Research Center, Learning Research Group, 1975. A Xerox PARC Orange Book.

19 Kahn, T. M. (1981*). An analysis of strategic thinking using a computer-based game.* Unpublished doctoral dissertation, Dept of Psychology, University of California, Berkeley.

20 Galas, C., & Freudenberg, R. (2010). Learning with Squeak Etoys. Constructionism 2010, Paris. VPRI Research Note RN-2010_002, Viewpoints Research Institute, http://www.vpri.org/writings.php.

21 Kahn, T. M., & Master, D. (1992). Multimedia Literacy at Rowland: "A good story, well told." *The Journal* (Technological Horizons in Education), 19, 7, 77–83.

22 Kohl, H., Kahn, T. M., & Lindsay, L., with Cleland, P. (1982). *Atari Games and Recreations.* Western Fairfax: Reston.

23 Malone, T. W. (1980). *What makes things fun to learn? A study of intrinsically motivating computer games.* Xerox Palo Alto Research Center. Cognitive and Instructional Science Series, Report CIS-7 (SSL-80-11).

24 Devlin, K. (2013). The Music of Math Games. *American Scientist,* 101(2), 87–91.

25 Kalogiannakis, M., Papadakis, S. & Zourmpakis, A-I (2021). Gamification in Science Education. A Systematic Review of the Literature. *Education Sciences*, 11(1), https://doi.org/10.3390/educsci11010022.

26 TechQuest (2022): Roblox & Computer History Museum, www.roblox.com/games/10905680506/Welcome-to-TechQuest-by-Computer-History-Museum (last accessed 10/30/22).

27 Werner, L., Denner, J., & Campe, S. (2014). Using Computer Game Programming to Teach Computational Thinking. In Schrier, K. (Ed). *Learning, Education and Games: Volume One: Curricular and Design Considerations.* Pittsburgh: ETC Press, 37–53.

28 Nebel, S., Schneider, S., & Rey, G. D. (2016). Mining Learning and Crafting Sci-entific Experiments: A Literature Review on the Use of Minecraft in Education and Research. *Journal of Educational Technology & Society,* 19, 2, *Intelligent and Affec-tive Learning Environments: New Trends and Challenges* (April 2016), pp. 355–366.

29 Zhang, S., Wong, G. K., & Chan, P.C. (2022). Playing Coding Games to Learn Computational Thinking: What Motivates Students to Use This Tool at Home? *Education and Information Technologies,* 28, 193–216. https://doi.org/10.1007/s10639-022-11181-7. https://link.springer.com/article/10.1007/s10639-022-11181-7.

30 Scistarter: Science we can do together. https://scistarter.org/

31 Nelson, B. C., & Erlandson, B. E. (2012). *Design for Learning in Virtual Worlds. Interdisciplinary Approaches to Educational Technology.* New York: Routledge.

32 Olsen, S. (2006). Are virtual worlds the future of the classroom? CNETNews.com, June 12, 2006.

33 Kahn, T. M. (2021). MINDSEUM: Magic Theater, Glass Bead Game...or a New Kind of Global Crowd-Sourced & Networked Museum Collaboratory? Webinar presented with support from NextNow online community, https://www.youtube.com/watch?v=L8pkgUFbx2Y and http://mindseum-next.herokuapp.com

34 15 Global Challenges. The Millennium Project: Global Future Studies and Research, https://www.millennium-project.org/projects/challenges/.

35 Clubhouse Network, The Clubhouse Learning Model: https://theclubhousenetwork.org/about/model

25

OUTDOOR AND OUTREACH

Informal Science Education Outside the Four Walls of Science Centers

Ganga S. Rautela

Research findings suggest that beyond the four walls of schools and science centers, the opportunities for science learning are abundant. A considerable part of the population in many countries, young and old alike, explore, experience, and learn science in informal learning environments by visiting outdoor and outreach science venues, such as outdoor science parks, zoos, nature parks, nature trails, camp sites, science circuses, mobile science exhibitions, agriculture science centers, and rural science centers in villages and clubs, use these learning resources to satisfy their curiosity and pursue their interests.

What characterizes these programs? How do these programs operate? How effective are they? This essay will address the above questions as they relate to various countries, especially in India.

After having spent 40 years in the science center field, I feel nostalgic when I look back to my first engagement as a science center professional in the Nehru Science Centre, Mumbai, India, that opened its outdoor science park, perhaps the first of its kind in the world, during the International Year of the Child in 1979. The outdoor science park was full of interactive exhibits that offered a unique and non-coercive learning setting to engage audiences and provide science experiences, unlike those available in an indoor setting. I witnessed public participation in this outdoor science park very closely and remained an active part of it and other rural science centers for almost 20 years, developing them and their outreach activities, as well as observing their impact on students and the general public. I still admire this concept. Since then, outdoor science parks are part of almost every science center in India and have opened in science centers around the world.

DOI: 10.4324/9781003145387-28

Developing nations in the world that cannot afford large and capital-intensive science centers in every part of their countries have introduced mobile science exhibition units, demonstration vans, science circuses, science expresses (trains carrying large exhibitions and science activity modules), mobile science labs with experiments, rural science centers, discovery centers, and science camps, even low-cost science parks, in order to offer non-formal STEM learning to their public, especially students in remote areas who are deprived of such resources. Since outdoor science centers and parks rarely require buildings, their implementation is speedy and economical.

The interrelationships of learning with different settings and learners have been explored in several studies. The relationship of science content with the natural landscape and its life forms, connecting science content with nature, may not be an intellectual exercise, but learning does take place from experience, action, and reflection. Science park-based activities show promise for supporting students' engagement and learning in science classes and in fostering students' interest in pursuing science.

The centrality of play in outdoor settings

The outdoor environment provides a variety of unique play-and-learn opportunities. A growing body of research suggests that play activities in natural environments are beneficial for children's development and learning in many areas. "Play is characterized by intrinsic motivation, active engagement, attention to means rather than ends, non-literal behavior, and freedom from external rules, a means for acquiring information about and experiencing the environment. Successful exhibitions for children and families share these qualities and outdoor exhibitions encourage a degree of exploration and full-body experience often not possible or appropriate inside a museum."[1]

Play and learning in outdoor settings are associated with changing conditions and unpredictability, as well as an abundance of available space and possibilities for open-ended activities.[2] At an early age, movement is children's primary method of action, expression, and learning.[3] Environmental complexity and diversity in nature provide a variety of opportunities for children to become familiar with the natural world through direct sensory experiences, various opportunities for adventurous play, and exploration.[4] Giving children the possibility to experience nature may also encourage their appreciation of nature.[2] Several researchers consider children's play in natural environments as an essential element for early childhood sustainability education, since play provides children with opportunities to build personal and meaningful relationships with nature and to strengthen their environmental consciousness.[5] Likewise, childhood nature experiences are considered to be important points of departure for the development of ecological ideas and embodied environmental understanding.[4]

The varieties of outdoor and outreach science experiences

Both indoor and outdoor learning environments can complement each other, in order to improve students' academic performance in science. Outdoor learning contributes to effective and influential impacts on students' academic performance in understanding science. In addition, students are often more enthusiastic about participating in the outdoors than about staying indoors, as they are provided with wide-ranging opportunities to make connections of their science learning in the classroom through observing and exploring in the outdoors.[2]

What are these different forms of outdoor science centers, outreach learning resources, and settings? We review some of them below.

Outdoor science parks. A science park or science garden is an innovative concept of teaching science in an informal way. These are excellent places for kids to learn new skills and ideas while having fun at the same time. In addition to imagination, creativity, physical movement, and social interactions, science playgrounds are perfect places for kids to explore science concepts. Children engage in a wide variety of outdoor experiences that lead to learning.[2]

When children gain knowledge, via hands-on experience in interactive outdoor settings, they can become familiar with their environment as well as apply what they have learned in the classroom. "The basic characteristic of a science park is that it promotes learning from hands-on experience, offering an ideal free-choice learning environment."[6] For example, research on the outdoor Science Learning Garden Activities (SciL) showed that they helped diverse students not only engage more productively in science class but also to think of themselves as individuals who could be successful and valued as contributors to the scientific community. These findings also lend support to a motivational model, based on self-determination theory, as a means for capturing the active ingredients of pedagogy, curriculum, and social relationships (Figure 25.1).[7]

The popularity of outdoor science parks with hands-on interactive exhibits has grown in India and throughout the world. The growth of science playgrounds is an international trend among museums. Their overall goal is to create safe and challenging outdoor environments that use play to explore the foundations of science. The outdoor exhibitions encourage a high degree of exploration and physical experience that is often not possible in indoor settings. However, challenges regarding such outdoor science parks exist, e.g., providing a safe environment, interpreting the exhibits and maintaining them.

Science Circus. Transported in a large van, a Science Circus is a multi-activity mobile outreach project that aims to share the wonders of STEM

FIGURE 25.1 A science park, the District Science Center, in Tirunelvelli, India.

with students, teachers, and the wider community, in order to generate their curiosity, skills, knowledge, and motivation. These outcomes are necessary both for the development of individual STEM careers as well for as a country's development. Science Circus uses low-cost everyday materials to showcase STEM through science shows for schools and public venues, workshops for teachers, and interactive exhibits for the community. It uses a rare blend of science, comedy, and circus arts. Student can learn basic science principles in unusual ways, such as gravity through bowling ball juggling, gyroscopic stability through glass bowl spinning, centripetal force via cowboy lariats, the center of balance through a unicycle and inertia through the old tablecloth pull.[8]

Science Circus has a positive impact on teachers. The founder of Science Circus Africa noted that "teachers were interested in the approach and eager to see more examples of fascinating demonstrations using simple everyday items, but weren't always so willing to try it out with their own hands. You have teachers who have been using blackboards and rote learning for their whole careers, so the idea of students doing science using common stuff from a supermarket is a paradigm shift."[8]

Mobile Science Exhibitions/Science Express. Known by various names – such as museo buses, museum on wheels, mobile science exhibitions, Mobile Museum, Museum on the Move, and Science Express – these vans or trains are mounted with interactive exhibits and have the aim of creating awareness on important issues of science and technology, while improving the quality of STEM education and providing hands-on learning opportunities to students in remote areas. For example, Science Express is a 13-compartment train designed to increase the coverage and reach of informal science education in India.

These vehicles are a popular mode of science communication in the developing world.[9,10] They can carry science shows, portable planetariums, science

FIGURE 25.2 Students exploring the transfer of momentum exhibit in a science bus.

films, telescopes for sky observation, and hands-on science kits for rural students and their teachers. This mode of science outreach for science centers in developing countries is not new (Figure 25.2). For many years they have been quick and viable alternatives for countries that cannot develop or rapidly expand their infrastructure for physical science centers or for countries that have economic or physical accessibility issues. As one of the pioneers of science centers in India wrote: "Artifacts and four walls are no longer pre-requisites of a museum. Today museum is a generic term for which museo buses could well be named as mobile science museums."[9]

With a mission of instilling a scientific culture, while fighting superstitions and obscurantism, in low-literacy areas, mobile museums have been an effective platform for non-formal science education.[9] The impact of these programs has been tremendous in India: for over 50 years they have operated in the country, with 50 units that have reached nearly 5 million people annually.[10]

The mobile exhibition program initially posed a few challenges, such as designing a bus that could negotiate narrow rural roads and accommodate large exhibits, maintaining the exhibits, and designing exhibits that could be interchangeable with those of other science centers in the country. Eventually a bus was designed with fixed-sized exhibit boxes that could easily be removed and exchanged with other exhibits that focused on different themes. This design allowed the mobile exhibition to bring exhibits with different themes to the same rural area, year after year. However, the program continues to face other problems, such as a shortage of staff members and volunteers, who avoid traveling to rural areas for long periods of time, due to a lack of creature comforts and difficult working conditions.

Despite these challenges, the mobile exhibition program has been quite successful. A manager of such a program said, "When the exhibition bus reached the village school, it was a festival for entire village and the neighboring villages too."[11] A report on the impact of mobile science exhibitions in India concludes that "the success of the Mobile Units is based on the team members, their ability to communicate with the rural population, their attitude to serve society in rural areas, far away from cities where science centers do not exist, and their proficiency in local language. The best way to integrate science and technology with the lives of the people in India is by taking science to the doorsteps of the rural or unreached population." In addition, impact studies indicate substantial learning by students in schools that periodically received the van.[6]

Mobile Science Labs/Lab in a Lorry

This program is known by various names: Lab in a Lorry, Mobile Lab, Science Exploratory, and Floating Classroom. Its basic purpose is to provide the conditions and props necessary to conduct hands-on experiments and thus supplement the curriculum in schools where laboratory resources are scarce or missing. "Science is better learned by props" is the motto of this program.

The interactive mobile science laboratory is often staffed by practicing science communicators or scientists and lab technicians. Young people are given the opportunity to do experimental science the way it actually occurs in laboratories, via authentic and exploratory experiences, informed by curiosity and intuition, but also bounded and guided by the experience and insight of the staff members. This program has been very effective in rural schools that have very little or no laboratory activity to supplement the STEM curriculum.

Rural science center. Since building full-size science centers is capital-intensive and requires a great deal of time, a quick, feasible, and alternative solution is to build smaller and more economic versions that can be multiplied easily and can be built with locally available materials. Rural science centers work on low budgets. Relatively small in size, they are meant to provide hands-on learning experiences to rural students and to the public. First started in India in the 1970s, these centers were developed in regions that had no informal science education facilities. Local governments had the agenda of social development for the poor population living in those areas. "What science should be presented in these areas?" was a big question. India's answer was to present exhibits that dealt mostly with topics of local interest, such as water, sanitation, agriculture, food, energy, and health. Moreover, motivating the local population to participate in the science center was difficult. Therefore, extensive outreach activities with local volunteers were arranged, to win the trust and support of the local population and community leaders. Gradually, the public realized the importance of science centers and started visiting them. Today they are extremely popular and viable.

Expansion of this initiative to more rural areas is always influenced by financial considerations. Therefore, the cost of building and operating these centers has to be kept low, with minimum staff members who are trained in multiple skills. A standard-building design, common and interchangeable exhibits, and mobile units have been adopted as common procedures. Due to the low rural economy, the rural public cannot bear the cost of these centers; rather, these centers must be supported by the government of the local municipality.

Discovery centers. Discovery centers are small but effective places to provide non-formal and hands-on STEM education to students beyond school hours. These facilities work on a different model than rural science centers and are part of a school or another institution. Discovery centers stimulate curiosity in science across all age groups of students.

In their core program called "science through experiments" (once or twice a week, with one-and-a-half-hour lab sessions, for a yearly program from June to March), children from various schools are encouraged to experience science by performing experiments and activities in well-equipped and state-of- the-art laboratories instead of in their formal science classrooms. Here the children learn to explore, experiment, invent, and innovate. This program is highly beneficial in developing students' conceptual understanding of the basic concepts of the various scientific disciplines.

Students interested in science participate in the program after school or during holidays. In this way, these students can pursue their passion for science. Evidence indicates that students participating in the activities of these facilities have shown deep interest in science and pursued careers in science or technology. Interest and motivation are linked to future career choices.[12]

Science camps. Many science centers have science camp facilities outside the four walls of the centers in forests, riversides, or in mountains rich in biodiversity and geological formations. These camps are popular for science enthusiasts and help in motivation and interest development in science and technology. The camp activities are specialized programs rich in hands-on experiments or activities. Scientists working in different disciplines help students in various activities to nurture their interests in science and technology and have encouraged many young participants to choose STEM careers. The informal environment provides opportunities for youth to connect with science in personally meaningful ways, develop their own science identities, and consider pursuing science careers through exhibits, activities, and extended programs.[13] Though rigorous longitudinal studies have not been conducted that link STEM careers with informal science education experiences, numerous scientists have credited early experiences specifically with museums as influential to their choice of a career in science.[14] In a study by the Cosmos corporation for NSF, visits to museums were cited by surveyed scientists as their most memorable informal science experiences as youth.[15] They also cited these experiences as the most influential source of ideas still used in the present.[16] In 2014, the Pew Research Center found that 8% and 7% of responding scientists attributed "childhood experiences in natural parks, science museums, star gazing, chemistry sets" and "books, movies, TV on science e.g., Cosmos series, biographies of scientists, and science fiction" were critical childhood experiences that initiated their science paths.[16]

The benefits of outdoor and outreach science programs

Science outreach activities are wonderful in so many ways. They can generate much-needed excitement and interest in science with students and the public. They can generate appreciation in the community for a variety of STEM-related professions and institutions. They can also provide a sense of accomplishment and community well-being to those who provide them; providing science outreach activities is also a great way for students and adult volunteers to gain a deeper understanding of science and its applications, as well as to develop valuable communication skills. For example, if a sixth grader learns to successfully explain an aspect of science or technology to a third grader, her own level of understanding is greatly enhanced. The same principle is true for scientists: it is comparatively easier for them to communicate scientific ideas to other scientists, but if they use the same approach with nonscientists, it may not work as effectively. Scientists who communicate with the general public in science outreach programs can improve their own science communication, for the benefit of themselves, their profession, and the general public.

Effective science outreach is not only fun and rewarding; it is also greatly needed now in our society. Science education is of great importance, due to its effect on the development of future citizens. Due to the continuous advancement of technology throughout the world, scientific literacy is an essential requirement for the citizens of every modern country. Due to limitations in school budgets and teacher training, in many places students rarely get to do more than read about the more exciting aspects of science. As a response to this situation, science outreach programs can help reverse negative attitudes, expose students to more exciting aspects of science, spark interest and enthusiasm, and encourage communities to support science education.[17]

Science has been in the core of societal development and people appreciate it. However, an exhaustive study on general public and scientists of AAAS has shown that both are critical on the quality of science, technology, engineering, and math (STEM subjects) in grades K-12. A total of 75% AAAS scientists say that weak STEM education for grade K-12 is a major factor in the public's limited knowledge about science. Here science centers or other informal education facilities can bring major change. Young adolescents who expected to have a career in science were more likely to graduate from college with a science degree, emphasizing the importance of early encouragement.[16]

STEM outreach and outdoor activities can be characterized by free-choice, informal, and non-coercive learning environments that offer a higher degree of exploration than can usually be found in formal science classrooms. Outdoor informal science facilities add an extra dimension to outreach activities. They offer unique physical and learning experiences and enhance one's connection to nature. They offer authentic natural settings that can be more authentic than indoor settings, for example, by presenting real solar energy and wind energy demonstrations, as well as real exploration of flora, fauna, land forms, weather conditions, and rivers. Moreover, making informal STEM education accessible to underserved learners in remote locations can increase their curiosity, interest, attentiveness, achievement, and skills.

Notes

1 Chermayeff, J.C., Blandford, R.J., and Losos, C.M. (2010). Working at play: Informal science education on museum playgrounds. *Curator*, 44(1): 47–60.
2 Wells, N.M. (2000). At home with nature: Effects of 'greenness' on children's cognitive functioning. *Environment and Behavior*, 32(6): 775–795.
3 Trevlas, E., Matsouka, O., and Zachopoulou, E. (2003). Relationship between playfulness and motor creativity in preschool children. *Early Child Development and Care*, 173(5): 535–543.
4 Stephenson, A. (2002). Opening up the outdoors: Exploring the relationship between the indoor and outdoor environments of a centre. *European Early Childhood Education Research Journal*, 10(1): 29–38.
5 Beery, T., and Jørgensen, K.A. (2018). Children in nature: Sensory engagement and the experience of biodiversity. *Environmental Education Research*, 24(1): 13–25.
6 Impact studies on MSE, *MSE – Science on Wheels (1965–2015): 50 Glorious Years of taking science to rural India* – a publication of National Council of Science Museums (India), pp. 62–68.
7 Williams, D.A., Brule, H., Kelly, S.S., and Skinner, E.A. (2018). Science in the learning gardens (SciLG): A study of students' motivation, achievements and science identity in low-income middle schools. *International Journal of STEM Education*, 5(1): 8.
8 Walker, G. (2017). A science circus for Southeast Asia. https://www.newmandala. org/science-circus-southeast-asia/. Retrieved November 6, 2022.
9 Ghose, S. (2015). Science on wheels – a perspective. *MSE – Science on Wheels (1965–2015): 50 Glorious Years of taking science to rural India* – A publication of National Council of Science Museums (India), pp. 1–8.
10 Rautela, G.S. (2015). In *MSE – science on wheels (1965–2015): 50 glorious years of taking science to rural India* – a publication of National Council of Science Museums (India), Foreword.
11 Kasar, N.T. (2015). Mobile science exhibition: An unforgettable and exciting lifetime experience. In *MSE – science on wheels (1965–2015): 50 Glorious years of taking science to rural India* – a publication of National Council of Science Museums (India), pp. 92–93.
12 Gorghiu, G., and Santi, E.A. (2016). Applications of experiential learning in science education non-formal contexts. In *Proceedings of 7th International Conference on Education and Educational Psychology* – ICEEPSY 2016 (eISSN: 2357-1330).
13 Adams, J.D., Gupta, P., and Ctumaccio, A. (2014). Long-term participants: A museum programme enhances girls' STEM interest, motivation and persistence. *Afterschool Matters*, 20: 13–20.

14 McCreedy, D., and Dierking, L.D. (2014). Cascading influences: Long-term impacts of informal STEM experiences for girls. In Proceedings of the 27th Annual Visitor Studies Association Conference. Albuguerque, New Mexico, pp. 44–46. https://visitorstudies.org/images/conference/2014/vsa_2014_abstracts_-_full. pdf#page=44.

15 Cosmos Corporation. (1998). A report on the evaluation of the National Science Foundation's Informal Science Education program. NSF Publication No. 98-65. Arlington, VA: National Science Foundation.

16 Tai, R.H., Liu, C.Q., Maltese, A.V., and Fan, X. (2006). Planning early for careers in science. *Science*, 312(5777), 1143–1144.

17 Dhanapal, S., and Lim, C.C.Y. (2013). A comparative study of the impacts and students' perceptions of indoor and outdoor learning in the science classroom. *Asia-pacific Forum on Science Learning and Teaching*, 14(2): Article 2, p. 1. https://www.eduhk.hk/apfslt/download/v14_issue2_files/dhanapal.pdf.

PART IV
Rethinking Informal Science Learning

26

CREATING AWARENESS OF THE CHALLENGES OF CROSS-CULTURAL COMMUNICATION IN INFORMAL SCIENCE SETTINGS THROUGH AN INDIGENOUS LENS

David Begay and Nancy C. Maryboy

Creating an awareness of the challenges of cross-cultural communication can be a complex process. Going from one language to another, going from an Indigenous perspective, in this case, to a non-Indigenous perspective is challenging. Explaining the differences is challenging. In this chapter we will be discussing some of the more basic components underlying the differences and commonalities of each perspective. The language and words used in many cases determine the world views. It is erroneous to think that all Indigenous languages are the same, but they may contain some significant commonalities.

The authors have extensive backgrounds in understanding Native American knowledge, particularly Diné. They have worked with many tribes in US, Canada, as well as Maori, Australian Aborigines, Nigeria, Norway Saami, Columbia, Hawaii navigators, Yucatan, Cherokee Eastern Band, Cherokee Nation, Alaska Athabaskan, and Pueblos. They have strong relationships and shared knowledge. They have years of experience in formal and informal education, having taught and administered at the K-12 level and undergraduate levels and graduate levels. Some of the universities with which we have been affiliated include University of Washington, University of New Mexico, Oregon State University, Diné College, Northern Arizona University, University of Arizona, Northwest Indian College, and University of Colorado. We have provided informal professional development for NASA, for National Science Foundation (NSF) officers, and over 110 science centers and museums. We work closely with Indigenous K-12 schools through formal and informal settings. We ourselves are Indigenous (Dr. Begay is Diné with ancestral connections to Jemez Pueblo, and Dr. Maryboy is Diné/Cherokee/ with ancestral connections to Santa Clara Pueblo). Our PhDs are in Indigenous Science.

DOI: 10.4324/9781003145387-30

In this essay we are speaking from the Navajo culture, language, and world-view, which is the one in which we are grounded. Although we are speaking from a Navajo perspective, we know these concepts are widely shared among many Indigenous people around the world. The word Navajo is more commonly known but the actual name of the people, and the one we prefer is Diné, meaning The People.

The languages of Indigenous people are primarily structured through a unique relationship with their environment and ecology within their place in the cosmic universe. This provides an ontological perspective and a holistic world view in which everything in the ecosystem is interrelated, though an elaborate system of knowing. This system, or systems, is more commonly referred to as traditional ecological knowing/knowledge. There are several other names and acronyms that have evolved over the years: Indigenous science, TKN (Traditional Knowledge), IKN (Indigenous Knowledge), TEK (Traditional Ecological Knowledge), and several others. We choose to use the word "knowing" to emphasize that this is a process, a verb, a dynamic perspective inherent in the language itself, rather than use the word "knowledge" which refers to a noun or more static situation. Thus we often use the term "Indigenous Knowing" Among Indigenous people when one uses the word "knowing," it just feels like the right word, something that is culturally correct and appropriate to their perspective.

In order to discuss the challenges of authentic and meaningful communication between Indigenous and non-Indigenous people, we wish first to create an awareness of the worldviews of the Diné which are much lesser known in the US than the scientific/academic worldviews which are taught in most schools and universities.

One of the main challenges of cross-cultural communication is the significant difference in education and learning between Indigenous and non-Indigenous settings. The academic definition of education draws a hard line between different disciplines and different ways of learning and knowing. In the Native American world education is based on holistically based knowing, where all things are interconnected, and it is very difficult for Indigenous scholars to define education in the breaking down of all knowledge into departments, disciplines, silos, institutional organizations, with non-specific meaning. The Diné language is very precise, built on interrelationships, and dynamic processes. Native people can speak in discrete pieces but they also know that there is an implied knowledge that is holistic, and they keep the holism in mind when they speak about the pieces.

In order to get to the crux of the challenges of cross-cultural communication it may be useful to discuss the language, in this case the Diné language, in order to illustrate the differences and challenges of communication. There is little verbatim translation from English to Diné and Diné to English. In addition, the Diné language, similar to many other Indigenous languages, is place-based.

Everything is interconnected to place, in a holistic way, foundational to life itself. You can talk about the pieces but you know everything is interconnected.

There are many commonalities as much as differences among Indigenous people and much diversity among Indigenous people with many different world views. These discrete Indigenous differences usually become clear through the languages, especially with the use of ancient words and expressions related to a strong sense of place.

Like many Indigenous languages, the Diné language is a holistic and verb-based language, unlike Euro-American languages that are largely noun-dependent and rooted in non-holistic Cartesian and Newtonian methodology and reductionism. In order to better understand the differences between the two distinct ways of knowing, it may be helpful to first define what is meant by Cartesian/Newtonian/reductionism, and holism as understood through traditional Indigenous perspectives.

Cartesian is a term that refers to the 17th-century French philosopher, mathematician, and scientist René Descartes. He influenced the development of the Euro-American education system, including modern philosophy, mathematics and physics. In the US we are all educated explicitly and implicitly, through his scholarly work and influences. Cartesian thinking has been described as a form of rationalism because it holds that scientific knowledge can be derived from innate ideas through deductive reasoning, very similar to the reductionist method of inquiry. Cartesian thinking is emphasized here because Descartes' work is significantly different from the way that traditional Diné people think. Descartes created an ontological dualism by writing about the separation of mind and matter, mind and nature, which, in the Diné consciousness is not separated, being thus a polar opposite to Cartesian thinking, if the two are juxtaposed.

Sir Isaac Newton, from England, was a mathematician, physicist (studies of motion), and natural philosopher who was recognized as one of the most influential scientists of all time, and in many academic circles, even today. His work on objective reality was important in the development of western science, and he was a key figure in the scientific revolution.

Newton's thinking is quite different from the Diné concept of objectivity. According to Diné thinking there is no way to be an independent observer apart from the physical and holistic reality. In other words, the human is always a part of the motion of things because of the interconnection of all things, a concept that David Bohm, another physicist, would go on to develop, years later, and name it holomovement.

These differences between reductionism and holism are significant. Holism, as understood by traditional Indigenous people, is not easily understood. It may be clarified by the juxtaposition of the two systems. It is not only native people who are conscious of the inherent differences between the two ways of thinking. Jerry Mander, author of In the Absence of The Sacred: The Failure of Technology and the Survival of the Indian Nations,[1] asks: Is it possible, then, for the two societies to coexist?

Mander has provided a meaningful look at the differences. The more detailed the comparisons, the more obvious it becomes, that in almost every category Indian and Western societies are at virtually opposite poles. Beyond "opposite," they are in contradiction.

To clearly articulate the differences, we developed Table 26.1 as a juxtaposition of significant factors of incoherencies between native ways of knowing and the western world view. Unless specifically indicated, the scientific method is representative of the western world view, in accordance with the Cartesian-Newtonian paradigm. We have juxtaposed key concepts to provide some distinction between them and to illuminate the differences. To clarify these inherent differences, we have selected, in some cases, extreme positions.

Classic science, wholeness sciences and indigenous ways of knowing

In the previous section we have discussed major differences between western science, defined through Cartesian/Newtonian concepts and Indigenous science.

Now we would like to add in a discussion of wholeness (quantum) sciences.

Western science contains a basic ontological assumption that the universe is made up, ultimately, of fundamental particles and quanta which are separate from one another except insofar as there are specifiable connections (Cartesian and Newtonian based).

The wholeness sciences, often called Quantum Mechanics or Quantum Sciences, are often based on an ontological assumption that the universe is basically a single whole within which every part is connected to each other. The wholeness includes every aspect accessible to human awareness and the physical world is discerned through our senses, and all the contents of consciousness.

Traditional Indigenous Knowledge is based on concepts that the universe is holistic, ordered, and all is interrelated, including consciousness and the *spiritual essence* of all things.

Going back to western science, or what may be termed *Classic Science*, a scientific explanation of a phenomenon (understanding of cause and effect in a scientific sense) consists of relating the phenomenon to more general relationships or "scientific laws." The ultimate scientific explanation would be in terms of the motions and interactions of the fundamental particles and quanta involved. There are causes and effects.

In terms of *Wholeness Science* or the Quantum Sciences, pragmatically useful scientific explanations enhance understanding of phenomena by relating them to other phenomena and relationships.

TABLE 26.1 Juxtaposition of Western Science and Indigenous Ways of Knowing

Western science	*Indigenous ways of knowing*
Ecological Context: Scientific Knowledge	**Ecological Context: Traditional Knowledge**
Separation of man from nature	Inseparable organic interconnection between man and nature
View of the universe as a place to explore and understand	View of cosmos as a system of interrelationships as participants
Primarily linear thinking	Primarily cyclical thinking
Planet resources are never-ending and commercially exploitable. Resources become economic, to be extracted and exploited for material gain	Ecological planetary responsibility established through kinship and a sense of stewardship and reciprocal responsibility
View of self as independent with right to experiment with nature and impact outcomes	View of self as interdependent with nature, no sense of exploitation
Organization of the mind: monoculture valued	Organization of the mind: diversity and equity are valued
Technologically and economically oriented society based on self-interest	Spiritually oriented society based primarily on community interest
Mass production-based industrial society with unlimited expansionist goals	Consume only what you need with attention to conservation
Competition and materialism are positive values of life	Sharing is a positive value
Rational, empirical faculties overrule intuitive faculties	Intuitive, spiritual faculties can overrule rational faculties
Political and economic logic and values guide social decisions	Holistic, spiritual values can advance human growth
Transmission of Scientific Knowledge	**Transmission of Traditional Knowledge**
Objectivity is valued, subjectivity is devalued	Subjectivity and objectivity are both equally valued
Primarily written	Primarily oral, but not diminished for being non-written
Library as repository of knowledge	Nature and elders are repository of knowledge
Primarily uses English language as well as language of mathematics	Holistic cultural and spiritual language transmission
Value of objective research. Emphasis on individuality	Participative orientation as integral part of cosmic whole, emphasis on value of collaboration
Primarily quantitative	Primarily qualitative
Validation through pre-agreed upon set of criteria	Self-validation comes from individual intuition, community recognition, application to relevant way of life

(Continued)

TABLE 26.1 Juxtaposition of Western Science and Indigenous Ways of Knowing *(Continued)*

Western science	*Indigenous ways of knowing*
Western World View	**Indigenous World View**
Use of scientific method includes hypothesis, experimentation, and replication	The use of hypothesis, experimentation, and replication is used but often is of lesser value
Reductionist, breaking down to smallest common denominator	Reductionist process not used, instead there is a conscious awareness of a unity of all things
Relationships are evolutionary and hierarchical, progressing to a higher intellectual plane	Interrelationship of everything in universe, nonhierarchical
Separation of matter and spirit	Unity of matter and spirit. Unity of body and mind
Separation of body and mind	
Separation of inanimate and animate	Everything in cosmos is animate
Spirituality not often a way of life	Spirituality as integral part of life
Religion is separate from state and profession	Spirituality is all encompassing, not separate from religion or profession

Any accounting for cause is within a specific context for a specific purpose. The search for ultimate reductionist cause is futile. There are no causes and effects, but rather the evolution of a whole system from the cause. In comparison to classic science and wholeness sciences, *Indigenous people* think in terms of relationships, historically, spiritually, multi-faceted. Relationship encompasses the whole universe. Causality is complex and includes a universal system of relationship.[2]

While reductionism can be seen as an approach to the study of complex systems or breaking down things into a set of simpler components, holism is a theory or belief that the whole is greater than the sum of its parts. The Diné emphasize awareness of parts as they relate to the whole dynamic system, without divisions. The reductionist process reduces everything down to the smallest common denominator to better understand the parts apart from the whole. While reductionism is a good approach for studying complex systems for research and scientific inquiry, it has its limitations, limitations that some Diné elders have come to realize that reductionism allowed one to focus only on the parts and not the entire system. Using the Diné perspective, climate change is a global phenomenon. It calls for a comprehensive examination of a whole system, a system of multiple, interrelated complex cause and effects.

Scientific reductionism is breaking down the building blocks of life, which begins with cells and molecules. From there it goes deeper all the way to neutrons, electrons and protons. Beyond these elementary particles, there is little known substructure and therefore little comprehensive language has been developed for it, in English. Some scholars have used words like quarks and string theory in an attempt to explain how things appear beyond the elementary particles. Others believe that everything becomes unified at this level, interconnected and boundless.

The Diné have always maintained that everything is interconnected, and everything works together, holistically affecting each other. The Diné phrase, *alchi' naazla*, articulates the idea that all is related in a state of flux and interaction that takes up time and space with many dimensions in a unified field of energy. One should be aware that much is often lost through the translation process. This is true of most of the communication challenges in cross-culture communication.

In juxtaposition, the Diné language is holistically based which makes it very difficult in English to single out a word or circumscribe meaning of individual Diné terms. Unlike the western scientific method, Diné elders do not usually single out a subject in order to understand it, but rather comprehensively discuss it in terms of how things interrelate systematically in relation of the implicit understanding of the whole. Although one may be talking about the pieces, the consciousness is always related to the whole. All the pieces enfold the whole of everything. This epistemological and ontological consciousness is fundamental and this organization of the Diné mind is articulated in the holistic construction of the Diné language. This holistic existence is not static, it is rather perceived to be in constant, unified motion expressed through an ancient Diné term, *Nidanit'á*.

This emphasis on holistic movement has been discussed by David Bohm, an American physicist, through the use of his word holomovement. According to Bohm, holomovement brings together the principles of undivided wholeness with an idea that everything is in a state of process or becoming. This idea, *hogáalth*, in a universal sense is also fundamental to Diné consciousness. Bohm also discussed this ontological matter working through quantum theory with his colleagues. He called this second consciousness the "implicate order" in contrast to the explicate order.

The word "science" is used by many people, in many societies, in many ways, countless times every day. Yet few words in the English language are used so much with such imprecise definition. At the same time, many people who use the word "science" do so with the conviction that they are being understood and that their definition is the only definition. To further compound matters, the word "science" has come to stand for a kind of authority in our current western (Euro-American) society, in fact it has become a gold standard, an arbitrator, for what is real and what is not real.

Many people around the world still live in accordance with a world view that is remarkably different from the Euro-American world view. Indigenous[1] peoples have rich and complex ways of knowing and ways of being and living in this universe. They have observed the natural processes and interrelationships of all things over millennia and can talk about the natural order through languages of process and relationship that allow for a natural expression of holistic systems through complex dimensions and intricate organizations of time and space. These ways of knowing are based on very different world views than that of the Euro-American immigrants who seized and colonized Indigenous lands around the world.

To return to the word "science," it may be helpful to examine the use and root of the word. Today "science" can mean the seal of approval, sometimes arbitrary, sometimes definite, given to a body of knowledge that fits within the parameters of the Euro-American worldview or, to be more definitive and expansive, the Euro-American consciousness. It can also be appropriated by Indigenous peoples to mean a way of observing and being in the world. One of the actual roots of the word "science" is the Latin word sciencia, which simply (and of course complexly) meant "to know." Interestingly, there is no word for "science" per se (in relation to the western definition), in most Indigenous languages. Nor are there words for western constructs such as "art" or "religion" in most Indigenous languages. The actual concepts are deeply and culturally embedded in interconnections of holistic world views.

So, if "science" means "to know" does it not follow that Indigenous or native ways of knowing the cosmos would be science? However, if science is only defined through the lens of Euro-American consciousness, as a rigid Scientific Method process that excludes ways of coming to knowing, such as intuition, spirit, and any such related processes which cannot be seen, proved, or manipulated in replicable experiments, then Indigenous science cannot be called science.

Interestingly, the new, modern or wholeness sciences (as they are often referred to) have strong relationships with Indigenous sciences and in some aspects even with early western civilizations, such as the Greek civilization (pre-5th-century BC, where science and philosophy were holistically united) and with many of the Eastern wisdom traditions. An example is a foundational process of complementarity which is seen through the Yin/Yang of the Chinese and the *Alkee Na'ashi* of the Navajo. The dynamic movement that is intrinsic to the being of these energies emerges and flows continuously through the seemingly oppositional attributes of Yin and Yang, of Male and Female, of Negative and Positive, vibrations. Actually they are not opposites or polarities at all, rather they are related through a multi-dimensional continuum and are vitally necessary to the unending process of cosmic movement, described in Navajo as *nanit'a*.[2]

In many native communities the concept of balance is of paramount importance. For the Navajo, a sense of balance, *as'aa naaghai*, pervades all thinking. The Navajo language embeds the concept of balance. Navajo language can be described as a language of paradox, where seemingly contradictory concepts are held together as a unified whole, culminating in a sense of philosophical and cultural balance. Paradox itself is a manifestation of complementary energies.

We have worked together for many years, and it was Dr. Begay's extensive knowledge of everyday and ceremonial Navajo language that enabled us to conduct much of our research. This often involved layers of translation. One example of our research process began with our study of early writings of Franciscan Fathers who came to convert the Navajo, learned the language and studied the culture. Father Berard Haile was one scholar-priest who left clues written in the old orthography, which we were able to use as a knowledge source of Navajo astronomy. Haile in turn got his information over a hundred and fifty years ago, from a renowned Diné knowledge holder, *Hastiin Bigishi Biye*, who drew the entire Diné universe in the sand. This was an incredible feat as he included dynamic movement in time and space to cover an entire year and place. Dr. Maryboy had to learn and read the old Navajo orthography and then pronounce it in perfectly accordance with today's Navajo language. Dr. Begay then had to translate what he heard into the high ceremonial language that was being used, to which he had been exposed to from birth. Few people today still know and speak that high spiritual/ceremonial language and from there into today's Navajo and then into literal English and finally we finished with an English translation that could be understood today. This involved translations from Navajo to science or from science to Navajo. It is somewhat hard to translate, as there are no direct verbatim translations. It has to be translated by description and sometimes metaphors. One of the main difficulties in these kinds of translations is that concepts of spirituality are implied in Navajo words and in the translation from Navajo to science, spirituality is taken out. It is possible that this is the only generation that could make these translations correctly. It takes someone who has the Navajo knowledge and holistic language as well as can understand and speak the language of western science and/or quantum science.[3]

Today, traditional Navajo people will say as you take out the spirituality you lose significant meaning, there is a real cost in these kinds of translations. With every level you are losing holistic content and meaning.

In addition, the implicit significance of the interrelationship and interconnectedness of all things can be lost. In Navajo thinking the position of the human is implicitly interconnected with all things. This is most often not carried over into Navajo to English translations. In fact, in strictly English language thinking the human is mostly separated from all other entities, in an effort to provide a kind of objective thinking. To a traditional Navajo this is not possible.

Another difficulty in cross-cultural communication lies in the realm of informal learning as a designated space where learning can take place. This pre-supposes that there are different learning spaces and styles, most often markedly separate. Formal learning is supposed to take place in a school setting while informal learning is supposed to take place in an informal, out of school, setting. To the Navajo mind there is no differentiation. In fact, almost all traditional learning would take place out of the school setting, among family and community. When Indigenous scholars and institutions that rely on federal funding, such as the National Science Foundation or NASA, go through the maze of regulations and administrative paperwork, it is difficult to answer the required questions of whether or not this or that activity will take place in a formal or informal setting. To the Navajo mind, the settings are interconnected, contiguous, for all knowledge transmission. For the most part there is no third box to check which brings the two together. Interestingly the Covid pandemic has allowed for some change to take place in areas where teaching and learning occur. This is now being called "hybrid" and describes learning transmission that takes place on site in a classroom and/or off site, at home.

There are so many examples of learning transmission that occur in informal or formal settings. Indigenous people are aware of the differences, significantly more than the non-Indigenous educators. There are so many examples of difficulties in pursuing cross-cultural work whether it be teaching, learning, evaluation, writing, speaking, etc. Building awareness of the differences and commonalities is one good way to bridge the gap and build authentic collaborations that can lead to real transformative change.

Notes

1 Mander, J. (1992). *In the Absence of the Sacred: The Failure of Technology and the Survival of the Indian Nations*. San Francisco, CA: Sierra Club Books, p. 214.
2 Harman, W. W. (1991). *A Re-Examination of the Metaphysical Foundations of Modern Science*. Sausalito, CA: Institute of Noetic Sciences.
3 For a detailed description of *nanit'a*, see Begay, David H. and Nancy C. Maryboy (1998). *Nanit'a Sa'ah Naaghai, Nanit'a Bikeh Hozhoon – Living the Order: Dynamic Cosmic Process of Diné Cosmology*. UMI Dissertation Services.

27

ON-RAMPS TO WHERE?

John H. Falk and Lynn D. Dierking

For generations educators have supported children and youth's free-choice science learning through informal education experiences, such as visits to museums, science centers, zoos and aquariums, both school visits, as well as those with family/friends; various online programs, summer camps and scouting, and a growing array of increasingly targeted science/STEM programs in afterschool, on weekends, and over the summer months. A recent UK-wide survey of science educator goals, both in school and outside,[1] found that despite the diversity of study participants, there was widespread convergence on programmatic goals. Informal science education leaders, as well as formal education leaders, all agreed that their top two goals were: "Make science enjoyable and interesting" (91%) and "Inspire a general interest in, and long-term engagement with science" (89%). Although these survey data are from a single nation, there is no reason to question the generalizability of these findings. Formal and informal educators consistently espouse these two goals. The question is, how successful are they at achieving these goals?

As we say in the US, there is good news and not-so-good news. The good news is that there is a considerable and growing body of research, showing that individually and collectively, free-choice science learning experiences do contribute to children and youth perceiving science as both enjoyable and interesting. A variety of studies have demonstrated that both individually and collectively, informal experiences result in positive attitudes toward science and its enjoyment.[2] However, evidence that free-choice learning experiences in a variety of informal science settings influence children/youths' long-term science interests and participation in science is less certain and more equivocal. This is the focus of this paper.

DOI: 10.4324/9781003145387-31

Fostering persistent, long-term interest, and participation in science

Although there is evidence that informal science education experiences significantly contribute to children and youth's long-term interest and participation in science,[3] these data also suggest that such contributions are far from universal and often specific to particular programs and youth. By contrast, substantial research suggests that youth persistence in science is typically more a consequence of affective, socioemotional factors, such as identity, interest, and motivation, and sociocultural/physical factors, such as social/cultural capital, income, education, and geography.[4]

Data from a decade-long research project, SYNERGIES, in a diverse, under-resourced Portland community reinforces these findings. Over the length of this effort, we have conducted long-term investigations of the science learning pathways of 11–14-year-old youth living in this community.[5] These data suggest that the conditions required for children/youth to move from initial, situated interest and participation, to continued in-depth engagement (interest and participation) that could lead to long-term engagement and mastery of a particular science topic/practice are much more involved and complicated than most informal science education practitioners have assumed. A recent in-depth study of three youth over five years who were interested in STEM[6] showed that informal science experiences themselves only marginally contributed to youths' long-term science interest and engagement. Much more significant were each of the youth's habitus,[7] in particular, a family's social, cultural, and financial capital. In the presence of family social, cultural, and financial capital, youth persisted in their interests, including being able to access informal education experiences. However, when families because of race/ethnicity, income and other factors did not have social, cultural, and financial capital, external resources – such as informal education, or for that matter schools – failed to be sufficient to ensure long-term persistence in science interest and/or engagement.

Of course, one might conclude that this is the (unfortunate) nature of things – with children/youth born into privilege having benefits and opportunities while those less fortunate not having those benefits and opportunities – but in fact, this need not be the case. There are many other non-science-related fields/ disciplines, in which organizations & institutions regularly provide supports that enable families to surmount their social, cultural, and financial capital.

Fostering persistent, long-term interest, and participation in sports and music education experiences

Organized sports are among the most popular free-choice learning activities for children and youth worldwide,[8] involving billions of children/youth in programs associated with everything from soccer/football, to martial arts, to swimming, and more. Such programs are available for children as young as 3 or 4, with tiered programs offered to all children to continuously participate from these early pre-school ages on through childhood and adolescence. Programs are specifically designed to support age-appropriate skill development, with programs for early childhood directly connected to programs for primary school-aged children, and primary school-aged programs designed to support and feed into secondary school-aged programs. At every level, children not only learn physical and social skills, but are encouraged to proceed to the next level, particularly those who demonstrate interest and talent. Although childhood participation in such programs still requires a degree of parental social, cultural, and financial capital, these kinds of free-choice sports opportunities are designed in ways that even children of parents lacking social, cultural, and financial capital are made aware of the opportunities and encouraged to have their children participate. It is not until the ages of 11–14 that youth engage in sports through school. Perhaps most notably, when children in these school-based programs exhibit specific interest or skill, the leaders of these programs work diligently to communicate with parents and guardians, strongly encouraging and supporting continued participation, often connecting them with out-of-school opportunities.

An analogous situation, although perhaps not as widespread or culturally supported, exists in the performing arts, particularly music programs.[9] As with sports, organized music programs outside of school exist for young children, as well as adolescents; most performance-oriented music programs in schools do not begin until age 11. Like sports programs, the leaders of these programs reach out to the parents/guardians of children who exhibit promise, and/or interest, and then strongly support and encourage continued participation. In both sports and music, youth with talent are keenly sought and often their continued participation is free or subsidized.

In this way, sports and music programs create clear pathways for children/youth (whether underrepresented or under-resourced), to move from novice to increasing levels of expertise. They provide an abundance of entry level programs, clear and well-signed opportunities, and scaffolding for children and youth to grow and progress at every level of expertise. Thus, long-term interest and persistence in both sports and music are not disproportionately dependent upon parental social, cultural, and financial capital, or if there are challenges, there is intentional effort to ameliorate such differences. As a consequence, unlike in the field of science, it is rare to find a professional athlete or classical musician, or for that matter, a highly skilled hobbyist in these domains, who did not come through the ranks of organized, informal education programs, since they mindfully and systemically provide support for children and youth to progress from novice to mastery.[10]

Towards a more systematic approach to long-term, in-depth engagement in science

The current informal science education model is clearly deficient when it comes to supporting one of its key goals – creating opportunities for children/youth to remain interested and engaged with science on a long-term basis. Sadly, this need not be the case. If informal/free-choice science learning experiences were more thoughtfully and systemically conceptualized and organized, perhaps modeled after comparable sports and music programs, more children and youth would be able and willing to move from interested novices to deeply engaged masters – whether vocationally or through leisure pursuits. For this change to happen, we believe three conditions should be in place. In particular, the science learning ecosystem needs to be better customized, coordinated, and connected.[11]

> *Customize opportunities*: SYNERGIES research shows that a major constraint for children/youth in science was the fact that there were too few opportunities to engage in their specific interests. By customizing informal science resources in the ecosystem, that is considering the interests of individual youth, rather than providing one-size-fits-all programs. Typically, most informal STEM programming is designed as a generic introduction to a particular topic, which piques interest, but leaves children and youth seeking the next step, which is more difficult to identify or may not exist. One of the youths in SYNERGIES research attended a community-based program on coding. The program succeeded in triggering his interest in coding, but the program curriculum was essentially the same every year, minimizing the opportunity for him, and other youth, to extend their knowledge to other coding languages (e.g., Python and JavaScript). If such programs offered opportunities for repeat-attendees to learn new skills and continue to be challenged in their learning, such programs might better be able to provide the support youth need to persist in their interest.
>
> *Coordinate resources*: A major constraint for youth when forming (and trying to sustain) science interests is the uncoordinated nature of various science offerings within different settings and contexts. This makes it extremely difficult for youth to find "the next thing," that might be aligned with their interests. The lack of coordination, and consistent "signposting" of experiences and opportunities, is one of the reasons why only children/youth with parents who have social, cultural, and financial capital persist in their interests, since these parents are able to navigate the uncharted waters of the ecosystem, while parents with less capital in these areas find it difficult to support their

children. SYNERGIES findings suggest that successful ecosystem coordination requires all science providers in an ecosystem to commit to ongoing and continuous communication among and between themselves, ensuring that opportunities and options are clearly and continuously signposted for children/youth and their families.

Connect learners to resources: As SYNERGIES data so strongly show,[12] the social networks and connections that an individual has can strongly influence his/her ability to access resources and opportunities in a learning ecosystem.[13] The results of differing levels of social, cultural, and financial capital were quite apparent in the differing pathways of the youth in this study. A major challenge in the future will be to develop effective supports to enhance "Navigational Knowledge" for families and other mentors. Possible solutions include the cultivation of science-specific mentors,[14] in which adult volunteers were hired specifically to serve as brokers between youth and the science learning resources in a community. Such brokering entails engaging in practices that connect youth to "events, programs, internships, individuals and institutions related to their interests to support them beyond the window of a specific program or event."[15]

If the free-choice learning organizations/institutions within the science learning ecosystem, along with schools, can implement these kinds of customized, coordinated, and connected mechanisms, it seems possible that they could not only continue to achieve their goal of fostering enhanced short-term interest and enthusiasm for science, but equally meet the goal of stimulating and supporting long-term interest and participation in science, which was Mac Laetsch's vision and dream.

Notes

1 Falk, J.H., Dierking, L.D., Osborne, J., Wenger, M., Dawson, E., & Wong, B. (2015). Analyzing science education in the U.K.: Taking a system-wide approach. *Science Education*, *99*(1), 145–173.

2 Bonnette, R. Crowley, K., & Schunn, C. (2019). Falling in love and staying in love with science: Ongoing informal science experiences support fascination for all children. *International Journal of Science Education*, *41*, 1626–1643. Bevan, B., Dillon, J., Hein, G.E., Macdonald, M., Michalchik, V., Miller, D., Root, D., Rudder, L., Xanthoudaki, M., & Yoon, S. (2010). *Making science matter: Collaborations between informal science education organizations and schools*. Washington, DC: Center for Advancement of Informal Science Education. Falk, J.H., Dierking, L.D., Swanger, L., Staus, N., Back, M., Barriault, C., Catalao, C., Chambers, C., Chew, L.-L., Dahl, S.A., Falla, S., Gorecki, B., Lau, T.C., Lloyd, A., Martin, J., Santer, J., Singer, S., Solli, A., Trepanier, G., Tyystjärvi, K., & Verheyden, P. (2016). Correlating science center use with adult science literacy: An international, cross-institutional study. *Science Education*, *100*(5), 849–876. Falk, J.H., Pattison, S., Meier, D., Livingston, K., & Bibas, D. (2018). The contribution of science-rich

resources to public science interest. *Journal of Research in Science Teaching*, *55*(3), 422–445. National Research Council. (2009). *Learning science in informal environments*. Washington, DC: National Academies Press. National Research Council. (2015). *Identifying and supporting productive STEM programs in out-of-school settings*. Washington, DC: National Academies Press. Stocklmayer, S.M., Rennie, L.J., & Gilbert, J.K. (2010). The roles of the formal and informal sectors in the provision of effective science education. *Studies in Science Education*, *46*(1), 1–44.

3 Crowley, K., Barron, B.J., Knutson, K., & Martin, C. (2015). Interest and the development of pathways to science. In K. A. Renninger, M. Nieswandt, & S. Hidi (Eds.), Interest in Mathematics and Science Learning (pp 297–313). Washington, DC: AERA. Falk, J. H., & Needham, M. D. (2013). Factors contributing to adult knowledge of science and technology. *Journal of Research in Science Teaching*, *50*(4), 431–452. Jones, M. G., Childers, G., Corin, E., Chesnutt, K., & Andre, T. (2019). Free choice science learning and STEM career choice. *International Journal of Science Education, Part B*, *9*(1), 29–39. Maltese, A. & Tai, R. (2010). Eyeballs in the fridge: Sources of early interest in *science*. *International Journal of Science Education*, *32*(5), 669–685. Rahm, J., & Moore, J. C. (2016). A case study of long-term engagement and identity-in-practice: Insights into the STEM pathways of four underrepresented youths. *Journal of Research in Science Teaching*, *53*(5), 768–801. Tai, R. H., Liu, C. Q., Maltese, A. V., & Fan, X. (2006). Planning early for careers in science. *Science*, *312*, 1143–1144. Venville, G., Rennie, L., Hanbury, C., & Longnecker, N. (2013). Scientists reflect on why they chose to study science. *Research in Science Education*, *43*(6), 2207–2233.

4 Archer, L., DeWitt, J., Osborne, J., Dillon, J., Willis, B., & Wong, B. (2012). Science aspirations, capital, and family habitus: How families shape children's engagement and identification with science. *American Educational Research Journal*, *49*(5), 881–908. Bell, P., Bricker, L., Reeve, S., Zimmerman, H. T., & Tzou, C. (2013). Discovering and supporting successful learning pathways of youth in and out of school: Accounting for the development of everyday expertise across settings. In *LOST opportunities* (pp. 119–140). Springer, Dordrecht. Shaby, N., Staus, N., Dierking, L., & Falk, J. (2021). Pathways of interest and participation: How STEM-interested youth navigate a learning ecosystem. *Science Education*, *105*(4), 628–652. https://doi.org/10.1002/sce.21621.

5 Falk, J.H., Dierking, L.D., Staus, N., Penuel, W., Wyld, J., & Bailey, D. (2016). Understanding youth STEM interest pathways within a single community: The Synergies Project. *International Journal of Science Education, Part B*, *6*(4), 369–384. Staus, N.L., Falk, J.H., Penuel, W., Dierking, L., Wyld, J., & Bailey, D. (2020). Interested, disinterested, or neutral: Exploring STEM interest pathways in a low income urban community. *EURASIA Journal of Mathematics, Science and Technology Education*, *16*(6). https://doi.org/10.29333/ejmste/7927.

6 Shaby, N., Staus, N., Dierking, L., & Falk, J. (2021). Pathways of interest and participation: How STEM-interested youth navigate a learning ecosystem. *Science Education*, *105*(4), 628-652. https://doi.org/10.1002/sce.21621.

7 Archer, L., DeWitt, J., Osborne, J., Dillon, J., Willis, B., & Wong, B. (2012). Science aspirations, capital, and family habitus: How families shape children's engagement and identification with science. *American Educational Research Journal*, *49*(5), 881–908. Bourdieu, P. (1986). The forms of capital. In J. Richardson (Ed.), *Handbook of Theory of Research for the Sociology of Education* (pp. 241–258). Westport, CT: Greenwood.

8 Kjonniksen, L., Anderssen, N., & Wold, B. (2009). Organized youth sport as a predictor of physical activity in adulthood. *Scandinavian Journal of Medicine & Science in Sports*, *19*(5), 646–654. Vertonghen, J., & Theeboom, M. (2010). The social-psychological outcomes of martial arts practice among youth: A review. *Journal of Sports Science & Medicine*, *9*(4), 528–537.

9 Hesser, B., & Bartleet, B.L. (Eds.). (2020). *Music as a Global Resource: Solutions for Cultural, Social, Health, Educational, Environmental, and Economic Issues (5th Edition)*. New York, NY: Music as a Global Resource.

10 Kjonniksen, L., Anderssen, N., & Wold, B. (2009). Organized youth sport as a predictor of physical activity in adulthood. *Scandinavian journal of Medicine & Science in Sports, 19*(5), 646–654. Tunstall, T. (2012). *Changing Lives: Gustavo Dudamel, El Sistema, and the Transformative Power of Music*. New York, NY: WW Norton & Company.

11 Falk, J.H., & Dierking, L.D. (2018). Viewing science learning through an ecosystem lens: A story in two parts. In R.D. Corrigan, C. Buntting, A. Jones, & J. Loughran (Eds.), *Navigating the Changing Landscape of Formal and Informal Science Learning Opportunities* pp. 9–29. Dordrecht: Springer Netherlands.

12 Shaby, N., Staus, N., Dierking, L., & Falk, J. (2021). Pathways of interest and participation: How STEM-interested youth navigate a learning ecosystem. *Science Education, 105*(4), 628–652. https://doi.org/10.1002/sce.21621.

13 Bourdieu, P. (1986). The forms of capital. In J. Richardson (Ed.), *Handbook of Theory of Research for the Sociology of Education* pp. 241–258. Westport, CT: Greenwood.

14 Allen, S., Kastelein, K., Mokros, J., Atkinson, J., & Byrd, S. (2020). STEM Guides: Professional brokers in rural STEM ecosystems. *International Journal of Science Education, Part B, 10*(1), 17–35. Falk, J.H., & Griesmer, R. (2019). Future trajectories for STEM education at Virginia Air and Space Center. *Dimensions, 20*(1), 31–36.

15 Ching, D., Santo, R., Hoadley, C., & Peppler, K. (2016). Not just a blip in someone's life: Integrating brokering practices into out-of-school programming as a means of supporting and expanding youth futures. *On the Horizon, 24*(3), 296.

28

FOSTERING YOUTH STEM IDENTITIES THROUGH SOCIAL NETWORK CONNECTIONS IN INFORMAL SCIENCE SETTINGS

Julia McQuillan, Trish Wonch Hill, Meghan Leadabrand, and Amy N. Spiegel

In this essay we provide a sociological perspective on creating informal science education to support lifelong interest in and use of STEM knowledge and practices. In informal STEM learning environments (e.g., museums, zoos, summer camps, clubs), youth can maintain interest in STEM by developing STEM identities through social connections and a sense of belonging in STEM-focused communities. We describe insights from our experiences with youth in informal settings and research on youth STEM identity formation. We also highlight how informal STEM settings offer the potential to broaden participation to groups historically marginalized and excluded from conventional, formal STEM settings. Finally, we provide examples of opportunities for intentional inclusion, accessibility, and identity development in designing informal STEM programming.

What is STEM identity?

Science identity is the extent to which people see themselves as scientists or "science kinds of people."[1] STEM identity development may depend less on academic achievement and STEM interest and more on ability to see oneself as a scientist and incorporate STEM into one's concept of self. Because pervasive STEM stereotypes promote a limited prototype of the typical scientist, for example, some youth self-select out of STEM interests. Focusing on sense of self in addition to STEM interest, the concept of STEM identity makes relevant how youth engage in STEM and how their engagement is related to their sense of who they are and where they belong.[2]

Identity is a conception of the self that is formed and maintained through developmental, interactional, and reflective cognitive and social processes.[3] In

DOI: 10.4324/9781003145387-32

childhood and adolescence, considerable identity development occurs through play and interaction, which leads to an understanding of "self" as separate from the "other," often referred to as the "generalized other" or as the abstraction of the prevailing societal view. Stories and images shared in informal settings, including information about who makes STEM discoveries, give youth ideas about how generalized members of society perceive them, thus bringing "society" into youth imaginations.[4] Informal STEM materials that help youth develop broader ideas about who can do STEM are valuable particularly for youth who rarely see "people like them" in STEM fields (e.g., minoritized racial/ethnic communities, youth who experience misogyny, youth with lower economic status).[5]

How does identity connect to broadening participation through STEM capital?

STEM identity reflects access to and is shaped by STEM capital. The theory of "science capital"[6] explains how social location (e.g., economic class, parent educational and occupational prestige) influences how families nurture science activities, identities, and future career aspirations.[7] People with high STEM capital understand STEM well enough to make daily, science-informed decisions about their lives and to hold those with power accountable for decisions that impact us all. Unequal access to STEM enrichment activities and low STEM identity can lead to disengagement in STEM and, in early adulthood, divergence from those with higher science identity in academic and career outcomes.[8] On the other hand, extracurricular STEM opportunities enhance youth awareness of and interaction with scientific terms, concepts, ways of knowing, activities, and scientists themselves – all elements of STEM capital – which in turn increase STEM identity and sustained STEM engagement.

Why does STEM capital matter for individuals and communities?

High STEM capital matters for individual and collective survival because of the dramatic ways that STEM innovations can benefit or harm humans, animals, and environments. The impacts of applications of STEM discoveries have dramatically shaped life on earth, yet some groups and communities have garnered more or fewer advantages across places and populations.[9] Some science-guided changes – such as vaccines and sanitation systems – have had more general benefits whereas others benefit producers/owners in the short run but have disproportionate negative impacts on others – such as toxic waste dumps that pollute low-income neighborhoods. Without consideration of long-term potential consequences for all populations, the tradeoffs of short-term, lucrative, for-profit STEM innovation are more likely to have unintended negative consequences, including for many species, catastrophic climate change.[10] These consequences may eventually harm everyone, but in the short run, historically exploited groups will likely suffer more. Many of the Millennial Developmental Goals[11] call for new approaches to STEM-based interventions to understand how to undo or modify the negative consequences of former interventions. Broadening access to STEM capital allows for more creative STEM solutions to societal problems, thereby demonstrating the value of seeking broader inclusion in STEM enterprises.[12]

Inequalities in STEM: Economic, racial/ethnic, and gender-based barriers to participation

Even though representation in STEM fields usually differs by gender and race/ethnicity, with over-representation of white men, propensity to engage in science (i.e., "discovery orientation"), science interest, and science ability vary neither by gender nor by race/ethnicity among youth in the United States.[13] Cross-national comparison data suggests that many girls in the United States are likely implicitly "indulging in [their] gendered selves" by conforming to notions of femininity that involve avoiding STEM careers.[14] Nevertheless, variations by country in the proportion of scientists who are women provide evidence that nothing inherent to being a woman leads to avoiding an STEM career. Instead, the variation suggests that social determinants (e.g., cultural practices, prevailing images of scientists, and social interactions) matter much more for participation in STEM fields.[15]

Gender and race/ethnicity are often of concern in the United States because of implicit biases and historical exclusion, but other axes of structural barriers beg for attention to increase inclusion. For example, youth in suburban school districts have better access to extracurricular activities than youth in impoverished rural and urban neighborhoods.[16] Furthermore, youth from lower-income families may face barriers to participation in informal STEM settings. More affluent families, by comparison, can and do invest time, energy, and financial resources to enable their children's participation in extracurricular activities from young ages.[17] Free and low-cost informal STEM opportunities (e.g., science museums, websites, afterschool programs) have the potential to reduce structural barriers to STEM participation. Because of implicit associations of gender, race/ethnicity, and social class with STEM engagement and careers, however, just having access does not mean that youth will choose to engage or will see STEM as for them. It is therefore important to be intentional about recognizing potential social and identity-based barriers to inclusion in STEM in informal opportunities. By considering implicit and explicit images, messages, and narratives, informal STEM engagement has the potential to provide all youth with opportunities to forge social connections, build senses of belonging, and form STEM identities.

How can informal STEM opportunities help increase STEM inclusion?

Before youth even begin school, access to science capital is unequal.[18] While public education systems provide opportunities for reducing inequalities in academic achievement, public schools are limited in their ability to reduce societal inequalities. Efforts to offset the limits of formal education include providing access to valuable STEM capital in informal learning settings, such as summer camps, library programs, museums, and science centers, or within the home or community. For example, 21st Century Community Learning Center clubs were created to meet the needs of youth living in poverty in the United States.[19] Participating in 21st Century Community Learning Center programs gives youth enriching experiences doing STEM during non-school hours that can reduce inequities in STEM capital created by social inequality.

People who design informal STEM programs, like those who design any programs, have implicit and explicit ideas about who the audience will be. Many of us imagine our younger selves when we are creating informal STEM activities, yet we all have specific social locations (such as gender, race/ethnicity, social class) that others do not share. If we occupy normative statuses in STEM (e.g., male, white, middle class, highly educated), we are unlikely to think about how our social location shapes our worldview. Therefore, seeking feedback and insights from people with diverse social locations will maximize effective reach. Those designing informal STEM programs might test activities and stories with youth from various backgrounds to discover what messages youth take away from the programs and if the programs help youth envision themselves participating in STEM fields. Some might focus on the cost of attending as a major barrier to participation, and it may be, but considering other social psychological dynamics may also help create culturally comfortable, relevant, and easily accessible STEM programs for diverse audiences, particularly those with lower science capital.[20]

Location in social hierarchies matters for what people find relevant; therefore, to increase access to and engagement in STEM experiences for youth from historically marginalized groups, conventional STEM education practices (e.g., lecturing, mandatory memorization of content, charging tuition for informal programs) should be reconsidered. If school science examples and applications are based upon the experiences of boys who are white, those from groups historically excluded from science are less likely to see science as relevant, interesting, and enjoyable.[21] Research suggests that individuals from marginalized communities may be more attracted to science knowledge and practices as a means to help others and to strengthen their communities.[22] Youth from these communities may find meaning in engaging in STEM to reduce harm and/or benefit the good of others. When program providers frame STEM activities and practices as cooperative, youth build community with one another around engaging in meaningful STEM practices.

Informal STEM learning settings also provide youth with space to explore STEM interests and aspirations. Sociological studies of aspirations about the future or "future possible selves" in science suggest that they can emerge through role-playing (e.g., wearing lab coats) and engaging in science practices in settings such as science classes, zoos, museums, or clubs.[23] Because adolescent development is in large part a process of transitioning group memberships and roles,[24] joining informal STEM learning groups gives youth the chance to explore roles as scientists. People negotiate and construct how they view themselves based upon feedback from group memberships that inform the self-verification process,[25] therefore youth also benefit when adults in informal STEM settings acknowledge them explicitly.[26]

How do social interactional dimensions of adolescence influence STEM identity and sustained STEM engagement?

Social connections are vital to human thriving.[27] The "eureka" image of a lone scientist does not match lived reality because collaboration is a key feature in scientific practices.[28] In contrast to the image of scientists as socially marginalized geniuses, many successful scientists belong to research teams and active science communities.[29] Therefore, those of us who create engaging informal STEM activities and resources should consider the social dimensions of science in addition to other evidence-based criteria for effective informal STEM outreach identified in "Dimensions of Success."[30]

Peers can inhibit or facilitate STEM identity and behavior (as with any kind of identity or behavior) through interaction that explicitly or implicitly indicates approval or disapproval.[31] Peer influence matters beyond academic environments because youth interact in social networks in many settings. Network studies show that youth often select friends who have similar behaviors and attitudes (i.e., homophily), but also that individual attitudes and behaviors converge over time through peer socialization.[32] Evidence from social network research[33] demonstrates the crucial role of social relationships for creating and maintaining identities, including STEM identities,[13] yet social interaction can also reinscribe stereotypes. Observational survey data of middle school youth in the United States found that both boys and girls were more likely to assume that among their friends, the boys more than the girls were science kinds of people, even adjusted for self-identification with and grades in science.[34]

Compared to formal education settings, informal settings provide youth with space to play, choose who to interact with, and to form friendships in ways that support STEM identity development.[35] In addition to designing STEM programs with accurate science and engaging content, explicitly incorporating opportunities for youth to build group identification with STEM fields increases the likelihood of sustained engagement after the program ends.[36] For example, youth who created more relationships around science during a science-related summer program also reported higher self-identification with behaviors in those fields.[37] In addition, having social network connections with friends in science settings was associated with higher STEM career interest.[38] Following youth from seventh grade past college showed that youth who participate in informal science programs were more likely to pursue STEM majors and careers. The association was due in part to interpersonal connections which helped youth build STEM-related social networks and shared STEM identities.[39]

Are informal STEM opportunities relevant for youth STEM identity development?

A report by the Afterschool Alliance[40] argued that after-school science can be a particularly important component for youth science identity development. Informal STEM exploration in after-school clubs can potentially strengthen STEM identity in several ways. Youth experience how science functions "in the real world" and discover STEM career possibilities. Engaging in STEM outside of school helps youth feel like they are doing "real" science, thus increasing their likelihood of identifying as a science kind of person.[41] Doing science activities outside of classrooms seems more like "real science" to many youth.[42] Youth who participate in after-school science clubs have higher odds of continuing in STEM education and occupational aspirations,[43] but it is hard to know if the association reflects selection or causation because youth are not randomized to after school clubs. Beyond the direct activities, youth who participate in after-school science tend to increase and maintain student engagement in science compared to youth who have not partaken in such clubs.[44]

In conclusion: From a sociological perspective, what is special about informal compared to formal STEM opportunities?

Informal STEM opportunities can be leveraged to allow for exploration of STEM identity and social relationships in a free-choice setting. People recognize themselves as members of groups and develop group identities (e.g., "scientists") in part through group membership. Therefore, creating informal settings that involve STEM-related groups (e.g., clubs, camps, robotics teams, math leagues) does more than introduce or reinforce field-specific content – it also creates opportunities for discovering STEM identity. Creating STEM opportunities that signal invitations to belong and support STEM identity development for youth from groups often marginalized by STEM could happen with intentional efforts in informal programs. In free-choice environments youth have more agency to express who they are (their "selves"). In addition, creating social relationships and a sense of belonging to an STEM group allows for deeper and longer-lasting connections to STEM, plus greater consideration of future possible selves in STEM careers. Free from the stressful assessments ubiquitous in formal education (e.g., tests or assignments), youth can "try on" and affirm STEM identities, even if they never thought that "people like them" could participate in STEM.

Funding source acknowledgement

This work is supported by the *Worlds of Connections* SEPA [R25GM129836] at the University of Nebraska–Lincoln, funded by the National Institute of General Medical Sciences of the National Institutes of Health. This content is solely the responsibility of the creators and does not necessarily represent the official views of the National Institutes of Health or the University of Nebraska.

Notes

1 Archer L, DeWitt J, Osborne J, Dillon J, Willis B, Wong B. (2012). Science aspirations, capital, and family habitus: How families shape children's engagement and identification with science. *American Educational Research Journal, 49*(5), 881–908.

2 Carlone HB, Scott CM, Lowder C. (2014). Becoming (less) scientific: A longitudinal study of students' identity work from elementary to middle school science. *Journal of Research in Science Teaching, 51*(7), 36–869.

3 Howard JA. (2000). Social psychology of identities. *Annual Review of Sociology, 26*(1), 367–393.

4 Martin J, Sokol B. (2011). Generalized others and imaginary audiences: A neo-Meadian approach to adolescent egocentrism. *New Ideas in Psychology, 29*(3), 364–375.

5 Schinske JN, Perkins H, Snyder A, Wyer M. (2016). Scientist spotlight homework assignments shift students' stereotypes of scientists and enhance science identity in a diverse introductory science class. *CBE—Life Sciences Education, 15*(3), ar47.

6 Jones MG, Chesnutt K, Ennes M, Macher D, Paechter M. (2022). Measuring science capital, science attitudes, and science experiences in elementary and middle school students. *Studies in Educational Evaluation, 74*, 101180. https://doi.org/10.1016/j.stueduc.2022.101180.

7 Jenkins TM. (2020.) Doctors' Orders. In: *Doctors' Orders*. New York: Columbia University Press.

8 Dufur MJ, Parcel TL, Troutman KP. (2013). Does capital at home matter more than capital at school? Social capital effects on academic achievement. *Research in Social Stratification and Mobility, 31*, 1–21. https://doi.org/10.1016/j.rssm.2012.08.002.

9 Temper L, Del Bene D, Martinez-Alier J. (2015). Mapping the frontiers and front lines of global environmental justice: The EJ Atlas. *Journal of Political Ecology, 22*(1), 255–278.

10 Díaz S, Settele J, Brondízio ES, et al. (2019). Pervasive human-driven decline of life on Earth points to the need for transformative change. *Science, 366*(6471), eaax3100.

11 Sachs JD, McArthur JW. (2005). The millennium project: A plan for meeting the millennium development goals. *The Lancet, 365*(9456), 347–353.

12 Dasgupta N, Stout JG. (2014). Girls and women in science, technology, engineering, and mathematics: STEMing the tide and broadening participation in STEM careers. *Policy Insights from the Behavioral and Brain Sciences*. https://doi.org/10.1177/2372732214549471.

13 Hill PW, McQuillan J, Spiegel AN, Diamond J. (2018). Discovery orientation, cognitive schemas, and disparities in science identity in early adolescence. *Sociological Perspectives*, *61*(1), 99–125. https://doi.org/10.1177/0731121417724774.

14 Charles M, Bradley K. (2009). Indulging our gendered selves? Sex segregation by field of study in 44 countries. *American Journal of Sociology*, *114*(4), 924–976.

15 Kim Y, Sinatra GM. (2018). Science identity development: An interactionist approach. *International Journal of STEM Education*, *5*(1), 51. https://doi.org/10.1186/s40594-018-0149-9.

16 Afterschool Alliance. (2016). *The Growing Importance of Afterschool for Rural Communities*. America after 3 PM Special Report.

17 Lareau A. (2011.) *Unequal Childhoods: Class, Race, and Family Life*. Berkeley, CA: University of California Press. Accessed October 19, 2016. https://books.google.com/books?hl=en&lr=&id=JuQuoPKMPF4C&oi=fnd&pg=PP1&d-q=lareau+unequal+childhoods&ots=up4bwM2BqP&sig=KlMRYCZ_kBF_N5sdovQo7tD-ws4.

18 von Hippel PT, Workman J, Downey DB. (2018). Inequality in reading and math skills forms mainly before kindergarten: A replication, and partial correction, of "Are schools the great equalizer?" *Sociology of Education*, *91*(4), 323–357.

19 U.S. Department of Education. (2018). *21st Century Community Learning Centers (21st CCLC) Analytic Support for Evaluation and Program Monitoring: An Overview of the 21st CCLC Performance Data: 2016–17 (13th Report)*. Washington, DC: U.S. Department of Education.

20 Feinstein NW, Meshoulam D. (2014). Science for what public? Addressing equity in American science museums and science centers. *Journal of Research in Science Teaching*, *51*(3), 368–394.

21 Saw G, Chang N, Chan HY. (2018). Cross-sectional and longitudinal disparities in STEM career aspirations at the intersection of gender, race/ethnicity, and socioeconomic status. *Educational Researcher*, *47*(8), 525–531. https://doi.org/10.3102/0013189X18787818.

22 Smith JL, Cech EA, Metz A, Huntoon M, Moyer C. (2014). Giving back or giving up: Native American student experiences in science and engineering. *Cultural Diversity & Ethnic Minority Psychology*. https://doi.org/10.1037/a0036945.

23 Buday SK, Stake JE, Peterson ZD. (2012). Gender and the choice of a science career: The impact of social support and possible selves. *Sex Roles*, *66*(3–4), 197–209. https://doi.org/10.1007/s11199-011-0015-4.

24 Williams LS. (2002). Trying on gender, gender regimes, and the process of becoming women. *Gender & Society*, *16*(1), 29–52.

25 Dasgupta N. (2011). Ingroup experts and peers as social vaccines who inoculate the self-concept: The stereotype inoculation model. *Psychological Inquiry*, *22*(4), 31–246. https://doi.org/10.1080/1047840X.2011.607313.

26 Jackson MC, Leal CC, Zambrano J, Thoman DB. (2019). Talking about science interests: the importance of social recognition when students talk about their interests in STEM. *Social Psychology of Education*, *22*(1), 149–167.

27 Graeber D, Wengrow D. (2021). *The Dawn of Everything: A New History of Humanity*. New York: Farrar, Straus and Giroux.

28 Atkin E. (2018). The sexism of "genius." *The New Republic*. Published online March 15. Accessed July 6, 2022. https://newrepublic.com/article/147463/sexism-genius-death-stephen-hawking.

29 Cooke NJ, Hilton ML, eds. (2015). *Enhancing the Effectiveness of Team Science*. Washington, DC: National Academies. https://nap.nationalacademies.org/catalog/19007/enhancing-the-effectiveness-of-team-science.

30 Noam GG, Allen PJ, Sonnert G, Sadler PM. (2020). The common instrument: An assessment to measure and communicate youth science engagement in out-of-school time. *International Journal of Science Education, Part B*, *10*(4), 295–318.

31 Crosnoe R, Riegle-Crumb C, Field S, Frank K., Muller C. (2008). Peer group contexts of girls' and boys' academic experiences. *Child Development, 79*(1), 139–155.

32 Brechwald WA, Prinstein MJ. (2011). Beyond homophily: A decade of advances in understanding peer influence processes. *Journal of Research on Adolescence, 21*(1), 166–179. https://doi.org/10.1111/j.1532-7795.2010.00721.x.

33 McPherson M, Smith-Lovin L, Cook JM. (2001). Birds of a feather: Homophily in social networks. *Annual Review of Sociology, 27*(1), 415–444.

34 Gauthier GR, Hill PW, McQuillan J, Spiegel AN, Diamond J. (2017). The potential scientist's dilemma: How the masculine framing of science shapes friendships and science job aspirations. *Social Sciences, 6*(1), 14. https://doi.org/10.3390/socsci6010014.

35 Afterschool Alliance. (2014). *America After 3PM: Afterschool Programs in Demand.* Washington, DC: Afterschool Alliance. Published online. https://www.wallacefoundation.org/knowledge-center/documents/America-After-3PM-Afterschool-Programs-in-Demand.pdf.

36 Stevenson N, Sommers AS, Grandgenett N, et al. (2022). Replicating or franchising a STEM afterschool program model: Core elements of programmatic integrity. *International Journal of STEM Education, 9*(1), 10. https://doi.org/10.1186/s40594-021-00320-0.

37 Lee JD. (2002). More than ability: Gender and personal relationships influence science and technology involvement. *Sociology of Education, 75*(4), 349–373. https://doi.org/10.2307/3090283.

38 Robnett RD, Leaper C. (2013). Friendship groups, personal motivation, and gender in relation to high school students' STEM career interest. *Journal of Adolescent Research, 23*(4), 652–664. https://doi.org/10.1111/jora.12013.

39 Habig B, Gupta P, Levine B, Adams J. (2018). An informal science education program's impact on STEM major and STEM career outcomes. *Research in Science Education, 50*, 1051–1074.

40 Afterschool Alliance. (2013). *Defining Youth Outcomes for STEM Learning in Afterschool.* Washington, DC: Afterschool Alliance. http://www.afterschoolalliance.org/stem_outcomes_2013.pdf.

41 Allen PJ, Noam G. (2016). *STEM Learning Ecosystems: Evaluation and Assessment Findings.* Cambridge, MA: The PEAR Institute: Partnerships in Education and Resilience. http://stemecosystems.org/wp-content/uploads/2017/01/STEM Ecosystems_Final_120616.pdf.

42 Wade-Jaimes K, King NS, Schwartz R. (2021). "You could like science and not be a science person": Black girls' negotiation of space and identity in science. *Science Education, 105*(5), 855–879. https://doi.org/10.1002/sce.21664.

43 Cohen MD, Therriault S, Scala J, Lavinson R, Brand B. (2019). *Afterschool Programming as a Lever to Enhance and Provide Career Readiness Opportunities.* Arlington, VA: College & Career Readiness & Success Center at American Institutes for Research.

44 Dabney KP, Tai RH, Almarode JT, et al. (2012). Out-of-school time science activities and their association with career interest in STEM. *International Journal of Science Education, Part B, 2*(1), 63–79. https://doi.org/10.1080/21548455.2011.629455.

29

SPECULATIVE DESIGN FOR STEM LEARNING

Camillia Matuk

Centering learning within youth's concerns

During an out-of-school STEM workshop, my team of facilitators and I asked our participants – about a dozen seventh-grade youth – for their thoughts on the future of STEM. The youth worked silently, and eventually, one wall of the university classroom was covered in colorful sticky notes expressing their hopes and concerns. These expressed misgivings about ambitious STEM innovations that go unchecked: That artificial intelligence might "get too smart and take jobs and overpower humans;" that genetically "modifying people to be perfect could go wrong in many ways;" and that our careless demands on natural resources would cause the Earth to "no longer exist." The stickies also reflected their hopefulness in the potential for STEM to offer bodily agency (e.g., using genetic engineering "to [have] superpowers" or at least "a little say on how I look") and to solve such global challenges as disease and overpopulation.

These youths' ideas – both fantastical and apocalyptic – reflect the public's broader hopes and concerns about the future, as well as our increasingly complex relationship with STEM. The anxiety associated with climate change[1] and the coronavirus pandemic,[2] together with advances in data science and artificial intelligence into morally gray territory,[3] have roused both confidence and mistrust in the ways that scientific institutions serve the public.[4] These concerns are also reminders of the tightly knit connections among science, individuals, and societies. They highlight the need for education that prepares youth with the multidisciplinary tools to identify and address challenges, now and in the future, as social challenges as well as science ones.

Traditional science education falls short of this goal, as it tends to focus learners on studying already solved problems, such as how plants get energy,

DOI: 10.4324/9781003145387-33

and the reason for the seasons. While there is no doubt that these topics are important to learn, a sole focus on them ignores youth's more urgent concerns and risks them failing to recognize the relevance of STEM to their lives. These approaches moreover neglect the creative and imaginative dimensions of STEM that allow us to envision and realize more hopeful futures.

As Australian futurist Peter Ellyard stated, "Humans can only work to build a future if they can first imagine it."[5] Building on this sentiment, this chapter describes speculative fiction as an approach to prepare students to tackle current and future socioscientific issues.

Speculative design and futures thinking

Speculative design is a process of envisioning plausible futures – whether probable, possible, or preferable, which results in the creation of an artifact representing some aspect of that future.[6] The ultimate goal is to provoke reflection on the links between present actions and future outcomes and to inspire action. The process involves playful improvisation, experimentation, and creativity and results in works – ranging from tangible objects, to performances, to stories, to games – that are as much feats of engineering as they are of art. For example, Meydan Levy designed *Neo-Fruits*, printed from cellulose and filled with a vitamin-enriched substance, to mimic the aesthetic and sensory qualities of natural fruit. Levy's work comments on the food industry's processing practices that have widened the gap between what our bodies need and what we consume, by envisioning a future in which all food is engineered. In another example, Lucy McCrae's *Solitary Survival Raft* is an inflatable structure that embraces and comforts the human body, a reaction to the absence of touch spurred by our increasing turn toward digitalization, and the social distancing practices introduced by the COVID-19 pandemic.

Speculative design differs from traditional approaches to science and design in that, rather than solving problems and answering questions about the world as it is, it seeks to raise questions and to address the world as it could be.[3] In this manner, speculative design can be an approach to preparing youth to tackle future socioscientific issues by developing *futures thinking*, sometimes referred to as *futures consciousness*[7] or *futures literacy*[8]: the ability to understand how today's trends may contribute to likely futures and to imagine plausible alternatives. Futures thinking has four dispositional characteristics: (1) *Agency beliefs*, which comprise the conviction that one's actions can have an impact; (2) *systems perceptions*, the awareness of how everything is interconnected;[9] (3) *openness to alternatives*, the curiosity and foresightedness that allows speculation on probable, possible, and preferable; and (3) *empathy*, consideration of the well-being of others.[10]

Speculative design can be a way to build learners' futures thinking via its grounding in several paradigms in the learning sciences. For example, as a constructionist, design-based approach to learning,[11] speculative design engages learners in creating, critiquing, refining, and presenting tangible artifacts. These activities can also build self-efficacy[12] and a sense of agency by providing learners with authentic tools and practices to participate meaningfully in social discourses. The artifacts they create become objects *through* which to learn, as learners use them to externalize their understanding; and objects *from* which to learn, as they engage audiences in reflecting on the ideas made visible by the designer.[13] Finally, as design is in service of others, speculative design can also build empathy as designers place themselves in others' situations to understand their experiences, needs, and desires.

Second, speculative design draws on features of *narrative* to support reasoning about complex socioscientific issues. By developing future scenarios, designers engage in storytelling practices such as worldbuilding and perspective-taking,[14] which create relatable contexts in which to identify meaningful questions and to reason about the significance of abstract ideas and their surrounding moral considerations.[15] Stories help build systems perceptions as they provide intuitive frameworks for recognizing the interconnectedness of things and for making predictions about the outcomes of actions and events. Stories also help build empathy, as they can allow one to experience situations vicariously through characters.

Third, speculative design bridges learners' formal and informal learning experiences, which broaden the possible ways to engage with STEM.[16] By involving creative writing, research, art and design, and popular culture, speculative design allows youth to develop disciplinary agency and identity through their own STEM and non-STEM interests. Speculative design also encourages systems perceptions: Its cross-domain nature highlights such domains as science, art, social studies, creative writing, and engineering design as valid and complementary ways of knowing.

Examples of youths' STEM learning through speculative design

Research on speculative design with youth has demonstrated the various opportunities it creates for youth's STEM engagement. For example, in one out-of-school program, youth of color use Wakanda, the fictional world from the *Black Panther* movie,[17] as a starting point for their own Afrofuturist designs.[18] The artifacts they created – made from such materials as fabric, cardboard, microcontrollers, and LEDs, and including a cloak that guards against stereotyping, a trash can that converts garbage into energy, a city seamlessly integrated with nature – reflect these youths' social and environmental concerns, and their desires to live harmoniously and sustainably. They moreover demonstrate how tinkering and building with materials could be a means by which youth could reflect upon and critique their relationships with their past, their environment, and with others. This also provided a medium with which to articulate solutions to issues that concerned them.

In another example, my own team facilitated a series of five-day out-of-school speculative design workshops for middle school youth, during which we guided them in designing artifacts to represent their futuristic visions of STEM. Early in one workshop on the future of genetic engineering, participants explored the concept of inheritance and the scientific and technological advances in the field. Through imaginative "what if" questions, such as "What if you could clone Einstein and raise him as your child?" and "What if you could enhance your body to achieve superhuman powers?" youth explored the debate surrounding nature vs. nurture and discussed issues of equity in a possible world in which these abilities would be available only to some.

Approaching these topics through imagination rather than through content spurred youth to ask questions that at once revealed their curiosity and their reasoning ("Would your clone have the same memories as you? Would they be the same age as you?") and allowed facilitators to adapt their guidance accordingly. We prompted youth to use what they had learned to imagine scenarios representing their desired futures for genetic engineering. One group staged a *tableau vivant*, picturing an operating room 200 years in the future. In it, they dressed as surgeons surrounding a patient on an operating table, and next to them, an organ cloning machine constructed from painted cardboard. In this future, they described, transplanting organs cloned during one's youth would be a common life-extending procedure. This group's work reflects humankind's reckoning with mortality, and our desire to control the time we have to live, and the quality of our lives. Through their design, these youth were able to articulate a vision for science and technology to grant this agency over life.

Our research also shows learning opportunities at the convergence of science, design, and storytelling, particularly as speculative design requires coordinating between science and imagination to achieve plausibility.[19] For instance, a participant in one of our workshops on the future of viruses sought to design a science fiction game about HIV. Following advice he received from facilitators to develop a backstory that would explain how his "guardian angel" character came to have wings, he conducted internet research on bird wing evolution, and on T-cell transfer, a current cancer treatment. The backstory he ultimately created described how, in his game's world, T-cell transfer was discovered as a cure for HIV in humans; however, due to a lack of human T-cells at the time of treating one patient, scientists transferred T-cells from a bird to him, leading to a mutation that resulted in him developing wings.

This example shows how this youth's motivation to create a believable story drove his independent research into relevant science content. As a story is *told* to others, and as my team had structured the workshop to support progressive refinements of milestones, this youth's story also served as a public artifact to be critiqued and improved through cycles of design, feedback, and revision.

Ultimately, the story offered a framework within which this youth could build evidence-supported claims that made both scientific and narrative sense.

Challenges with speculative design for STEM learning

The interdisciplinary nature of speculative design focused on socioscientific issues offers numerous informal STEM learning opportunities. While it is conceivable that a teacher with sufficient flexibility in their schedule and curriculum may find ways to incorporate aspects of speculative design into their classroom instruction, the genre poses numerous challenges in traditional school-based science learning contexts. For instance, there is the challenge of facilitating open-ended, student-driven projects, and of assessing standardized outcomes given the diversity of individual learners' experiences. Moreover, speculative design treads a narrow path between science fiction and science fact: Without proper guidance, learners of the genre may find it more natural to base their speculations on fantasy rather than on scientific understanding and to prioritize design choices based on aesthetic preferences rather than on scientific evidence and argument coherence.[20] For example, we observed youth in our workshops to superficially add "fun science facts" to their speculative game designs rather than meaningfully integrate science into the game mechanics and narrative. We also observed students preferring to retain design ideas for their aesthetic appeal (e.g., a "shadowstepper" character, who travels by stepping between others' shadows) and to reject advice to find more scientifically plausible explanations (e.g., a futuristic version of teleportation that builds on the present-day concept of quantum teleportation). Additionally, while the narrative aspects of speculative design offer an intuitive and concrete context for reasoning about abstract ideas, the linear nature of narrative can also conflict with the complex systems that are characteristic of socioscientific issues, creating possibilities for oversimplification and misunderstanding of science concepts.

However, attempting to achieve more "scientific rigor" in speculative design may be at the expense of allowing learners to follow their interest-driven lines of inquiry and to find connections between science and their various personal interests and experiences. Ultimately, the learning potential in speculative design is in building futures thinking, which comprises a broader set of skills and dispositions than is encompassed by science alone. The brief examples above show how speculative design can promote the development of agency beliefs (through youth's creating solutions to anticipated problems and worlds according to their visions of preferable futures); systems perceptions (by working out specific causes and effects through their design of stories and games, which are themselves systems); openness to alternatives (by developing potential future outcomes based on their assessment of today's trends); and empathy (through their design of objects that would help people in the future). These examples also show the affordances of speculative design in engaging learners in a design process of progressively refining their ideas; in using storytelling as a frame for building explanations; and of encouraging interest-driven inquiry into issues that concern them.

While ongoing research in this area might further explore ways to support learners in balancing content understanding with imagination in STEM contexts, it is within this limitless space that speculative design may allow youth to explore their passions, form identities, and build relationships with disciplinary ideas.

Notes

1 Léger-Goodes, T., Malboeuf-Hurtubise, C., Mastine, T., Généreux, M., Paradis, P.-O., & Camden, C. (2022). Eco-Anxiety in children: A scoping review of the mental health impacts of the awareness of climate change. *Frontiers in Psychology, 13*. https://doi.org/10.3389/fpsyg.2022.872544.

2 Blanchflower, D. G., & Bryson, A. (2022). Covid and mental health in America. *PloS One, 17*(7), e0269855.

3 e.g., Köbis, N., Bonnefon, J.-F., & Rahwan, I. (2021). Bad machines corrupt good morals. *Nature Human Behaviour, 5*(6), 679–685.

4 Anderson, J., Rainie, L., & Vogels, E. A. (2021). Experts say the 'new normal' in 2025 will be far more tech-driven, presenting more big challenges. *Retrieved from PEW*: https://www.Pewresearch.org/Internet/2021/02/18/Experts-Say-the-New-Normal-in-2025-Will-Be-Far-More-Tech-Driven-Presenting-More-Big-Challenges. https://www.pewresearch.org/internet/2021/02/18/worries-about-life-in-2025/. Gundersen, T., Alinejad, D., Branch, T. Y., Duffy, B., Hewlett, K., Holst, C., Owens, S., Panizza, F., Tellmann, S. M., van Dijck, J., & Baghramian, M. (2022). A New Dark Age? Truth, trust, and environmental science. *Annual Review of Environment and Resources.* https://doi.org/10.1146/annurev-environ-120920-015909.

5 Ellyard, P. (1992, July). *Education for the 21st Century.* New Zealand Principals Federation conference, Christchurch, New Zealand. Jones, A., Buntting, C., Hipkins, R., McKim, A., Conner, L., & Saunders, K. (2012). Developing students' futures thinking in science education. *Research in Science Education, 42*(4), 687–708.

6 Dunne, A., & Raby, F. (2013). *Speculative Everything: Design, Fiction, and Social Dreaming.* Cambridge and London. MIT Press.

7 Ahvenharju, S., Minkkinen, M., & Lalot, F. (2018). The five dimensions of Futures Consciousness. *Futures, 104*, 1–13.

8 Wilenius, M., & Pouru, L. (2020). Developing futures literacy as a tool to navigate an uncertain world. In UNESCO (ed.) *Humanistic Futures of Learning: Perspectives from UNESCO Chairs and UNITWIN Networks,* UNESCO, *5*, 207–210.

9 Liu, S.-Y., Lin, C.-S., & Tsai, C.-C. (2011). College students' scientific epistemological views and thinking patterns in socioscientific decision making. *Science Education, 95*(3), 497–517. Sadler, T. D., Barab, S. A., & Scott, B. (2007). What do students gain by engaging in socioscientific inquiry? *Research in Science Education, 37*(4), 371–391. Yang, F. (2005). Student views concerning evidence and the expert in reasoning a socio-scientific issue and personal epistemology. *Educational Studies, 31*(1), 65–84.

10 Heinonen, S., & Balcom Raleigh, N. (2015). Continuous transformation and neo-carbon energy scenarios. *FFRC EBOOK, 10*, 2015. Lombardo, T., & Cornish, E. (2010, October). Wisdom facing forward: What it means to have heightened future consciousness. *The Futurist; Washington, 44*(5), 34–42. Miller, R. (2007). Futures literacy: A hybrid strategic scenario method. *Futures, 39*(4), 341–362.

11 Papert, S., & Harel, I. (1991). Situating constructionism. *Constructionism, 36*(2), 1–11.
12 Schunk, D. H., Meece, J. L., & Pintrich, P. R. (2013). *Motivation in Education: Theory, Research, and Applications*. Pearson College Division.
13 Clark, D., Nelson, B., Sengupta, P., & D'Angelo, C. (2009). Science learning through digital games and simulations: Genres, examples, and evidence. *National Academy of Sciences Workshop on Learning Science, Computer Games, Simulations and Education, Washington, DC.*
14 Liveley, G., Slocombe, W., & Spiers, E. (2021). Futures literacy through narrative. *Futures, 125*, 102663.
15 Cohn, D. (2000). *The Distinction of Fiction*. Baltimore and London:" JHU Press. Zeidler, D. L., Herman, B. C., & Sadler, T. D. (2019). New directions in socio-scientific issues research. *Disciplinary and Interdisciplinary Science Education Research, 1*(1), 1–9.
16 Ito, M., Gutiérrez, K., Livingstone, S., Penuel, B., Rhodes, J., Salen, K., Schor, J., Sefton-Green, J., & Craig Watkins, S. (2013). *Connected Learning: An Agenda for Research and Design*. Digital Media and Learning Research Hub, Irvine, CA, USA. National Research Council. (1996). National science education standards. *Washington, DC: National Academy of Sciences. National Research Council. (2012). A Framework for K–12 Science Education: Practices, Cross-Cutting.*
17 Coogler, R. 2018. Black Panther. http://www.imdb.com/title/tt1825683/
18 Holbert, N., Dando, M., & Correa, I. (2020). Afrofuturism as critical constructionist design: Building futures from the past and present. *Learning, Media and Technology, 45*(4), 328–344.
19 Matuk, C., Hurwich, T., & Amato, A. (2019). How Science Fiction Worldbuilding Supports Students' Scientific Explanation. In *Proceedings of FabLearn 2019 on – FL2019*. https://doi.org/10.1145/3311890.3311925.Matuk, C., Hurwich, T., Prosperi, J., & Ezer, Y. (2020). Iterations on a transmedia game design experience for youth's autonomous, collaborative learning. *International Journal of Designs for Learning, 11*(1), 108–139.

30

LIMITING CLAIMS ABOUT MUSEUMS, STEM, AND SOCIAL ISSUES

John Fraser, Rebecca Joy Norlander, and Kathryn Nock

Museum workers believe that museums are critical vectors for social change. The 2022 ICOM definition of museums claimed that museums are necessary for fixing social wrongs, paths for cultural diplomacy, and venues for advancing a sustainable future. Unfortunately, there seems to be a scarcity of evidence to back up these social impact claims. An effort to synthesize research in the USA, published in the first two decades of the 21st century sought to describe what can be considered common understanding in the museum field about how social issues and science, technology, engineering, and math (STEM) come together in museum practice.

Our study focused on methods and data reporting; we examined where claims may overshoot what should be considered generalizable fact. To do that, we analyzed a subset of papers assembled through a configurative review of the museums, STEM, and social issues domains in the USA.[1] The initial review described the topics and types of research related to our focal subject. Here, we focus on the choices made about the research methods. By selecting only those papers that assessed the intersection of STEM and social issues in museums, we were able to look across three primary sources of knowledge: peer-reviewed journals, grey literature from a national online repository, and academic dissertations or theses in the ProQuest database. We used these reports to understand whether there is sufficient evidence to make claims about the museum sector or museums as a class capable of supporting the many claims about their impacts. In this case, we focused only on museums' capacity to use STEM to engage audiences with social issues and acknowledge the exclusion of humanities content as a path for social change.

DOI: 10.4324/9781003145387-34

Background

The scholarly research into museum efforts make programs through community engagement has increased substantially since the turn of the millennium. These trends are evident in peer-reviewed journals and grant funded research. The topic was also in the forefront of the contentious debate over the definition of what constitutes a museum at the 2019 and 2022 International Council on Museums congresses. The 2022 international definition demonstrated that there is a consensus that museums should be engaging communities with humanities issues of representation, human rights, and moral decision-making. But this concept of community engagement around social issues may not pertain to all types of museums, including those that focus on the sciences, technology, engineering, or math (STEM).

We questioned the nature of what constitutes social science evidence in support of these claims about the museum enterprise and whether STEM content might be a useful lens for examining this contentious idea. We firmly believe that the museum enterprise can fulfill these aspirations with STEM content, but are concerned that much of the writing about what museums are capable of doing may be based on limited evidence and may even be a function of confirmation bias. Museums, as a type of intervention, cannot be easily subjected to long-term randomized controlled studies due to variation in visitor knowledge, museum types, and the nature of the learning that occurs, but that limitation does not preclude examination of any research that could suggest whether the form can have the social impacts it claims.

We took an appreciative position in our study, aiming to confirm claims of impact that could be attributed to museum form rather than a specific intervention. We value the hybrid knowledge that emerges when different epistemic traditions and methods examine their truths about a phenomenon. We explored how methodological choices afford comparison, contrast, additive, or explanatory value when brought together. Despite the breadth of disciplines that represent museum studies, we found the data to be sparse. This finding is notable, given that museum studies have been highly promoted for decades and there are more than 35,000 museums in the USA alone.[2]

In 2015, Fu and colleagues categorized STEM learning evaluation methods and found a remarkably narrow set of designs, suggesting the need for more empirical research designs that could quantify the impact, scale, and statistical models that would support generalizability.[3] In parallel, Morrissey and colleagues[4] examined an online repository of grey literature to understand what can be said about STEM and social issues. Even with an expansive sample frame, they found that many of the projects tended to focus on temporal conditions during an experience, rather than the impact that might accrue over time. Together, the two studies demonstrated a lack of statistical rigor, compounded by variation in cultural conditions, complexity, and specificity that challenge any synthesis or claims about the form.

These two studies' findings did suggest that museum visiting has long-term impacts. The research demonstrates that museum experiences are memorable as significant life events[5] and have a mnemonic function in memory reconstruction.[6,7] But, museums also face challenges with equal representation in exhibition display, staffing, and visitors[8–10] and as both studies acknowledged, participant self-selection may exclude some voices.

Knowledge production

Methods are a purposeful choice, driven by research aims rather than any need to justify the museum sector's work. Research and evaluation operate at the intersection of understanding specific conditions for monitoring or improvement, and the efforts to generalize knowledge based on aggregate learning. When research or evaluation is undertaken to support improvement, design, change, or manage risk, they fulfill their goals. But when considered as a body of knowledge, we must acknowledge that methods preference some truths while suppressing awareness of others.

Museum studies, like many other fields, rely extensively on case studies, comparative assessments, and non-randomized studies. Using a quantitative approach can explain how common something is across a population, or to what extent something is occurring. Qualitative approaches give us a better understanding of what happens and why, often reflecting the deeper and unique lived experience of individuals, but frequently without the ability to make broad claims about representativeness.

Methodologists advancing mixed methods theory, the use of qualitative and quantitative methods in combination, reveal a series of crises that challenge claims. They have demonstrated a crisis of validity because of a dichotomization of samples as either qualitative or quantitative, leading to reductive assessments that can be misleading. They describe these challenges as self-limiting crises of representation, legitimation, integration, and epistemic politics.[11]

Museum studies is also a victim of non-representativeness in attendance exacerbated by opt-out bias, and the challenge of integration when qualitative and quantitative methodologies are used in isolation, the relative affluence of museum visitors, and historical defaults to heteronormativity.[12–15] These conditions lead to challenges when researchers focus on a preconceived normative visitor when attempting to address a community level social issue.

Common methodologies and sample sizes in the data

We identified 127 separate studies that reported sample size in 71 papers (Figure 30.1). We consider a sample size to be small if the margin of error in statistical tests would not generalize beyond 5,000 people or effectively

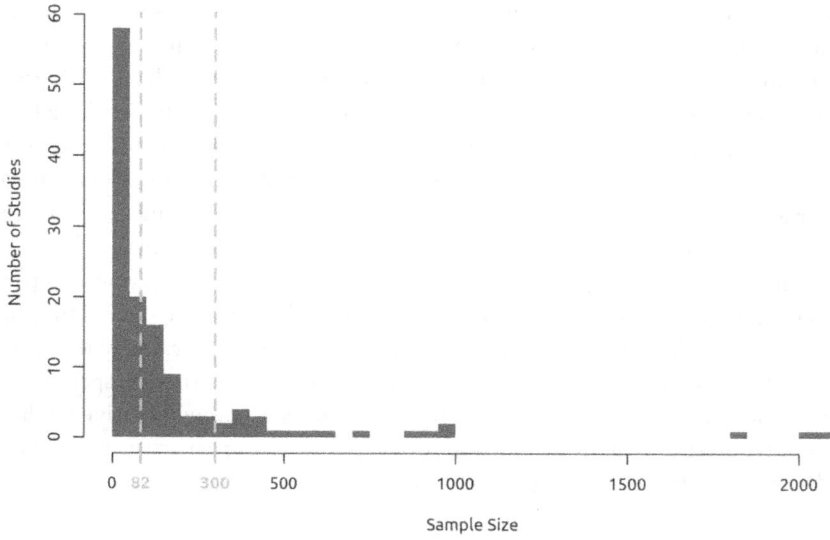

FIGURE 30.1 Sample sizes compared to the number of studies relating to social issues and STEM in museums, with break-points at 82 participants, suitable for two-tailed tests of significance, and 300 participants, where margin of error may be large enough to make our inventory.

$N = 300$. We found that median sample sizes in peer review were $N = 116$, theses and dissertations $N = 28.6$, and grey literature hovering near $N = 82$, the minimum required for a two-tailed test. We conclude that studies focusing on STEM and social issues in museums relied on non-representative, convenience sampling with responses from samples too small for quantitative analysis, or sample sizes that prevent validity testing. There were a few notable exceptions where large sample sizes allowed a robust statistical analysis of the findings which we explore in the Synthesis section below. Across our review, the majority garnered at least 30 responses, considered the upper limit of the minimum range recommended for experimental testing[16] or grounded theory[17] but below the minimums recommended by methodologists for correlational one-tailed testing (64 participants), two-tailed testing (82 participants), or causal-comparative study (51 participants). That is to say, the topic of STEM and social issues was most commonly represented by qualitative research, even when researchers were reporting quantitative data.

Sampling strategies

Studies that use truly randomized sampling strategies enable researchers to more easily conduct meta-analysis because they can be combined to provide a more generalizable understanding of a phenomenon. Studies in our dataset tended to focus on surveys, many with small convenience samples collected at a single institution or a small set of institutions. Criterion or convenience sampling seemed to be the most widely used with visitor audiences. A great many articles did not describe a sampling strategy, so our analysis inferred the methodology from how the data were described. While criterion and convenience were the most frequently employed, we also found explorations of variation, including stratified, representative, purposive, extreme case, critical case, snowball, maximum variation, or the use of key informants.

In a few notable cases, deliberate criterion selection or invitation of a participant or participant group based on a specified characteristic offered insights into some social issues. These choices may offer explanatory value when interpreting findings in larger quantitative studies because the criterion specifically illustrates minority concerns that might be concealed in quantitative studies. Unfortunately, without a companion quantitative representative study, these data remain limited in their utility to the museum phenomenon itself.

Analytic approaches

Some papers in our dataset were clearly grounded in methodological theory, including detail about the steps involved and their decision-making. This clarity was more common in evaluations, theses, and dissertations, while peer review was notably lacking detail. The most frequently reported techniques were descriptive statistical measures, such as counts, percentages, and averages.

Analyses tended to be either quantitative or qualitative, with few references to triangulation between types of data. The qualitative analysis consisted of coding with predefined codes, pattern emergence, *post hoc* code books, inductive coding, or thematic coding. In most cases, projects involving more than one coder included data did report inter-coder reliability and the use of well-known software. Only 31 papers had samples that were large enough to be used in a synthesis study, and few offered benchmarking useful for connecting to larger studies.

We also uncovered some serious concerns in the studies involving STEM and social issues that may extend into the larger museum studies field. We observed studies that employ data summaries and tools without describing how that analytic approach was appropriate to the data. We found calculations of mean and standard deviation or statistical comparison of paired ratings for sample sizes that fall below recommended minimums, or reporting proportions without N values, making it hard to know whether the results were statistically significant. We commend the work by Fife and others[18] who employed strategies to resolve the majority of statistical traps researchers may fall into, but note that our data included quite a few examples that replicate misconceptions or lack of training in applying statistical methods for small samples.

Synthesis

Even though we uncovered methodological issues, we observed a handful of projects that stand as methodological models, and may offer some generalized understanding of the affordances of museums. These models illustrate some of the latent potentials for museums to contribute to public understanding, attitudes, and behavior. They also offer some strategies for addressing methodological critiques revolving around representation, legitimation, integration, and politics.

Representation: Museums, viruses, and vaccines

Papers by Diamond and various colleagues[19,20] described a three-phase sequential mixed methods research strategy to identify the root causes of public debate or confusion about viruses and vaccines and to test interventions to address these gaps. The project exemplifies an effective approach to addressing representation because the team gathered data that reflected beliefs and then compared specific populations.

They presented a foundational descriptive quantitative study conducted through a set of questions added to a biannual state survey of residents to establish deficits and a rationale for their intervention with specific target audiences. The second phase employed stratified, purposeful, and semi-structured interviews of key informants, followed by a third phase randomized comparison test of their intervention with their target audiences.

These authors demonstrated that museums can identify a social issue that requires STEM reasoning to understand and make decisions about, critical audiences for that intervention, and testing of specific tools to address that deficit. Their approach led to an intervention useful for a general population and avoided the self-selection crisis.

We conclude that museums are well equipped to create learning experiences that can build essential STEM literacies to help a population navigate a social issue. Their methods address the challenge of representativeness by expanding their source data beyond visitor populations. By sampling in controlled settings and using key informants, they were able to overcome potential bias found in relying on visitors alone for data. These studies also demonstrated that museums represent a node in the social process of learning, that complements other settings and materials like formal school, comic books, and public media.

Legitimation: Participatory research, reflexivity, and controversy

Ostman, Zirulnik, and McCullough Cosgrove[21] undertook one of the few studies that anchored itself in the general movement to address societal issues at museums. Their study of museums' use of controversial topics affords an examination of the legitimation challenge – specifically, single methods studies that attempt to represent the diversity of audiences. They used the apparent conflicts that emerge between religious theologies and science enterprise as their prime focus, with the intention of synthesizing claims about museum social experiences and how STEM content can align with audiences' needs. More importantly, they recognized that analyzing a sample of projects around a controversial topic would offer broad theory detection for museum practice.

They noted that despite the promise of museums to provide integrative experiences that bring STEM to social issues, the literature was without compelling evidence beyond small topical studies that we reaffirmed in this study. Therefore, they employed participatory research methods, a strategy that acknowledges researchers own experience and knowledge as central to theory detection, a constructivist acknowledgement of the presence of cultural histories, potential biases, and values of the museum staff as part of the knowledge construction, and the importance a reflexive process to theory construction.

Mixed-methods research theorists assert that the legitimation crisis results from any research paradigm taking precedence over another as a path to truth. Through five unique interventions, they developed a quantitative dataset triangulated to qualitative reporting from participants, program leaders, and the research team. Their instruments focused on the construct level they labeled engagement, rather than any focus on specific content learned. This approach enabled them to compare across subject matter to build generalizable knowledge that could speak more generally to a phenomenon. Although the final dataset ($N = 315$) offered a limited degree of certainty and was weighted toward two of the five settings, the model remains instructive. It revealed a set of critical practices that support using social issues as a motivational tool for engaging with STEM content in a museum setting. In particular, they identify four principles for museum experience, which represent foundational constructs for future research that we paraphrase as:

- Neutral space, through creating a space for users; personal reflection, idea-sharing ideas, and encouraging open-minded attention to the perspectives of others
- Presenting visitor's contrasting perspectives to aid affective visitor experiences
- Creating space for personal reflection and documentation to enhance learning outcomes
- Providing opportunities for engagement beyond the experience

The affordances of setting would likely vary based on the content area being pursued. But more importantly, they addressed the legitimation concern by describing their findings as tentative, and calling for future exploration to challenge their theory in service of advancing generalizable knowledge.

From a methodological perspective, Ostman, Zirulnik, and Cosgrove demonstrated that STEM and social issues in the museum sector will likely continue to be in the theory development phase, relying on more qualitative approaches, and noting that theory requires sufficient agreement on constructs across topics and types before it can be pursued through large-scale quantitative study. Their work also offered a useful critique of the qualitative bent in museum studies, and the over-reliance on evaluation reports reporting frequencies and percent change for small populations irrespective of significance.

Given both the case study implications of Ostman and colleagues' work and our findings, we raise a concern about the post-positivist preference for numbers that may not have meaning.[22] The acknowledgment of the early phase of this field suggests that the interpretivist and critical theory paradigms may continue to be more useful until constructs related to social impact are broad enough to explore the principle of social impact that museums aspire to achieve.

Integration: Triangulating between knowledge worlds

A challenge with literature synthesis lies in the possibility of triangulating, comparing, expanding, or consolidating quantitative data when the source material is small and purpose-driven. Leblang and Osche[23] evaluated a multi-site film, lecture, and discussion program, employing parallel measures to understand how the intervention impacted individuals in different roles and settings. They gathered data from the public, students, and teachers at five sites, supplementing this work with online qualitative studies, and interviews with the program leaders.

Their methods demonstrated that visitors tend to be equipped with knowledge about the social issue they will encounter, and use museum assets to strengthen these users' understanding. They demonstrated that 80% of program participants expanded their knowledge of a topic, and most claimed they could apply their learning to different contexts and concerns. The integrative approach revealed a tension important for all questions of how STEM content and contentious social issues might be well suited to museum. Furthermore, their results demonstrate that small samples can be used to explore the larger thematic questions of museum affordances and opportunities if used in mixed-methods approaches.

Discussion

We found that the motivation for exploring informal STEM learning about social issues at museums remains an emerging field, relying heavily on qualitative inquiry. Unfortunately, there appears to be little cross-referencing between sources, creating a risk of under-representation of the full breadth of evidence that can demonstrate the value of museums bringing STEM to social issues, or where foundational precedent may be found.

The lack of cross-referencing creates a challenge for researchers, evaluators, and research consumers. For researchers, it suggests that each study exists in autonomy and undermines the scientific principle of building on prior knowledge through the use of frameworks and precedent research. For evaluators treating each project de novo is expensive, time-consuming, and may be less cost-effective than building on prior knowledge or triangulating data to precedent. For evaluation consumers, it suggests that synthesis and benchmarking before commissioning new studies can offer a more focused use of resources.

We acknowledge that this study was possible because self-published evaluation reports generously offered the most detailed information on methodology and more frequently illustrated the complexity of a mixed-methods study, while the peer-reviewed literature tended to focus on a single question or subset of data without the value that comes from triangulation between methods. This review suggests that peer-review journal editors should encourage the publication of methods to support the advancement of synthesis and reduced redundancy.

Theses and dissertations were more likely to acknowledge the positionality of the researcher, while peer-reviewed literature adhered more closely to a post-positivist paradigm that may not fully represent the impact of an intervention. More concerning, the noted absence of thesis and dissertation research from the references in the evaluation and peer-reviewed literature suggests a blind spot that could fill gaps in knowledge.

We were surprised by the widespread lack of validated and published instruments for research and evaluation. The tendency to construct targeted instruments for each study without reference to the literature or triangulation with other datasets related to social issues appears to be a deficit in the field. It is impossible to assess broad or generalizable findings.

We were similarly concerned about a sampling fallacy that seems to pervade descriptions of museum populations without caveats or benchmarking. It is well known that museum visitors cannot be considered representative of a general population. Without refusal rate reporting and remediation strategies for non-participants, "random" sampling may produce a false positive affirming the bias of museum professionals that an intervention is achieving desired goals. Although many articles acknowledge these challenges, most reports tended to claim representativeness of the sample to their attendees based on specious indicators like sex or age, a practice that can confound explorations of impact because aggregating data containing errors in source data can report on a majority of the majority, exacerbating exclusion.

Conclusion

This exploration of methods used for the study of STEM and social issues in museums reveals an overweighting toward case studies that should limit claims about the role of museums' ability to link STEM content with social issues.

This challenge can lead to a misrepresentation of the full depth and value of the museum sector as it supports its service populations' engagement with social issues. Our results offer a cautionary lesson that museums should limit their claims to impact until there is sufficient evidence to stand behind. We recommend that more effort be given to anchoring new case studies to precedent, using validated scales to help build enough data to support future meta-analysis and paying attention to minoritized voices to ensure that we do not continue to preference the majority of the majority. Through minimal additional effort in reporting, it may be possible to demonstrate the aspirations of the museum sector and the social purposes at the heart of the ICOM 2022 definition.

Acknowledgments

This manuscript was based upon work supported by the National Science Foundation under Grant No. DRL-1906556. Any opinions, findings, and conclusions or recommendations expressed in this material are those of the author(s) and do not necessarily reflect the views of the National Science Foundation.

Notes

1 Morrissey, K., Fraser, J., Norlander, R., Nock, K., Ball, T., & Flinner, K. (2021). A Research Synthesis: Addressing Societal Challenges through STEM. Knology Publication #NSF.022.499.01a. Knology.

2 Institute of Museum and Library Services. (2014). *Government doubles official estimate: There are 35,000 active museums in the U.S.* https://www.imls.gov/news/government-doubles-official-estimate-there-are-35000-active-museums-us#:~:-text=The%20Institute%20of%20Museum%20and,of%2017%2C500%20from%20the%201990s.

3 Fu, A. C., Kannan, A., Shavelson, R. J., Peterson, L., & Kurpius, A. (2016). Room for rigor: Designs and methods in informal science education evaluation. *Visitor Studies, 19*(1), 12–38.

4 Morrissey, K., Petrie, K., Canning, K., Windleharth, T.W., & Montano, P. (2014) *Museums & social issues: A research synthesis of an emerging trend.* Available at http://www.informalscience.org/museums-social-issues-research-synthesis-emerging-trend.

5 Fraser, J. (2009). *An examination of environmental collective identity development across three life stages: The contribution of social public experiences at zoos.* Doctoral dissertation, Antioch University New England, 2009. Dissertation Abstracts International. Fraser, J., & Sickler, J. (2009). *Why zoos and aquariums matter: Handbook of research key findings and results from national audience surveys.* Association of Zoos and Aquariums.

6 Anderson, D., Shimizu, H., & Iwasaki, S. (2021). Recollections of who we were: Nostalgic retrospective perceptions of Japanese Society following a visit to a Shōwa-era Museum. *Curator: The Museum Journal, 64*(1), 17–40 and others by these authors.

7 Dean, D. (2013). Museums as sites for historical understanding, peace, and social justice: Views from Canada. *Peace and Conflict: Journal of Peace Psychology, 19*(4), 325.

8 National Research Council. (2009). *Learning science in informal environments: People, places, and pursuits.* National Academies Press.

9 Isselhardt, T., & Cross, L. (2020). You love them, but you don't know them: Recognizing and welcoming lived experiences. *Curator: The Museum Journal, 63*(4), 571–578.

10 Martinez, N. (2020). Increasing museum capacities for serving non-white audiences. *Curator: The Museum Journal, 63*(4), 585–590.

11 Collins, K. M., Onwuegbuzie, A. J., & Jiao, Q. G. (2007). A mixed methods investigation of mixed methods sampling designs in social and health science research. *Journal of Mixed Methods Research, 1*(3), 267–294.

12 Coffee, K. (2008). Cultural inclusion, exclusion, and the formative roles of museums. *Museum Management and Curatorship, 23*(3), 261–279.

13 DiMaggio, P. (1996). Are art-museum visitors different from other people? The relationship between attendance and social and political attitudes in the United States. *Poetics, 24*(2–4), 161–180.

14 Dawson, E. (2014). "Not designed for us": How science museums and science centers socially exclude low-income, minority ethnic groups. *Science Education, 98*(6), 981–1008.

15 Brow, K., & Buckner, J. (2020). The "Rich Gay"? Small museums & funding "Difficult" history. *Curator: The Museum Journal, 63*(4), 605–610.

16 Onwuegbuzie, A. J., Jiao, Q. G., & Bostick, S. L. (2004). *Library anxiety: Theory, research, and applications.* Lanham, MD: Scarecrow Press.

17 Creswell, J. W. (2002). *Educational research: Planning, conducting, and evaluating quantitative and qualitative research.* London: Pearson Education.

18 Fife, D. (2020). The eight steps of Data analysis: A graphical framework to promote sound statistical analysis. *Perspectives on Psychological Science, 15*(4) 1054–1075.

19 Diamond, J., Jee, B., Matuk, C., McQuillan, J., Spiegel, A. N., & Uttal, D. (2015). Museum monsters and victorious viruses: Improving public understanding of emerging biomedical research. *Curator: The Museum Journal, 58*(3), 299–311.

20 Diamond, J., McQuillan, J., Spiegel, A.N., Hill, P.W., Smith, R., West, J., & Wood, C. (2016). Viruses, vaccines, and the public. *Museums & Social Issues, 11*(1), 9–16.

21 Ostman, R., Zirulnik, M.L., & Cosgrove, J.M. (2019). Storytelling, science, and religion: Promoting reflection and conversation about societal issues. *Curator: The Museum Journal, 62*(2), 117–134.

22 Treagust, D. F., Won, M., & Duit, R. (2014). Paradigms in science education research. In N. G. Lederman & S. K. Abell (Eds.), *Handbook of research on science education.* Volume II. Abingdon, Oxfordshire: Routledge, pp. 3–17.

23 Leblang, J. & Osche, E. (2011). *Ice Planet Earth: Summative evaluation report.* Lesley University, Program Evaluation & Research Group. https://www.informalscience. org/ice-planet-earth-summative-evaluation-report

31

THE VIRTUOUS CYCLE OF AFFECT, ENGAGEMENT, AND LEARNING

Martin Storksdieck and Nancy Staus

What leads a person to get better at something, to improve understanding or knowledge, to refine skills or abilities? Scholars have pondered these questions for decades to better understand "how people learn."[1] Much of the conversation about learning, understandably, has been focused on formal education (K-16).[2-4] In formal education contexts learning outcomes mostly refer to knowledge, understanding, and abilities. Outcomes related to so-called 21st-century skills, dispositional outcomes related to interest, motivation, growth mindset, or self-efficacy, and behavioral outcomes related to engagement with learning experience are increasingly valued.[5] Consequently, educational programs, primarily those in out-of-school settings, address them, either in addition to cognitive learning outcomes or, less commonly, as replacement.[6]

The outcomes described above are not completely distinct concepts, but ones that overlap and interrelate in ways that are consequential to our understanding of how people learn, persist, or disengage from science-related activities or endeavors. For instance, exposure to and continued engagement in particular topics or disciplines might explain how interest develops over time, and interest itself can then motivate continued engagement and the development of strong self-efficacy and identity beliefs.[7] Indeed, the US National Academy of Sciences recognized six interwoven cognitive, affective, and behavioral strands that, together, describe the elements of successful informal science learning environments, suggesting an interaction between disposition, engagement, and learning without explicitly stating the nature of this interaction.[5]

Empirical research has documented some of the connection between self-efficacy, affect, engagement, and motivation in school and out-of-school settings.[8-10] For instance, feelings of pleasure were significant direct predictors

DOI: 10.4324/9781003145387-35

of short-term learning in an Informal Science Learning (ISL) context where greater levels of arousal were associated with successful learning outcomes.[11]

In this chapter we offer the concept of the *Virtuous Cycle of Affect, Engagement and Learning* as a means to connect factors that alone and in combination contribute to learning success and are themselves influenced by one another. The Virtuous Cycle recognizes how factors that support learning pathways over time can serve as "on-ramps," and lack of them might create "off-ramps" at any point in the learner's journey, while serving as mediating or contributing factors that move a person along a unique learning pathway. Building on the idea that learning is a lifelong process of connected experiences, some shallow and some rather deep and impactful,[4,5] we offer a holistic perspective on how key factors that support learning fit together and interact over time. Figure 31.1 describes the Virtuous Cycle. In this model of personal growth related to learning and engagement we posit that interest, motivation, self-efficacy, learning, and identity are all components of a constantly repeating cycle of reinforcing connections between cognition, affect, and behavior. At any one stage the cycle can get interrupted, ushering learners to an off-ramp for that particular learning pathway. Conversely, learners who complete many repetitions of the cycle can

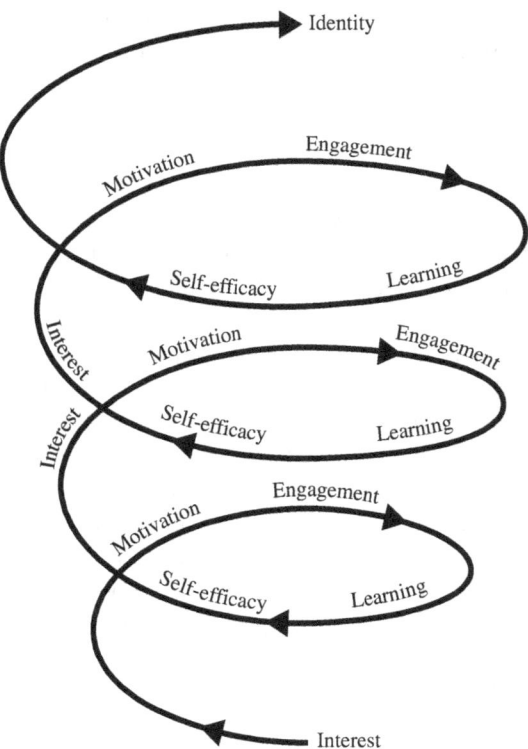

FIGURE 31.1 The reinforcing cycle of affect, engagement, and learning.

develop a sustained interest that may become part of a person's sense of self or their identity.

Illustrative examples of the Virtuous Cycle in action

The first example comes from the field of hobbyism. Multiple studies have investigated the culture, norms, and activities of amateur astronomy clubs in the intersection of hobbyism, volunteerism and group culture and the way in which they lead to astronomy expertise among amateur astronomers.[11–13] The research revealed that initial topical interest in astronomy, astrophotography, or telescope technology and the desire to share one's passion with like-minded individuals motivated many to join a club. Sustained participation in a club depended on a person's ability to find connections and purpose in the club and to get positive reinforcement on their hobby. An online knowledge test suggested that amateur astronomers who were active club members for at least three years and who were active in public outreach scored higher than faculty at a major US research university on a standard concept inventory test for astronomy.[14] In the context of the Cycle, this example can be understood in the following way: initial *interest* in astronomy or something related to it, such as photography, may lead to joining a club (on-ramp) where the interest is affirmed in a group of like-minded people, who are *motivated* to *engage* in their hobby and who challenge, correct, or teach each other, leading to greater *learning*. This system of positive reinforcement for "getting better at the hobby" (*self-efficacy*) leads to greater *interest* and *motivation* to continue to *engage* in the hobby.

The second example focuses on one adolescent youth with an interest in computer coding to show how the Cycle may be interrupted, leading learners to an off-ramp when they are unable to continue on their learning pathway.[15] Charlie's *interest* in computer coding was triggered at age 12 when he attended an after-school program called Pixel Arts Game Education (Pixel Arts) in which youth designed and created computer games. He attended the program for three years during which he *learned* a coding language called Scratch which he used to create video games collaboratively with other youth in the program. Charlie became very proficient in coding with Scratch, describing it as "a breeze" and rating his expertise in coding highly, indicating a high level of *self-efficacy* for this activity. As his knowledge, skill and confidence in coding grew, Charlie's *interest* in coding also evolved. He wanted to learn new coding languages such as JavaScript and Python which were not offered in Pixel Arts which *motivated* Charlie to look for other ways to *engage* in his interest. For example, Charlie watched videos on YouTube and searched for other coding classes, but neither he nor his parents were able to find relevant resources and opportunities in their community. [It did not help that his personal computer was broken for almost two years during which he was unable to pursue online coding options.] By age 16, Charlie's interest in coding was waning because of the lack of opportunities to engage in activities that would support this interest. Thus, Charlie's story illustrates how lack of *engagement* (options) can become a potential off-ramp from the Virtuous Cycle when access to activities of *interest* become limited or otherwise difficult to access.

Elements of the Virtuous Cycle

The Cycle consists of discrete cognitive, affective, and behavioral elements that, as we see it, interact in ways that shape the learning trajectory. Although the power of the Cycle as a conceptual model lies in the intersecting relationships between and among its components, we believe it is useful to also describe how we are conceptualizing individual components of the model. Here, we describe these components using examples from the above cases to further illustrate the entire system of internal and external factors that ultimately influence one's learning pathway.

Interest. In the context of informal STEM learning experiences which are usually chosen by the learner, interest is often (but not always) the on-ramp to the Cycle. Interest is a multi-dimensional variable consisting of both affective/emotional and cognitive components: early phases of interest are typically characterized by emotion, but later, more stable phases of interest are composed of emotion and a number of cognitive components including growing knowledge and increasing feelings of value for the topic, as well as greater self-efficacy and ability in association with the content.[16] That is, someone who is interested in astronomy will have positive feelings about astronomy (liking it, being fascinated by it, etc.) and will also have significant knowledge, value, and competence in the topic. It is becoming increasingly clear that interest is foundational for learning in all content areas in and out-of-school, regardless of the demographic characteristics of the learner.[17]

Interest development is often conceptualized as progressing in several phases, from early less-developed ones to more well-developed, enduring phases of interest. All interests begin as situational interests that are triggered by an event (or person) in the environment that captures the learner's attention. Many of these interests are quite transitory and dissipate quickly, but if the situational interest is maintained through repeated engagement in the activity of interest and through support from others, it may mature into a well-developed individual interest over time.[18] In the example of Charlie, his interest in computer coding was triggered when he attended an after-school program called Pixel Arts and later maintained through repeated engagement in the program over time.

However, interest is not static but rather constantly in flux. Therefore, the progression between phases of interest can be quite fluid with learners progressing to more well-developed phases or returning to earlier phases (or losing interest altogether) depending on factors such as how effectively the interest is supported through the efforts of others (e.g., adult/peer encouragement) and the availability of interest-related activities and opportunities that can be accessed by the learner.[19] Learning environments can facilitate the deepening of well-developed interest by providing numerous opportunities that challenge learners to build knowledge and competence.[7] However, this was not the case in Pixel Arts which taught the same curriculum each year and did not offer additional learning opportunities for repeat attendees, thus not serving Charlie's need for deeper engagement.

Motivation. In addition to its cognitive and affective components, interest is also a strong motivational variable directly linked to participation in STEM-related activities. That is, learners who are interested in topics related to STEM have a strong predisposition to seek out and engage with related content and activities both in- and out-of-school.[20] Self-determination theory distinguishes between two types of motivation: intrinsic motivation underlies activities that people do freely for personal reward or enjoyment, whereas extrinsically motivated behaviors are those that they feel coerced or pressured to do or for which they receive an external award such as money, grades, or competition ratings and rankings.[21] While interest has been found to be closely linked to the former, both extrinsic and intrinsic motivation are important elements in the Cycle. Especially in the case of younger learners, an external stimulus may be necessary in order to engage youth in experiences that may be of interest. Furthermore, earlier phases of interest are almost exclusively externally motivated and supported by others such as parents, peers, or teachers.[7]

Later phases of interest development are strongly associated with intrinsic motivation and self-regulated learning behaviors because the activity itself is rewarding. For example, Charlie was strongly motivated to re-engage with the Pixel Arts after-school program over several years because of his well-developed interest in coding, and despite the fact that his parents did not provide strong encouragement for this activity: "My family likes that I have something I'm interested in, but mostly they don't really encourage me to do it."[15]

Engagement. While some scholars use the terms engagement and motivation interchangeably,[22] we see engagement and motivation as distinct but related variables wherein motivation represents intention and engagement is related to cognitive and behavioral "action."[23] Engagement is essential for positive learning outcomes both in and out of school.[24] Like interest, engagement can be conceptualized as a multidimensional construct with behavioral, emotional, and cognitive dimensions that can be distinguished at different levels and time frames, from a short task engagement all the way to longer-term commitment to a topic or activity.[25] Behavioral engagement includes attendance and participation in learning activities and indicators of in-the-moment behavioral engagement, including displays of effort, persistence, and attention.[26] Emotional engagement includes learners' emotional reactions to topics or activities, development of a sense of identity or belonging and valuing of the learning experience. Cognitive engagement includes a willingness to participate in challenging tasks, use flexible problem-solving and self-regulation.[27] In general, youth who have more positive measures of engagement in these three dimensions have shown greater achievement in formal school environments. Although there has been little research on the effects of engagement in informal spaces, it is likely that engagement in this context will also lead to greater learning outcomes. Unfortunately, lack of engagement opportunities can ultimately derail the learning trajectories of even the most committed learners, as Charlie's example shows.

Learning. Definitions of learning can vary widely; we conceptualize learning as a process of constructing meaning based on one's past experiences, existing knowledge, interests, and motivations and the type and quality of the learning situation itself. Because everyone's mental foundations are unique to their life experiences, no two people will learn the same information in the same way.[28] Particularly in informal settings, learning can encompass procedural knowledge and skill development such as coding video games or setting up telescopes and finding objects in the night sky as learners engage in hands-on authentic activities. Such experiences can also lead to the learning of less tangible skills such as critical thinking, problem-solving or developing a keener and explicit awareness of oneself, i.e., identity development.[29]

As illustrated powerfully in the level of astronomy knowledge that club-based amateur astronomers who conduct public outreach achieve on a standard test, informal STEM learning is part of a continuous and cumulative process of learning that unfolds over time.[1] The principal benefits of informal STEM learning are the long-term and cumulative transformations that can occur in learners as they pursue a series of experiences across a diversity of learning contexts, including school, home, media, museums, and other designed spaces.[5] Collectively, these experiences and associated settings in which they occur comprise each person's (STEM) learning ecosystem which provides opportunities to acquire skills and knowledge throughout a lifetime.[30] Within the larger ecosystem, learners construct unique STEM learning pathways guided by their personal goals and interests, and mediated by the affective, cognitive, and behavioral elements illustrated in the Virtuous Cycle.[13]

Self-efficacy. At its most basic, self-efficacy is the belief that one is capable or competent in a certain skill or ability. Self-efficacy beliefs determine how people feel, think, motivate themselves, and behave. Learners with high levels of self-efficacy approach difficult tasks as challenges to be mastered rather than as threats to be avoided, set challenging goals and maintain a strong commitment to them, and quickly recover after failures or setbacks.[31] Generally, successful experiences increase self-efficacy beliefs, while experiences of failure lower them. However, too many early successes can actually decrease efficacy beliefs as the learner may be easily discouraged by later failure. Therefore, a resilient sense of efficacy requires experience in overcoming obstacles through persistent effort.[32] In other words, some setbacks and difficulties serve a useful purpose in teaching that success usually requires sustained effort.

Self-efficacy is strongly linked to the concept of interest: individuals with a well-developed individual interest are characterized as being able to persevere through frustration and challenge in order to meet their learning goals.[20] Thus, self-efficacy brings us full circle back to interest and the Virtuous Cycle continues unless or until a learner meets a challenge that ushers them toward an off-ramp.

The Virtuous Cycle as part of a learning trajectory

As illustrated in the examples of Charlie and the astronomy hobbyist, the Virtuous Cycle may repeat itself multiple times as learners engage in numerous activities related to their interest. During this time, elements of the Cycle – interest, motivation, engagement, learning, and self-efficacy – are reinforced in ways that may be consequential to one's learning trajectory. For example, a situated interest in coding could become a sustained individual interest characterized by self-regulated learning, increased knowledge about different coding languages, and a high level of self-efficacy at writing code. If the Cycle continues to repeat itself in positive ways, interest and engagement in the activity could become a part of a learner's permanent sense of self; that is, they may develop a strong science or STEM identity.

Many of the elements of the Virtuous Cycle are strongly influenced by the social context of the learner which may include parents and family members, friends, teachers, and others who are important for the development of the learning trajectory. For example, parental encouragement has been shown to play a strong role in supporting youth interest and engagement in STEM.[33-35] In addition, there is a strong social component in the construction of identity: in order to sustain a strong science identity, one must not only feel like a "science person" but must also be recognized as a "science person" by meaningful others within the learning environment.[36] Thus, each learner's unique social context will also affect how they experience the Virtuous Cycle. For example, the son of one of the authors (MS) developed a strong interest in dinosaurs at the age of three and engaged in dinosaur-related activities for the next four years during which his knowledge of dinosaurs grew rapidly, as did his identity as a future paleontologist. However, when his social focus shifted from his parents to same-aged friends who did not share his interest, his engagement with dinosaurs ended, and his knowledge quickly faded.

Conclusions

Informal STEM learning is often conceptualized as a linear process in which people become interested in a topic or activity and seek opportunities outside of school in which to engage. In contrast, the *Virtuous Cycle of Affect,*

Engagement and Learning interprets informal learning as a cyclical process, consisting of numerous cognitive and affective elements that may influence whether a learner remains interested and engaged in STEM or disengages and breaks the cycle. The Cycle illustrates that there are numerous on-ramps to STEM learning including but not limited to interest in a STEM topic or activity, as well as off-ramps that end the cycle, such as learners being unable to find the "next thing" that supports their interest and engagement or a change in the social context that leads a learner to align interest and engagement with new friends.

As with all models, the Virtuous Cycle has several limitations. First, although the model acknowledges the socio-cultural context within which learning occurs, it does not explicitly account for all the social aspects that shape a person's engagement in learning. Secondly, the model is conceptual in nature; that is, we recognize the complexity of the learning process and acknowledge that the relationships depicted in the Cycle could be expressed and visualized quite differently. We also recognize that the model is not designed to be predictive, but rather may serve as an analytical tool to help understand the learning process as it unfolds over time, which may suggest interventions that help learners successfully navigate the Cycle.

By attending to the numerous on- and off-ramps associated with STEM learning, ISL educators and researchers may be able to design more successful learning ecosystems that support STEM interest and engagement for more youth and adults alike. For example, Charlie's interest in coding decreased due to lack of engagement opportunities suggesting that his personal STEM ecosystem required more opportunities and resources related to technology in order to support his learning. In other cases, low self-efficacy may be the off-ramp, suggesting more options for low-pressure engagement may be needed to avoid frustration and low self-esteem. And sometimes the rise of other interests (rather than declining interest in a topic) may be the off-ramps, and we may have to accept that learners legitimately change course, choosing new adventures that better fit their personal and social circumstances. Understanding why and how different learners enter and exit the Cycle could inform the design of more customized ISL programs that better support STEM learning pathways over time, such as providing ways to strengthen motivational resilience when learners hit a snag or facilitating better communication among STEM providers so they can direct learners to other resources of interest.[13]

While there is no perfect representation of STEM learning and engagement, we are suggesting that the Virtuous Cycle is valuable for raising awareness of the entire "system" of internal and external factors that drive whether someone engages or disengages. In that sense, our model is meant to provide a framework for thinking about the ways in which learning trajectories develop. While the underlying model of the Cycle may not be perfect, especially when analyzed with deep scrutiny, we hope it might nonetheless be useful for stimulating discussion and further research.

Notes

1 National Academies of Sciences, Engineering, and Medicine (2018). *How People Learn II: Learners, Contexts, and Cultures.* The National Academies Press.
2 National Academies of Sciences, Engineering, and Medicine (2019). *Science and Engineering for Grades 6-12: Investigation and Design at the Center.* The National Academies Press.
3 National Academies of Sciences, Engineering, and Medicine (2022). *Science and Engineering in Preschool Through Elementary Grades: The Brilliance of Children and the Strengths of Educators.* The National Academies Press. https://doi.org/10.17226/26215.
4 National Research Council (2015). *Reaching Students: What Research Says About Effective Instruction in Undergraduate Science and Engineering.* The National Academies Press.
5 National Research Council (2009). *Learning Science in Informal Environments.* National Academies Press.
6 National Research Council (2015). *Identifying and Supporting Productive STEM Programs in Out-of-school Settings.* National Academies Press.
7 Hidi, S., & Renninger, K. A. (2006). The four-phase model of interest development. *Educational Psychologist, 41*(2), 113–127. Renninger, K. A., & Hidi, S. E. (2022). Interest development, self-related information processing, and practice. *Theory Into Practice, 61*(1), 23–34. https://doi.org/10.1080/00405841.2021.1932159.
8 Ben-Eliyahu, A., Moore, D., Dorph, R., & Schunn, C. D. (2018). Investigating the multidimensionality of engagement: Affective, behavioral, and cognitive engagement across science activities and contexts. *Contemporary Educational Psychology, 53*, 87–105. https://doi.org/10.1016/j.cedpsych.2018.01.002.
9 Price, C. A., Greenslit, J. N., Applebaum, L., Harris, N., Segovia, G., Quinn, K. A., & Krogh-Jespersen, S. (2021): Awe & memories of learning in science and art museums. *Visitor Studies.* https://doi.org/10.1080/10645578.2021.1907152.
10 Staus, N. L., & Falk, J. H. (2017). The role of emotion in informal science learning: Testing an exploratory model. *Mind, Brain, and Education, 11*(2), 45-53.
11 Storksdieck, M., & Berendsen, M. (2007). Attributes and practices of amateur astronomers who engage in education and public outreach. In Gibbs, M.G., Berendsen, M., & Storksdieck, M. (Eds.). *Science Educators under the Stars: Amateur Astronomers Engaged in Education and Public Outreach* (pp. 31–42). Astronomical Society of the Pacific.
12 Storksdieck, M., Stein, J. K., & Jones, E. C. (2012). Hobbyists in the Role of Environmental Educator: The Case of Amateur Astronomy Clubs. Chapter 5 in Kopnina, H. (Ed.). *Anthropology of Environmental Education.* Nova Science Publishers.
13 Yocco, V., Jones, E. C., & Storksdieck, M. (2012). Factors contributing to amateur astronomers' involvement in education and public outreach. *Astronomy Education Review, 11*(1). http://dx.doi.org/10.3847/AER2011040.
14 Berendsen, M., & Storksdieck, M. (2007). Knowledge of astronomy among amateur astronomers. In Gibbs, M.G.; Berendsen, M. & Storksdieck, M. (Eds.). *Science Educators under the Stars: Amateur Astronomers Engaged in Education and Public Outreach* (pp. 43–56). Astronomical Society of the Pacific.
15 Shaby, N., Staus, N. L., Dierking, L., & Falk, J. (2021). Pathways of interest and participation: How STEM-interested youth navigate a learning ecosystem. *Science Education, 105*(4), 628–652.
16 Renninger, K. A., & Hidi, S. E. (2016). *The Power of Interest for Motivation and Engagement.* Routledge.

17 Ito, M., Martin, C., Rafalow, M., Tekinbas, K. S., Wortman, A., & Pfister, R. C. (2019). Online affinity networks as contexts for connected learning. In K. A. Renninger & S. E. Hidi (Eds.), *The Cambridge Handbook of Motivation and Learning* (pp. 291–311). Cambridge University Press.

18 Renninger, K. A. & Hidi, S. E. (2011). Revisiting the conceptualization, measurement, and generation of interest. *Educational Psychologist, 46*(3), 168–184.

19 Renninger, K. A., & Su, S. (2012). Interest and its development. In R. Ryan (Ed.). *The Oxford Handbook of Human Motivation* (pp. 167–187). Oxford University Press.

20 Renninger, K. A. & Riley, K. R. (2013). Interest, cognition, and the case of L- and science. In S. Kreitler (Ed.). *Cognition and Motivation: Forging an Interdisciplinary Perspective* (pp. 352–382). Cambridge University Press.

21 Deci, E. (1992). The relation of interest to the motivation of behavior: A self-determination theory perspective. In K.A. Renninger, S. Hidi, & A. Krapp (Eds.). *The Role of Interest in Learning and Development* (pp. 43–70). Lawrence Earlbaum Associates, Inc.

22 National Research Council and the Institute of Medicine (2004). *Engaging Schools: Fostering High School Students' Motivation to Learn.* The National Academies Press.

23 Azevedo, R. (2015). Defining and measuring engagement and learning in Science: Conceptual, theoretical, methodological, and analytical issues. *Educational Psychologist, 50*(1): 84–94.

24 Finn, J. D. & Zimmer, K. S. (2012). Student engagement: What is it? Why does it matter? In S. L. Christenson et al. (Eds.). *Handbook of Research on Student Engagement* (pp. 3–20). Springer.

25 Fredricks, J. A., Blumenfeld, P. C. & Paris, A. H. (2004). School engagement: Potential of the concept, state of the evidence. *Review of Educational Research, 74,* 59–109.

26 Renninger, K. A., & Bachrach, J. E. (2015). Studying Triggers for Interest and Engagement Using Observational Methods. *Educational Psychologist, 50*(1), 58–69. http://dx.doi.org/10.1080/00461520.2014.999920.

27 Sinatra, G. M., Heddy, B. C., & Lombardi, D. (2015). The challenges of defining and measuring student engagement in science. *Educational Psychologist, 50*(1), 1–13.

28 Fosnot, C. T., & Perry, R. S. (2005) Constructivism: A psychological theory of learning. In C. T. Fosnot (Ed.), *Constructivism: Theory, perspective and practice,* 2nd ed. Teachers College Press.

29 Young, J. R., Ortiz, N. A., & Young, J. L. (2017). STEMulating interest: A meta-analysis of the effects of out-of-school time on student STEM interest. *International Journal of Education in Mathematics, Science and Technology, 5*(1), 62–74.

30 Traphagen, K., & Traill, S. (2014). *How Cross-sector Collaborations Are Advancing STEM Learning.* Noyce Foundation.

31 Bandura, A. (1994). Self-efficacy. In V. S. Ramachaudran (Ed.), *Encyclopedia of Human Behavior* (Vol. 4, pp. 71–81). New York: Academic Press.

32 Bandura, A. (1997). *Self-efficacy: The Exercise of Control.* Freeman.

33 Archer, L., DeWitt, J., Osborne, J., Dillon, J., Willis, B., & Wong, B. (2012). Science aspirations, capital, and family habitus: How families shape children's engagement and identification with science. *American Educational Research Journal, 49*(5), 881–908.

34 Harackiewicz, J. M., Rozek, C. S., Hulleman, C. S., & Hyde, J. S. (2012). Helping parents to motivate adolescents in mathematics and science: An experimental test of a utility-value intervention. *Psychological Science, 23*(8), 899–906.

35 Staus, N. L., Falk, J. H., Penuel, W., Dierking, L., Wyld, J., & Bailey, D. (2020). Interested, disinterested, or neutral: Exploring STEM interest profiles and pathways in a low-income urban community. *EURASIA Journal of Mathematics, Science and Technology Education,* 16.

36 Carlone, H. B., & Johnson, A. (2007). Understanding the science experiences of successful women of color: Science identity as an analytic lens. *Journal of Research in Science Teaching,* 44(8), 1187–1218.

32

DATA NARRATIVES FOR ACTION

Innovative Approaches to Data Collection and Reporting to Tell the Story of Informal Science Learning

Gil Noam, Dawn McDaniel, Bailey Triggs, and Abby Bergman

With three years of pandemic-related school disruptions, there is increased pressure on out-of-school time (OST) programs to leverage data to address the impacts of lost instructional time, particularly in the subject areas of science, technology, engineering, and math (STEM). OST programs are critical to filling these gaps in STEM learning as well as creating opportunities for youth harmed by the national racial achievement gap in STEM (e.g., racial/ethnic minorities, females, and children from low socioeconomic neighborhoods).[1] More than ever, it is important to identify the characteristics of OST STEM programming that are most likely to help youth overcome the barriers to access and address issues of learning loss driven by the pandemic. We need to take innovative approaches to how we help STEM OST programs collect, interpret, and share data to inform quality improvement decisions that lead youth to deeper interest and lifelong engagement in STEM fields.

Though OST program leaders, staff, and funders see the value of using data to inform decisions and strengthen their reach and quality, there is a misalignment between the high pressure to provide data and programs' low capacity for data collection and reporting. Despite the challenges, many programs are prioritizing data collection and reporting with the help of supportive partnerships in various industries. These partnerships have resulted in improvements in data collection, reporting, and evaluation that have made OST STEM programs an exemplar for the general out-school-time sector. Further, as these interorganizational relationships grow, the rigor and scale of the evaluations have grown with them. For instance, large-scale evaluations examining system-building efforts in STEM quality and youth outcomes in afterschool have been carried out in over 11 states.[2]

DOI: 10.4324/9781003145387-36

Partnerships in Education and Resilience (PEAR), incubated at Harvard, Mass General Brigham, and McLean Hospital, the main psychiatry teaching hospital of Harvard Medical School before becoming an independent organization in 2020, is leading a new approach to evaluation in OST STEM. Using a tripartite model of communicating data findings, PEAR uses the same aggregate data for three distinct purposes: program management and improvement, systematic program evaluation, and policy initiatives. This new system, called Data Narratives for Action (DNA), is designed for OST program evaluation. It has been tested and implemented with several PEAR partners, including large national youth-serving organizations and STEM ecosystems.

The DNA approach was built from PEAR's field-building efforts to study the efficacy of STEM programming. DNA represents a huge step forward for OST STEM programs in embracing data systems to quantify and monitor STEM program quality and outcomes. Integrating quantitative data enables programmatic successes to be more persuasively communicated to a wider audience, particularly funders and policymakers. Many stakeholders want to better understand the program models, practices, and characteristics that yield the greatest outcomes for students. DNA creates a data system that is simultaneously evaluative, immediate, usable for management purposes, and focused on data. It is both relevant for policymakers and useable to researchers alike.

DNA's data collection and reporting system

Beginning in 2014, with the support of the Noyce Foundation, PEAR developed a data system that has grown over time and is now used across the United States and internationally to collect data on student social emotional well-being, STEM engagement, educator facilitation experience, and observations of program quality. This system helps programs and schools across the country connect research to practice and speak a common language around data for continuous improvement. The assessments developed as part of this data system include the Common Instrument Suite (CIS), a validated measure of STEM engagement and social emotional/21st-century skills,[3] and the Holistic Student Assessment (HSA), a validated screening tool to understand the social emotional strengths and challenges of youth.[4] In addition to the student versions of each assessment, educator versions have been created that offer an additional perspective on the program or schools' experience to be incorporated into the data system. With National Science Foundation (NSF) funds, PEAR developed the Dimensions of Success (DoS), a STEM program observation tool and associated training materials to certify observers across the country. All three perspectives: student, educator, and observer are integrated into dynamic dashboards (see Figure 32.1).

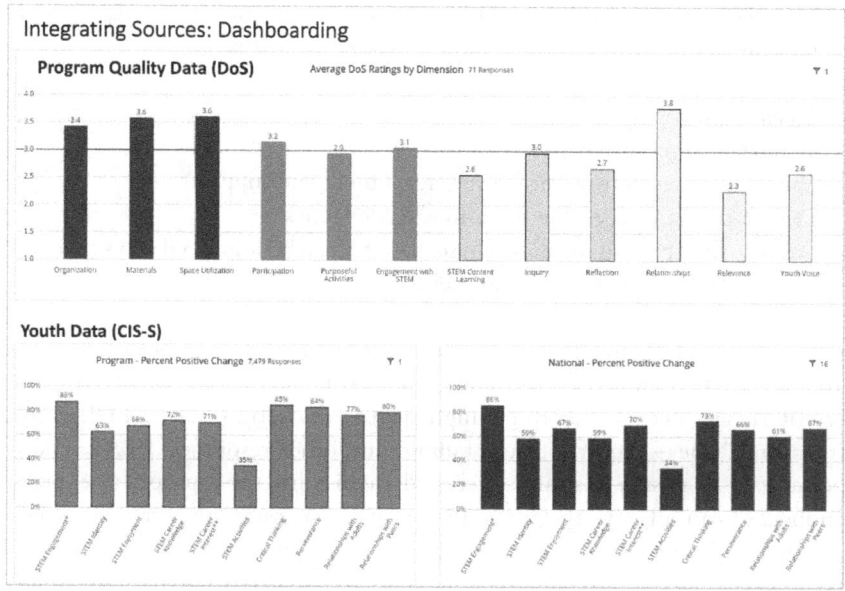

FIGURE 32.1   Interactive dashboard that integrates program observation data, program youth responses, and national youth responses.

Over the past decade, PEAR has used these tools to help partners improve program quality and youth outcomes across the country. With common measures at the foundation, norms have been created as an important method for easy comparisons. The volume of PEAR's data collection has also contributed to the success: PEAR's database includes over 300,000 students, educators, and parents, with student voice data comprising a considerable majority of responses. National norms are tied to the census and allow for comparisons across communities. Additionally, the investment included the buildout of a data visualization system using the widely used Qualtrics system. Qualtrics is compatible with most existing systems, and its results are highly useable for practitioners, evaluators, and policymakers. This system provides streamlined data collection and rapid reporting with easy-to-access findings that can be incorporated into materials for funding, reporting, and policymaking.

The data narratives for action (DNA) process

Building on a success story model that describes a proven solution to a challenge, DNA reports illustrate an evidence-based story of a program's achievements. In a four- to five-page brief, the reports integrate data collection, reporting, and interpretation in an easily digestible format. Integrated with

PEAR's data system, the DNA report merges data science, storytelling, and visualization. PEAR prides itself on producing reports that are:

Tailored: they cater to the needs and interests of a particular program and/or target group

Compelling: they activate program engagement and support

Concise: they are accessible to various reader types

Evolving: they present the program such that it can be improved and scaled

Empirical: they center on data at their core

In identifying a potential partner's readiness to undertake the DNA evaluation process, PEAR performs an informal assessment to verify if the DNA suitability criteria are met. Conducting a high-quality evaluation – even a relatively streamlined one – requires time, staffing, and other resources. PEAR takes special care to ensure that any interested partner has set appropriate expectations and can allocate the necessary resources to carry out the evaluation tasks that fall under their purview. The ideal evaluation partner is seasoned, accountable, engaged, collaborative, flexible, focused, and able to contribute. Nonetheless, PEAR can work with partners across the readiness spectrum to adapt the model based on their current needs and resources.

To capture a program's unique story of success, the DNA evaluation model is comprised of a menu of service options from which partners can select what is relevant to them given their evaluation goals, capacity, and the developmental stage of their program. When a partner expresses interest in partnering with PEAR to conduct an evaluation, PEAR works with them to collaboratively design an evaluation plan that aligns with their unique needs and capabilities. Designing an evaluation involves understanding a partner's reporting needs – clarifying for whom and for what purposes they need to communicate and contextualize their use of data. The PEAR team determines if the aims of the evaluation include:

A summative focus: such as an annual program summary for stakeholders and funders

Performance monitoring: such as the communication of program quality

A formative emphasis: such as highlighting a specific or new program's results

Regardless of the individual partner's evaluation purpose, PEAR's general DNA report describes the program in a way that builds an evidence base, articulates the quantitative aspects of its impact, and provides research that rationalizes its approach. Consequently, PEAR's DNA report can be adapted to varied partner needs while supporting a more global understanding and use of data to tell a program's story. A great deal of communication, collaboration, and intentional design is required to produce a captivating data narrative – and this planning phase is the beginning of a multistep process (see Table 32.1).

TABLE 32.1 DNA Development Process

1. **Planning Phase**		
Project Design	•	Report purpose
	•	Audience identification
	•	Tool selection
	•	Ongoing guidance
2. **Data Phase**		
Collection	•	Student (CIS-S) and educator (CIS-S) self-reports and program quality observations (DoS)
	•	Program and demographic data
	•	Assessment Tools in Informal Science (ATIS) tools
	•	Document archive review
Reporting	•	Data dashboard
	•	Statistical analysis
	•	Visualizations
Contextualization	•	Program background
	•	Existent research and rationale
	•	Data interpretation and implications of findings
3. **Writing Phase**		
Synthesis	•	Integration of data sources and other resources
	•	Collaborative revision process

The versatility of the DNA model means that it can communicate OST STEM impacts at various levels, spanning from an individual site to an entire organization, network, or ecosystem. Its flexibility gives partners informed agency over many evaluation decisions. The model's adaptability enables an evaluation to be performed that captures the uniqueness of an individual program while ensuring that it remains comparable to others. With the support of the PEAR team, partners make selections at multiple points in the planning process to tailor their report based on their needs (see Table 32.2).

Data Collection Plan. The first partner-directed decision is in selecting the data collection tools. After being presented with information on the PEAR

TABLE 32.2 DNA Model Decision Points

General project components	Data sources	Reporting products
1. Data Collection Plan	1. Common Instrument Suite – Student Survey (CIS-S)	1. Data Dashboard
2. Communication Plan		2. Tailored Statistical Report
3. Success Story Workshop *(Optional)*	2. Common Instrument Suite – Educator Survey (CIS-E)	3. Data Narratives for Action Report
4. Data Debrief and Celebration *(Optional)*	3. Dimensions of Success (DoS) Observation Tool	
	4. Assessment Tools in Informal STEM (ATIS)	
	5. Program Archive Documents	

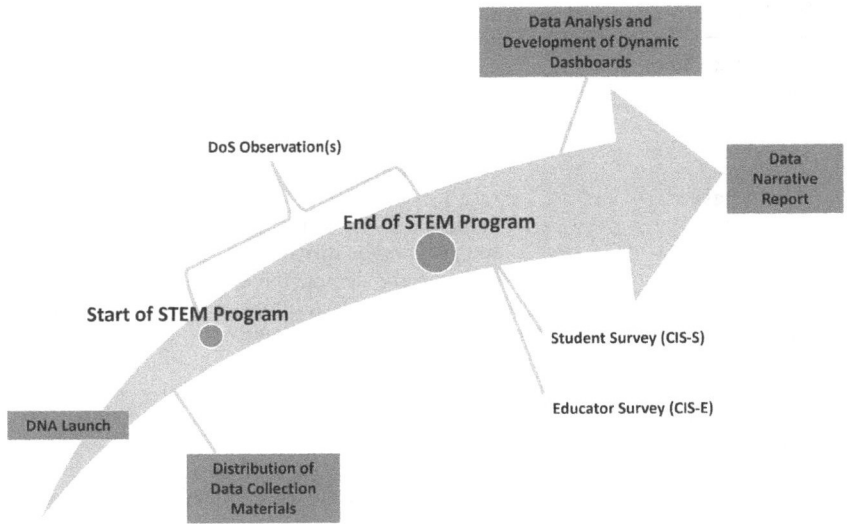

FIGURE 32.2 Dynamic DNA timeline aligns with partner's schedules and needs.

assessment tools integrated into the PEAR data system (see PEAR Measures below), partners choose the measures most relevant to their programming and suited to their needs. In addition to selecting the tools themselves, they can decide the frequency and format of each data collection tool. A data collection timeline will be created to summarize the data collection plan (see Figure 32.2).

The diversity of available tools means that there is a method to assess a variety of outcomes, perspectives, content areas, and program contexts. If there is a partner-specific assessment area interest that is not offered, PEAR's evaluation staff have the skills and training to develop, pretest, and administer custom scales of high psychometric quality. Regardless of data collection tools selected, PEAR will create and share a dashboard that integrates all data sources and reports a descriptive and graphic summary of results.

PEAR measures

The measures described below are reliable and valid assessment tools developed by PEAR that provide multiple perspectives to support continuous program improvement and are integrated into PEAR's data system for ease of collection and reporting. They are available for DNA partners to use as part of their data collection plan.

Common Instrument Suite – Student Survey (CIS-S). The Common Instrument Suite – Student is a rigorously validated student self-report survey,[5] built on an evidence-based youth development framework and designed to help programs tailor services to better support the STEM interest and engagement of youth. It measures four STEM-related constructs (STEM engagement, STEM career interest, STEM career knowledge, STEM activity participation), four social-emotional development (SED) constructs (perseverance, critical thinking, relationships with adults, relationships with peers), and one construct that integrates STEM and SED (STEM Identity).

Common Instrument Suite – Educator Survey (CIS-E). The CIS-E is a retrospective self-report survey designed to capture the unique qualities of STEM programs and the practitioners who lead them. Developed to complement both the CIS-S and the DoS observation tool, the CIS-E enables programs to capture key information about STEM activity facilitators, educators' levels of support, and their perceived efficacy in delivering STEM content. The survey was designed for any in- and out-of-school time staff leading or co-leading STEM activities.

Dimensions of Success Observation Tool (DoS). Dimensions of Success is a framework that defines key aspects of a quality STEM learning experience.[6] DoS forms the backbone of a suite of tools and guides designed to help OST programs improve the quality of their STEM offerings. It was developed and studied with funding from the NSF by PEAR and partnering organizations, Educational Testing Services (ETS) and Project Liftoff. The DoS suite includes an observation tool, a program planning tool, and a feedback & coaching guide. These tools allow researchers, practitioners, funders, and other stakeholders to track the quality of STEM learning opportunities and continuously improve their programs. DoS observation data can be integrated into a dashboard in addition to the youth and educator perspectives for a full picture of the strengths and challenges of the program.

Communication Plan. In the initial planning process, the partner and PEAR will co-create a project communication plan. This communication plan will set a cadence for both parties to regularly discuss programming and DNA progress. An external communication plan will also articulate the intended purposes, formats, and audiences for each evaluation product. These respective plans will establish a collaborative, accountable, and sustainable partnership and can still be modified to accommodate unexpected situations, developments, and opportunities as they arise in real-time.

Additional Training and Workshops. Depending on the nature and capacity of the program, the partner may choose to opt into optional trainings that complement the DNA model. The first training is PEAR's *Success Story Workshop*, a half-day event to launch a DNA project. Workshop participants will receive basic evaluation training; they will become familiar with report elements, learn to integrate data and stories into a report, and lay the groundwork for communicating their program's success with data. The second optional training is the *Data Debrief and Celebration*. In this half-day session, participants will learn the purpose and intention of data collection, to parse data trends, and to discuss potential programmatic supports that may maintain or improve trends. These sessions are helpful for partners with the capacity to and interest in developing their internal evaluation "muscles" and will help their staff become familiar with the creation and utility of DNA evaluation materials.

Tailored Statistical Analyses and Interim Reports. In evaluation planning meetings, PEAR staff work with partners to tailor statistical analyses that are best suited to the overall goals and focus of the evaluation. These analyses are included in an interim statistical report that forms a critical component of the final DNA report. These reports provide a written explanation of the performed analyses as well as a summary of quantitative findings from the program's data collection. They also include a visual presentation of key results in the form of bar graphs.

While the formal data collection process is underway, PEAR will work with the partner to discuss and share materials that will contextualize the quantitative findings and make them meaningful. DNA reports produce a holistic portrait of an OST STEM program. To do so effectively, the data findings cannot be presented in isolation. Existing documents like videos, testimonials, quotes, stories, photographs add a human element to the reports and make them appealing to a wider audience. Secondary research, including any previous studies of the program, existent theories justifying the program rationale, or frameworks underlying the program model provide a robust evidence base for describing the program's current impacts and positing its future effects.

DNA Reports. Once data collection has concluded, a member of PEAR's evaluation team analyzes the data. Frequency distributions of demographic data are summarized for student outcomes. Then, inferential analyses are conducted to note any significant differences in the scale averages across the overall dataset and by key demographic factors such as gender, grade, and race. These analyses are often reviewed with the partner organization in an interim statistical report. Frequently, partners have specific subpopulations of the student data that they want a deeper understanding of, such as the experiences of groups with intersecting minoritized identities (e.g., girls of color, English Language Learners, etc.) or participants of a newly implemented program type. An initial review of the overall student dataset helps partners better articulate the lens through which to view the data and shape their narrative.

Due to their likely smaller sample sizes, educator and observational data are reported at a broader level. Educator-report data compatible with student-report data, such as the educator perception of student outcomes, are compared, providing an additional perspective for the OST STEM program. Other educator data, such as their average rating of STEM identity and program characteristics, are reported at the group level. Finally, for observational data, the averages for each dimension are compared to a benchmark rating of program quality. Dimensions with averages equal to or greater than this threshold are seen as areas of high program quality, whereas dimensions beneath the threshold are potential areas of interest for future quality improvement efforts. The combined findings from these multiple data sources can provide a powerful picture of the entire OST STEM experience from the student, educator, and observer perspectives.

Insights from the three data sources are synthesized in the DNA report. Such triangulation provides confirmation for drawing strong conclusions about for the programmatic impact. Additional partner-provided data may be incorporated via a document review of the program, especially when existing materials, testimonials, images, and prior studies supplement the most recent data findings. Reports are also bolstered with secondary research that contextualizes the data and their implications for the larger field.

The report writing process is a sequenced and collaborative endeavor that takes place over a series of meetings with the partner. Different portions of the report development can occur in parallel, such as collection of data with the writing of the background section of the report. PEAR's evaluation team has created numerous guidance materials that simplify the writing process, making it more accessible and achievable for programs with a range of capacity considerations.

Data narratives for action (DNA): The use with STEM learning ecosystems (SLEs)

To see how the DNA process has been put into action, we share an example of a PEAR partnership with a cohort of STEM Learning Ecosystems (SLEs) in an initiative to create individual DNA reports and learn evaluation methods within a community of practice. Because DNA was modeled on PEAR's Tulsa, Oklahoma SLE's case study report[7] and it uses data and other factual information as its backbone, ecosystems were eager to participate.

Throughout the initiative, PEAR provided participating sites with guidelines and supports including best practices, reliable and valid data collection tools, and a data platform for collection and visualization. Each participating site met monthly with a member of PEAR's evaluation staff for technical assistance. During these meetings, they gradually co-created the report in an intentionally iterative, sequenced process. Different portions of the DNA process were happening in parallel – such as reviewing archival materials and collecting student data – so that the report could be built over time. As evaluation activities coincided with the start of program implementation, sites could also use their data for continuous improvement efforts. Delivering the DNA report in a cohort structure was a scalable, accessible, and empowering way for these ecosystems to conduct a brief evaluation.

Participating ecosystems joined the initiative with varying levels of readiness and comfort with evaluation activities. Many were familiar with more formal case study approaches that were perceived as impractical for their cost and time required. Further, many ecosystems lacked the capacity to conduct in-depth investigations with any frequency on their own. Though case studies provide valuable insights into transforming STEM education models and generating hypothesis to inform future studies on a larger scale, many of their functions can be more practically captured with DNA.

DNA's user-friendly evaluation model establishes a standard reporting process by which ecosystems collected data from multiple sources and answered unique research questions. This model enabled them to provide data and outcomes to funders, improved the quality of service, and plan for their future. This system was a much more feasible way for ecosystems to create accountability systems, communicate their outcomes, and craft a program narrative that centers data.

This initiative empowered SLEs to receive quality reports building off PEAR's knowledge of the ecosystem model, without the need for expensive external researchers or evaluators. There are plans to continue the initiative with additional cohorts, with the potential to expand to all 89 existing US and international ecosystems. These reports tell the stories of how communities are transforming STEM education models around the world.

Conclusion

We share PEAR's DNA as an example of how OST programs across the world can use data to tell the story of their impact and to inform continuous improvement. The system's shared name with genetic material is no accident: DNA emphasizes the importance of recognizing the underlying building blocks of an OST STEM program. The process of storytelling within the DNA model involves recognizing the program's connection to both its legacy and brighter future; it meaningfully captures the iterative process of organizational and program development. In addition to expressing a program's uniqueness, DNA also conveys its universality. The model is founded upon the importance of sharing a common language around data across the OST STEM community. Our research has found that youth participating in higher-quality OST STEM programming reported more growth than peers participating in lower-quality programs.[8] By better understanding the building blocks of impactful STEM OST programs, we tell the story to other programs to help young people across the world engage more deeply and fully with STEM throughout their lives.

To move forward, we must address the challenges that hinder evaluation and assessment of STEM learning in OST efforts. These deterrents include the burden of data management, dearth of common measures OST STEM learn-ing assessment, limited capacity to implement data collection and reporting processes, and feelings of suspicion toward collecting data on student academic performance.[9] The field must approach these challenges with innovation and creativity with a focus on leveraging existing resources and capacity for maxi-mum impact. With more simple, accessible, and flexible evaluation tools, data can be used for multiple purposes that garner funder and policymaker support, drive program improvement, and expand and scale opportunities for students.

Notes

1 Cunningham, B. C., Hoyer, K. M., & Sparks, D. (2015). *Gender Differences in Science, Technology, Engineering, and Mathematics (STEM) Interest, Credits Earned and NAEP Performance in the 12th Grade (NCES 2015-075)*. Institute of Education Sciences, National Center for Education Statistics, US Department of Education. https://nces.ed.gov/pubs2015/2015075.pdf.

2 Allen, P. J., Noam, G. G., & Little, T. D. (2017). Multi-state evaluation finds evidence that investment in afterschool STEM works. *STEM Ready America*. http://stemreadyamerica.org/multi-state-evaluation-finds-evidence-that-investment-in-afterschool-stem-works/.

3 Noam, G. G., Allen, P. J., Sonnert, G., & Sadler, P. M. (2020). The Common Instrument: An assessment to measure and communicate youth science engage-ment in out-of-school time. *International Journal of Science Education, Part B, 10*(4), 295–318. https://doi.org/10.1080/21548455.2020.1840644.

4 Malti, T., Zuffianò, A., & Noam, G. G. (2017). Knowing every child: Val-idation of the Holistic Student Assessment (HSA) as a measure of social-emotional development. *Prevention Science, 19*(3), 306–317. https://doi.org/10.1007/s11121-017-0794-0.

5 Noam et al. (2020). *Ibid*.

6 Shah, A. M., Wylie, C., Gitomer, D., & Noam, G. (2018). Improving STEM program quality in out-of-school-time: Tool development and validation. *Science Education, 102*(2), 238–259. https://doi.org/10.1002/sce.21327.

7 Lewis-Warner, K. M., Allen, P. J., & Noam, G. G. (2019). *Partnerships to Transform STEM Learning: A Case Study of Tulsa, Oklahoma's Ecosystem*. The PEAR Institute: Partnerships in Education and Resilience.

8 Allen, P. J., Chang, R., Gorrall, B. K., Waggenspack, L., Fukuda, E., Little, T. D., & Noam, G. G. (2019). From quality to outcomes: A national study of afterschool STEM programming. *International Journal of STEM Education, 6*(1), 37. https://doi.org/10.1186/s40594-019-0191-2.

9 Noam, G. G., Allen, P. J., Shah, A. M., & Triggs, B. B. (2017). Innovative use of data as game changer for afterschool: The example of STEM. In H. J. Malone & T. Donahue (Eds.), *The Growing Out-of-School Time Field: Past, Present, and Future*. Charlotte, North Carolina: Information Age Publishing.

33

SHOULD SCIENCE CENTRES EVICT STEM? THE 1960s DREAM FOR PEACE, BEAUTY, AND AWE VERSUS COLD WAR PRAGMATISM AND PROBLEM-SOLVING

Hooley McLaughlin

In 1969, the Exploratorium in San Francisco and the Ontario Science Centre in Toronto launched a new kind of museum. Using a play-oriented space, the visitor was guided towards an "Aha!" experience where she caught a glimpse into the ideas that have shaped the history of science and technology – the eureka moments. In recent years, the science centre has embraced a different style of learning: STEM and its closely associated Makerspaces, where visitors investigate 21st- century problems, in an open-ended playspace. We appear to be on a course to eliminating the original science centre experience in favour of the attractive STEM learning model. Of greater concern are the goals of the politicians and educators who promote a theory that has its roots in the Space Race of the Cold War and has been designed around the pragmatic need to produce workers who are focussed on "real-world problems" as defined by political leaders. The science centre of the late 1960s, in contrast, was born during the time of the *anti-war* movement when many of us were pursuing beauty and awe in all areas of study and life, in a non-political manner. The Exploratorium and the Ontario Science Centre were created by people who had seen the horrors of World War II from different sides of the conflict. They were devoted to the elusive elegance of science and mathematics and saw beyond the biased needs of the politicians. If the science centre is to survive, we may ask if it is time to analyse the impact of STEM learning spaces more closely and choose to move away from STEM to return to the original 1960s model.

The Cold War model for learning

I visited the Soviet Union during *Perestroika* [restructuring] during 1990 and 1991, when I was a member of the Ontario Science Centre exhibit team travelling under the guidance of the Adventure Club,[1] led by the Arctic explorer

DOI: 10.4324/9781003145387-37

Dmitry Sparo.[2] While in the northern region of the Yamal, I examined a collection of traditional children's toys, remarking how their playthings seemed to resemble small versions of tools used for adult chores. Came the reply, "No play! *Rabota* [work]! *Rabota, rabota, rabota!*"

Like many who have studied under teachers who had their own formative education before the height of the Cold War, my early professional years were underscored by apolitical scholarship,[3] a bias shared by many Russian scientists in the 1980s and 1990s.[4] Although pragmatic problem-solving may have been considered essential, it was considered as a sub-category that only occurred *after* scientists were guided by pure curiosity, and certainly, pragmatic concerns had no place in children's education in the pre-Cold War era. The Soviet project made me reflect on my personal educative trajectory to some degree, however.[5] Stalin's railroad, lying ruined on the heaving Arctic tundra, gave proof of the cruelty of GULAG labour camps, and I realized that, in practice, science itself must inevitably be shaped by the history surrounding it.[6] Those reflections do not, however, preclude my concern that the latest developments in science education have turned the natural sequence of scientific exploration upside-down.

At the Toronto opening of the *Siberia, spirit of survival* exhibition, in September 1991, Dmitry showed up with what looked like tank treads dug into his right cheek. In explanation he replied, "Barricady!" [barricades!]. He'd been at the demonstrations in Moscow on August 21, where unarmed citizens had stood at homemade barricades in the streets to confront a putsch led by Soviet hardliners.[7] I had been in Moscow myself earlier that August and I can attest there had been tension that could have easily ended up with someone like Dmitry stopping a tank with his face.

The Cold War began after WWII, but escalated exponentially after the USSR detonated its first atomic bomb in 1949. In the early 1990s the Cold War was starting to thaw, only to return a few years later. Attempts at disarmament that began in the 1950s have failed to this day when we are again at the brink of potential nuclear war in March of 2022.[8] This stark reality has shaped the trajectory for educational theory and practice on both sides of the conflict.

Dmitry Sparo's mother, Nina Gimers, despite her own father having been sent to the GULAG by Stalin, was, by 1956, working as a mathematician on the orbital calculations for the launches of both the world's first ICBM, and Sputnik 1. Dmitry personifies for me a dichotomy: He has what I would call Russian Sputnik-Pride, stemming from both personal and nationalist roots, while at the same time, similar to only a few brave Russians like himself, he actively expresses a hatred for the regime that launched Sputnik in the first place. In western countries, where harsh contrast is not as obvious, STEM education in the West can also be traced to the launch of Sputnik, which resulted in the passage of the US National Defense Education

FIGURE 33.1 The R-7 rocket undergoing erection before launch. The Soviet Union developed the R-7 in the 1950s. It became the engine for the world's first warhead-equipped intercontinental ballistic missile successfully launched on August 21, 1957. The same rocket model carried Sputnik 1, the earth's first human-made satellite, into orbit on October 4, 1957. The Cold War went beyond the space race, to include new commitments for an increased emphasis on science and mathematics education. Author's drawing based on photo from http://mentallandscape.com/S_R7.htm.

Act, based on the perceived need for greatly enhanced education in science, mathematics, and foreign languages.[9] And therefore, I would ask if the educative direction in western nations, without our being fully aware of it, hasn't been trending towards an outcome similar to the one seen in Russia, a recruitment of young "science-workers," rather than actual "scientists" – a Cold War army, without attention being given to the pursuit of personal, less political, goals that might have, in less warlike times, included the investigation of pure scientific and mathematical concepts that lie well outside the realm of short-term problems (Figure 33.1).

STEM learning spaces emphasize a pragmatic introduction to the world of scientific thinking, utilizing individual exploration and invention. The theory and practice of STEM have been under considerable discussion since the US Science Foundation began using first the term SMET, and then STEM in the early 2000s, but despite theoretical debates on some aspects of theory,[10] STEM definitions have consistently centred on a common interest in solving real-world problems.[11] One outcome of a focus on technology and engineering seems to be a de-emphasis on the more classical study of abstract mathematics and science. Particularly in mathematics teaching, the move away from classical teaching modes has been considered, by STEM proponents, as important for teaching methodologies.[12] Thus, "Process" begins to take over from "Content," and attention turns towards earlier 20th-century experiments in educational psychology.

"Construction" of knowledge

Drawing on the writings of Plato and Friedrich Froebel,[13] John Dewey said the child is the instigator, rather than the teacher. "…the initiative lies with the learner. The teacher is a guide and a director: he steers the boat, but the energy that propels it must come from those who are learning."[14] Jean Piaget postulated that children form theories of the world, much like "little scientists,"[15] a theoretical framework that has a current manifestation known as "Theory Theory." Alison Gopnik, a proponent of Theory Theory, describes a child's "theory-of-the-world" in constant evolution, and thus Dewey's "boat" is undergoing continuous reconstruction as it is in use on the water. Gopnik cites Otto Neurath, comparing a child's construction of knowledge to Neurath's metaphor of *science* itself as a ship that is underway at sea, a craft that must have its timbers and sails constantly restored, with no time to take the ship into drydock for a good looking-over.[16] This begs the question: does this much-touted approach to learning fail to produce boat-builders, but rather merely give us hardworking sailors who must take their direction from others?

Arguing against Gopnik, Katherine Nelson describes children not as immature scientists, but rather as creatures who think in a fashion entirely different from adults.[17] Nelson cites the need for more teacher or parental intervention for successful learning, ascribing her approach largely to Lev Vygotsky, who emphasized the child's being guided during formative years by people more knowledgeable than herself. STEM as delivered in the Makerspace owes much to thinkers like Gopnik, Piaget, and Dewey, and I would suggest the Aha! experience derives much of its philosophy from the tradition of Vygotsky. In contrast to the classic science tradition, STEM marries "play" to the problem-solving needs of the political and elite classes and thereby diverts children from exploring the contextual richness of their civilizations and history, thus turning the process of play into a training ground for work. No play, effectively, but simply *rabota*!

The science centre model of learning

There is an improvisational nature of a STEM learning space which is initially quite attractive to museums. The original science centre exhibit experience was a carefully constructed "adventure," the advantages of which take longer to establish, and their fragility in the face of STEM's juggernaut is all too obvious as we see the growing popularity of STEM Makerspaces.[18] The science centre experience can be seen in demonstrations that explore basic physics and chemistry phenomena such as Boyles Law,[19] or pulleys,[20] or the law of the conservation of angular momentum,[21] or perception,[22] or even the history of discrimination and prejudice,[23] or any subject, it would appear from the lexicon of standard science centre exhibits that can be presented using an exhibit method that leads the visitor towards a cathartic and focussed experience, an epiphany, the "Aha! moment."[24] Unlike what typically happens in a school classroom, the lesson is short and yet powerful enough to have lasting impact that inspires further thinking, research, and an increased willingness to tackle explorations into many other areas of personal interest.[25] The classic science centre's Aha! experience is but one small step removed from the deeper sense of *awe*,[26] and if it is working at its highest level of impact and effect, just like we see when we are engaged in scientific research, the science centre experience can feel awesome.[27]

STEM came out of the Cold War. The science centre was born from the *anti-war* movement in the 1960s. It is not irrelevant that Frank Oppenheimer and Raymond Moriyama came from opposite experiences of the horrors of WWII. The research career of Oppenheimer, through his work as a mathematician on the Manhattan Project, and the national and cultural heritage of Moriyama through his experience as a child prisoner in a Japanese internment camp in Canada have resulted in their both being tied indelibly to the making and utilization of the atom bomb.[28] Yet, they ended up devoted to peace and the promotion of children's learning through the exploration of awe and wonder (Figure 33.2).

A call for change

Freed to enter the history of ideas, we can become independent leaders. However, if the educational method is tied to a country's short-term political needs, the student is steered away from this deeper essential knowledge, and the liberated thinker is lost. Are we unwittingly supporting the state's need for the "well-trained" *blinkered* worker, who has not had a chance to be inspired by the higher ineffable questions of science and mathematics? James Baldwin said the purpose of education is "To ask questions of the universe," but he follows this statement of purpose with a description of the unfortunate reality of today's education approach, and a call for change: "What societies really, ideally, want is a citizenry which will simply obey the rules of society... If a society succeeds in this, that society is about to perish. The obligation of anyone who thinks of himself as responsible is to examine society and try to

FIGURE 33.2 Manhattan Project Trinity Test, 15 seconds after blast, the first atom bomb test, July 16, 1945. It is argued that the atom bomb was both an incredible feat of science, math, and engineering, and a monstrous event that triggered the eventual peace movements of the 1960s, an era that saw the development of the science centre. Author's drawing based on photo from https://www.atomicheritage.org/key-documents/trinity-test-eyewitnesses.

change it and to fight it – at no matter what risk. This is the only hope society has. This is the only way societies change."[29]

Barricady!

Notes

1 Adventure Club, Adventure Club, Russia: http://www.shparo.com/, retrieved November 25, 2021.
2 Dmitry Sparo led the Russian team in the joint 1988 Soviet-Canadian Polar Bridge Expedition from Siberia to Canada over the North Pole (Weber, Richard, Dexter, Laurie, Holloway, Christopher, and Burton, Max. (1992). *Polar Bridge: An Arctic odyssey.* Toronto: Key Porter Books).
3 I did my PhD with Richard A. Liversage at the University of Toronto Dept. of Zoology. Liversage believed that scientists' pursuits must be guided by curiosity and, to be successful, can serve no pragmatic purpose, a tradition he had inherited from his own graduate supervisors, Oscar Schotté at Amherst, and Elmer G. Butler at Princeton. Butler and Schotté had both studied with Hans Spemann in Freiburg. Spemann is considered the father of developmental biology, receiving the Nobel Prize in Physiology or Medicine in 1935. Spemann studied with Theodore Boveri and Wilhelm Röntgen in Würzburg, shortly before Röntgen won the Nobel Prize in Physics in 1901. With this heritage, I still struggle to allow that science has a cultural context. I find I can easily admit to its relativity in practice (the theme of *A Question of Truth* which I curated, and which opened at the Ontario Science Centre in 1996), but I still believe that the pursuit of the ineffable and the infinite *cannot start* with a practical, current problem that needs solving.
4 Kojevnikov, Alexei. (2002). Introduction: a new history of Russian science. In *Science in Context* 15(2). Cambridge, Cambridge University Press. "A historical thesis did not have to be philosophically oriented, but quite technical, focussed primarily on the development of scientific ideas and concepts…not the least important was that such scholarship was perceived as "apolitical" in the late Soviet mindset." 177.
5 McLaughlin, Hooley. (1999). *The Ends of Our Exploring: Ethical and Scientific Journeys to Remote Places.* Toronto: Malcolm Lester Books, Chapter One, Siberia, 7-36, Chapter Three, The Dark Wood, 77–105.
6 Applebaum, Anne. (2003). *GULAG: A History.* New York: Anchor Books, Random House; Ash, Lucy. (2012, June 7). Joseph Stalin's deadly railway to nowhere. *BBC News,* retrieved October 12, 2021: https://www.bbc.com/news/magazine-18116112; I owe a great deal to Lyudmila Lipatova, curator of the Salekhard Museum, who collaborated with us for the *Siberia* exhibition, cf McLaughlin, Hooley. (1998). The pursuit of memory: Museums and the denial of the fulfilling sensory experience. *The Journal of Museum Education,* Vol. 23, No. 3, Too Hot to Handle? Museums and Controversy (1998), 10–12.
7 Dobbs, Michael. (1991, August 21). Protesters confront tanks in Moscow. *The Washington Post,* retrieved October 9, 2021: https://www.washingtonpost.com/archive/politics/1991/08/21/protesters-confront-tanks-in-moscow/38d0772c-4276-4b00-8f3f-ee5b773131d8/; Sarotte, M.E. (2021) *Not One Inch: America, Russia, and the Making of Post-Cold War Stalemate,* New Haven and London. Yale University Press, 116–119.
8 Osnitskaya, G.A. (1960) Neutrality and Atomic Weapons, USSR. Moscow. *Sevelsoye Gosuderstvo I Pravo,* 2, 101–104, https://apps.dtic.mil/sti/pdfs/ADA362100.pdf., retrieved March 15, 2022; Stockton, Ron. (2022) The Ukraine War of 2022. https://deepblue.lib.umich.edu/bitstream/handle/2027.42/171764/UkraineWarOf2022TextOfAPodcastPDF.pdf?sequence=1&isAllowed=y, retrieved March 15, 2022.

9 Sputnik Spurs Passage of the National Defense Education Act. The United States Senate, retrieved October 11, 2021, https://www.senate.gov/artandhistory/history/minute/Sputnik_Spurs_Passage_of_National_Defense_Education_Act.htm; Sputnik 1 was launched by the Soviet Union on October 4, 1957, the first human-made space object in orbit around the Earth. Sputnik 3 launched on May 15, 1958, and stayed in orbit until April 1960; Yee, Gary & Kirst, Michael. (1994) Lessons from the New Science Curriculum of the 1950s and 1960s. *Education and Urban Society*, 26(2), 158–171; other western countries followed the US lead, cf as one example, Brittain, Pamela. (2021) Math curriculum in Ontario: A historical perspective. https://pambrittain.ca/wp-content/uploads/2021/04/CMS-Presentation-April-12-2021.pdf, retrieved March 17, 2022.

10 cf Chesky, Natalie Z. & Wolfmeyer, Mark R. (2015). *Philosophy of STEM Education: A Critical Investigation*. New York: Palgrave Macmillan.

11 Martín-Páez, Tobías, Aguilera, David, Perales-Palacios, Francisco Javier, & Vílchez-González, José Miguel. (2019). What are we talking about when we talk about STEM education? A review of literature. *Science Education*, 103(4), 799–822; Shaughnessy, J. Michael. (2013). Mathematics in a STEM context. *Mathematics Teaching in the Middle School*, 18(6), 324; Sanders, Mark. (2009). STEM, STEM education, STEMAnia. *The Technology Teacher*, 68(4), 20– 27; Bybee, Rodger. W. (2013). *The case for STEM education challenges and opportunities*. Washington, DC: National STEM Teachers Association.

12 Li, Yeping & Shoenfeld, Alan H. (2019). Problematizing teaching and learning mathematics as "given" in STEM education. *International Journal of STEM Education*, 6(44). https://doi.org/10.1186/s40594-019-0197-9, retrieved March 17, 2022.

13 Dewey, John. (1997). Activity and the training of thought. In *How We Think*. Mineola, New York: Dover, 162.

14 Dewey, John. (2008). Native resources in training thought. In J. A. Boydson, B. A. Walsh, & H. Furst Simon. (Eds) *John Dewey: the later works, 1925–1953*. Carbondale, Illinois. Southern Illinois University Press, 140.

15 Piaget, Jean (1954). *The Construction of Reality in the Child*. (M. Cook, Trans.). Basic Books. https://doi.org/10.1037/11168-000.

16 Gopnik, Alison. (2003). The theory theory as an alternative to the innateness hypothesis. In L. M. Antony & N. Hornstein. (Eds) *Chomsky and His Critics*. Oxford: Blackwell, p. 242; Neurath, Otto. 1973. Anti-Spengler. In M. Neurath & R.S. Cohen (Eds) *Empiricism and Sociology*. Vienna Circle Collection. Vol 1. Dordrecht: Springer, 199; The Ontario Science Centre had a demonstration of childhood-learning in *Psychology: Understanding Ourselves, Understanding Each Other*, a travelling exhibition created jointly by the Ontario Science Centre, the American Psychological Association, and the Association of Science and Technology Centers. Alison Gopnik was a primary advisor to us for this exhibit.

17 Nelson, Katherine. (2007). *Young Minds in Social Worlds: Experience, Meaning and Memory*. Cambridge, MA: Harvard University Press; Kozulin, A., Gindis, B., Ageyev, V. S. & Miller S.M. (2003) Introduction: Sociocultural theory and education: students, teachers, and knowledge. In A. Kozulin, B. Gindis, V. S. Ageyev, and Miller S.M. (Eds.) *Vygotsky's Educational Theory in Cultural Context*. Cambridge: Cambridge University Press, 1–11.

18 Brahms, Lisa & Werner, Jane. (2013). Designing makerspaces for family learning in museums and science centers. In M. Honey & D.E. Kanter (Eds). *Design, Make, Play*. New York: Routledge; Raup, Andrew B. (1019). STEM education's lost decade and tenor: Contemporary insights into a popular, global movement. Data Driven Investor, retrieved November 26, 2021: https://medium.datadriveninvestor.com/stem-educations-lost-decade-and-tenor-3f741bd728e6.

19 See, for example: The Sci Guys: Science at Home: Boyle's Law of Ideal Gases, July 13, 2014: https://www.youtube.com/watch?v=eR49g3ubTBg, retrieved October 9, 2021.

20 Pulleys! In See-Sciencecenter.org.: https://see-sciencecenter.org/pulleys/, retrieved October 13, 2021.

21 Physics Girl. Nov 22, 2014. Crazy Pool Vortex. https://www.youtube.com/watch?v=pnbJEg9r1o8, retrieved October 9, 2021.

22 Duensing, Sally & Miller, Bob. (1979). The Chesire Cat Effect. *Perception*, 8(3), 269–273; https://doi.org/10.1068%2Fp080269.

23 Hooley McLaughlin, (2002) Questioning Scientific Authority, a case study. In B. Lord and G. Dexter Lord (Eds) *The Manual of Museum Exhibitions*, Walnut Creek, CA: AltaMira Press.

24 Kounios, John, & Beeman, Mark. (2014). (First published 2009). The cognitive neuroscience of insight. In *Annual Review of Psychology*, 65, 71–93; Campbell, Stephen R. (2021). Chapter 14: The Aha! Moment at the nexus of brain and mind. In *Creativity of an Aha! Moment and Mathematics Education*. Brill, pp. 365–397. https://doi.org/10.1163/9789004446434_015.

25 Falk, John H. (2009). Memories. In *Identity and the Museum Visitor Experience*. London and New York: Routledge, chapter 6, 129–156.

26 The concept of 'awe' has a religious context, and experiences of awe in science can be linked to these. Cf Gottlieb, S. (2018). Awe as a Scientific Emotion. *Cognitive Science*, 42(6), 2081–2094; I would also point to Fyodor Dostoyevsky's *The Idiot*, in which his character Prince Mishkin describes Hans Holbein the Younger's *Portrait of the Body of the Dead Christ in the Tomb* as having an effect so profound that it threatens to take away one's religious faith. The combination of a soaring metaphysical experience of faith, combined with the sense of awe that is at once terrible and beautiful, is hard to come by. Dostoyevsky's description comes close.; cf Dostoevsky, Fyodor. (2003). *The Idiot*. Trans Richard Pevear and Larissa Volokhonsky. New York: Vintage Books, Random House. (Originally published in 1868–1869.)

27 Cuzzolino, Megan Powell. (2019). Scientists' experiences of awe and its relationship to learning and discovery. *American Educational Research Association*. Retrieved October 12, 2021. https://scholar.harvard.edu/megancuzzolino/publications/scientists%E2%80%99-experiences-awe-and-its-relation-learning-and-discovery; Nerantzaki, Katerina, Efklides, Anastasia, & Metallidou, Panayiota. (2021). Epistemic emotions: Cognitive underpinnings and relations with metacognitive feelings. *New Ideas in Psychology*, 63. https://doi.org/10.1016/j.newideapsych.2021.100904. It could also be noted that the sense of awe is closely related to the original search for the exotic that we saw in earlier forms of museums. Such orientialist concepts have been heavily criticized (cf Bryce, Derek & Carnegie, Elizabeth. (2013). Exhibiting the 'Orient': historicising theory and curatorial practice in UK museums and galleries. *Environment and Planning A: Economy and Space*,.45(7), 1734–1752. https://doi.org/10.1068/a45359), but it is worth looking at alternative evaluations that have looked at the exotic from different cultural perspectives and directions (cf Dening, Greg. (1996). Possessing Tahiti. In *Performances*. Chicago: University of Chicago Press, 128–167; Boon, James A. (1999). Why Museums Make Me Sad. In *Verging on Extra-Vagance: Anthropology, History, Religion, Literature, Arts ... Showbiz*. Princeton: Princeton University Press, Chapter 5, 126–142).

28 Frank Oppenheimer started the Exploratorium. He worked on the Manhattan Project with his brother J. Robert Oppenheimer (Frank Oppenheimer – from the Atomic Heritage Foundation. Retrieved October 13, 2021: https://www.atomicheritage.org/profile/frank-oppenheimer); Sally Duensing (personal communication, October 2014, Raleigh) told me that Frank had loved big science, and the ideas that inspired major scientific projects, the kind that lead you to wonder

about the origins of the universe, cf In Memoriam: Dr. Sally Duensing, Exploratorium: https://www.exploratorium.edu/people/sally-duensing; Raymond Moriyama was the architect for the Ontario Science Centre, and with others, like Taizo Miyake, developed the Ontario Science Centre's hands-on approach. He was a child prisoner in one of Canada's Japanese Internment Camps during WWII, Rinaldi, Luc. (2014, May 3). "We were immediately cast as enemy aliens." In Maclean's Retrieved October 13, 2021: https://www.macleans.ca/news/canada/architect-raymond-moriyama-on-internment/; with respect to the 1960s peace movement, the development of an increasing interest in Eastern religion was foreshadowed by J. Robert Oppenheimer's famous quote upon witnessing the first atomic explosion in 1945: from the Bhagavad Gita: "Now I Am Become Death, the Destroyer of Worlds." *Open Culture*, retrieved November 27, 2021: https://www.openculture.com/2020/09/j-robert-oppenheimer-explains-how-he-recited-a-line-from-bhagavad-gita.html.

29 Baldwin, James. (1963). A Talk to Teachers. Delivered October 16, 1963. https://www.spps.org/cms/lib010/MN01910242/Centricity/Domain/125/baldwin_atalktoteachers_1_2.pdf, retrieved online March 16, 2022. Baldwin's 1963 writings apply today as much as they did when he wrote them, in my opinion.

34

REFLECTIONS ON 50 YEARS IN THE SCIENCE CENTER PROFESSION

Dennis Schatz

When Judy Diamond and Sherman Rosenfeld talked with me about contributing to the volume, they suggested a reflection on my 50 years in the science center field. The profession has certainly changed since I was hired in mid-1973 by Alan Friedman at the Lawrence Hall of Science to teach a one-day-a-week afterschool astronomy class for elementary school-aged youth. By year's end, I had a full-time job working the planetarium and developing science experiences for visually impaired elementary school students. I became an informal science education professional when no one grew up saying, "I want be a science center professional when I grow up."

I've seen many changes over the years – moving to more interactive exhibitory, using formative evaluation during exhibit development, building IMAX theaters and using blockbuster exhibits as major program and financial drivers at large science centers, and sending outreach vans to deliver science programing to rural part of states and to unconventional settings (e.g., community fairs, shopping centers). I've also seen a lack of changes over the years – programs still primarily appeal to the science-enthused rather than the broader community. They focus on children/families at the expense of the adult audience and avoid dealing with difficult societal issues.

Not just my reflections

While I could provide a personal perspective on what science centers have done well and done poorly over the past 50 years, plus the implications for the future, I thought it would be more useful to survey people in the field to get a broader perspective. With this in mind, I sent email inquiries to 61 people in the science center community who represented an international perspective,

DOI: 10.4324/9781003145387-38

but mainly a North American focus. I did not seek new members to the field, who have not had time to develop a broader understanding of the field. But I did not just focus on old timers like me, so that the comments do include opinions from people who will be influencing the field 10 years from now. My questions to them were simple ones. What are the: Top three to five things science museums have done well over the past three to five years? Top three to five things science museums have done poorly over the past three to five years? Top three to five implications for the future of science museums?

Forty-two people responded to my request. You can see the list of respondents at the end of the chapter.

I developed a spreadsheet showing their responses and then found similarities among their comments. Synthesizing these categories allowed me to identify a number of areas regarding what science museums have done well, done poorly and the implications for the future. You will recognize that the reflections and recommendations are ones that have been discussed by many in the field over the years. A 3,000-word essay cannot do justice to the wealth of wisdom conveyed by the 42 people. My apologies to those individuals I did not quote. It is not a reflection of the quality of the comments.

If you want to "cut to the chase" to see the reflections and recommendations based on my 50 years of experience in the field and the comments from respondents, go to the end of this article to read the summary. You can then come back to the earlier sections to read the comments for more details.

Exhibits and programs

Kristin Leigh's (Explora) comment regarding what science museums have done well exemplifies the overwhelming responses in some form that I received from 30 of the 42 respondents:

Museums have provided important, alternative approaches to learning, making science concepts more understandable through fun, hands-on, manipulative experiences and "wow" factor in programs and events. Barry VanDeman (El Paso Children's Museum and Science Center) echoed this idea: Science centers created engaging experiences with science, increasing interactivity to a new level from previous science museums. This interactive way of engaging with the museum audience is now pervasive in all types of museums. Even art museums don't stage an exhibition without some sort of discovery area.

Sue Allen (Sue Allen and Associates) noted another major advance over the last 50 years: Integrating formative evaluation into a design process to create far more effective learning experiences. Today, I would not expect anyone in the field to develop an exhibition or program without first trial testing the elements before finalizing them. It is also routine to do front-end evaluation with visitors to identify topics of interest for an exhibit or program. Many people commented on how open the field has been with sharing learnings from their institutions and in developing collaborative exhibit-development projects.

We should be proud of the proliferation of opportunities for communities to offer science center experiences. As pointed out by Mikko Myllykoski (Heureka), the number of science museums globally has grown from 400 in 1989 to 3,000 today. The growth of science centers in small communities has been especially important for reaching individuals who live far from large population centers.

In addition to the proliferation of science museums, a number of people commented on the enhanced delivery of program via outreach efforts (e.g. science on wheels vans) to school and community event locations. While these are positive outcomes that influenced science museum visitors and the broader museum field, many comments clearly indicated there is much to improve in the future, especially related to inclusiveness of audiences not currently attentive to science.

Serving a more diverse audience

More than half of survey respondents made it clear that we need to diversify whom we serve. As Claire Pillsbury (Bay Area Discovery Center) pointed out: Science centers have been too comfortable preaching to the choir and teaching those who already identify as interested in science.

> Other people noted the lack of cultural responsiveness and inclusivity: Like most institutions across the arts and culture sector, science museums have been challenged by being limited in their cultural responsiveness and inclusiveness in how they create and deliver content and experiences (Darryl Williams, Franklin Institute). What science museums have done spectacularly poorly ... is engaging with individuals and communities from non-dominant backgrounds, or those who have traditionally not engaged with science centers.
>
> *(Jen DeWitt, Research and Evaluation Consultant)*

> With some exceptions, science museums have largely failed to make meaningful progress in shifting the demographics of visitors to more closely align with those of their communities or the nation at large. Efforts in outreach, free days, and community-focused programs have been insufficient in moving the broader needle.
>
> *(Rabiah Mayas, Museum of Science and Industry)*

> Not enough voices and points of view are involved ... They could be a place where we are learning about many different points of view. Cultural institutions could be the place where everyone feels welcome,

and all stories are told. They could provide an opportunity to share beliefs, build a common language, beliefs, and vision.

(Diane Miller, Detroit Zoo)

Kristin Leigh (Explora) nicely summed up what many people expressed as the problem: Museums have over-relied on their own expert knowledge and have relied less on public knowledge about what's important to local communities; museums have failed to recognize that "inspiring a love of inquiry" or "fostering scientifically-literate citizens" are goals that are much less important to families (especially historically underserved families) than they are to museum professionals.

Much as we tell teachers that they need to move from a teacher-centric classroom to a student-centric classroom, science museums need to move from an institution-centric approach to a community/guest-centric approach. Even as the science museum field asks for more front-end input from the community, we still focus on what we think is important for people to learn and still focus on what happens at our institutions.

More co-development of programs and exhibits needs to occur where the audience is engaged throughout the development process and has authentic influence on what is developed. In addition, more effort needs to be in serving audiences at locations they typically frequent (e.g. community centers, churches) (Figure 34.1).

FIGURE 34.1 Outreach van co-developed with indigenous population.

Staff/board diversity

Making science museum programs and exhibits more community\-centric with experiences that are more relevant to them will not easily occur until science museum staff and boards represent the diversity of the community that each science museum serves. As Rabiah Mayas (Museum of Science and Industry) indicated, this will require: ... internal [staff and board] development, goal-setting, commitment to and accountability on DEAI (Diversity, Equity, Accessibility, Inclusion) issues – from frontline staff all the way through senior leadership and the Board of Trustees.

While most science museums are cognizant of this need, more should be done so they regularly set goals and measure results to move any effort forward. Institutions not only need to look outside to find diverse staff members, but consider building a "career ladder" program, such as what exists at the New York Hall of Science and other science museums, where community youth are engaged as explainers, with steps to move into more responsible positions at the institution. The Discovery Corps program at Pacific Science Center was inspired by the Career Ladder effort at the New York Hall of Science (Figure 34.2a and b).

Serving adults and dealing with societal issues

The "bread and butter" audience for science museums has been families and children. I can't count the number of times I was told, "Yes, I used to go there when my children were young" when talking with people about Pacific Science Center. A significant number of survey respondents mentioned the need to broaden the audience served by having more programs for adults and programs/exhibits that deal with important societal issues.

> Science museums are seen as destinations for families with elementary-age kids. How do they design to signal that both adults and children should engage in the activities – or have adult issues?
>
> *(Judy Koke, Institute for Learning Innovation)*

> [Need to bridge] the gap between science museum and young adults – most youth tend to associate science museums with school or family outings, and they think visiting a science museum is too childish for them, and being seen among kids is not cool.
>
> *(TM Lim, Science Centre Singapore)*

> Science centers should move beyond their current focus on school children and families with children.
>
> *(David Ucko, Museums+more)*

(a)

(b)

FIGURE 34.2 a and b Discovery Corps youth in action at the Pacific Science Center.

[Science museums need to address] current problems, issues, controversies in science in a nimble and timely way that is responsive to public interest and concern.

(Marsha Semmel, Marsha Semmel Consulting)

New forms of science engagement in a democratic and pluralistic society [need to] become part of the menu of offerings, and museums don't shy away from tough conversations and from being involved in critical science-society conversations.

(Martin Storksdieck, STEM Research Center, Oregon State University)

There is some progress in these areas. In recent years, many science museums have offered evening events for adults, including serving alcohol. *The Race: Are We So Different?* exhibit from the Science Museum of Minnesota and the relatively new *Hall of Fossils* at the Smithsonian Natural History Museum, with its focus on climate change, are examples of exhibits that have taken on important societal issues.

A major challenge is that serving adult audiences and dealing effectively with important societal issues do not provide the same "return on investment" as serving families and children. To be sustainable in broadening our audiences and dealing with important issues to the community require new ways of funding science museums.

Finding a new financial model

"Science centers need to develop more effective business models, and diverse and stable bases for support." This statement by Cary Sneider (Portland State University) echoes what a large number of the respondents to the survey identified as a major issue. This is a critical need for science museums to be successful in the future, especially if these institutions want to diversify their audience and deal more with societal issues.

The success of science museums at the end of the last century led many institutions to expand their space, adding IMAX theaters and regularly featuring blockbuster traveling exhibits (e.g., Titanic, King Tut). This all added to greater operating expenses, which led to a greater need to focus on mission-supporting activities at the expense of mission-driven programs. Traveling exhibits often had limited connections to science and the IMAX theaters needed to give more showings of Hollywood films (e.g., Harry Potter, Star Wars) at the expense of science-based films, which was the original argument for the value of having IMAX theaters in a science museum.

This may be the greatest challenge facing science museums, and limited suggestions were made regarding solutions. Libraries are often mentioned as how science museums should be treated. They should be an essential resource to the community – see the chapter by Judy Diamond in this book. Claire Pillsbury (Bay Area Discovery Center) nicely stated what needs to occur for this to happen and shows how libraries often offer programs and facilities that compete with what science museums offer.

Science museums [will] remain vital when they commit to ongoing evolution, revision and diversifying of services in response to community interest and for relevance to new topics and interests. Fortunately, the model of the public library shows us how this can be done. Libraries have been impressive in their proliferation of services, responsive roles for their communities with teen centers, job search workshops, story time for toddlers and parents, multiple language resources, and computer and internet access.

> This need to be more embedded in the larger community is echoed by Antonio Gomes de Costa.
>
> *(Science Communication Consultant)*

[We need to] intentionally position museums as effective hubs in broader, community-wide learning ecosystems.

Charlie Walter (Mayborn Museum Complex, Baylor University) noted that "We need to build the strongest possible case and get state legislative support for the work we do." Whether through funding from the state or local municipality or private support, the need to make the case for the value of learning in science museums is a major need in the future.

Making the case

Much conversation focuses on developing and sharing the evidence for the value of learning in museums, and I would add learning at any venue outside of the formal classroom. Increased funding for science museums is not likely to occur without sharing funding with other out-of-school science learning experiences (e.g., afterschool programming, such as 4H, scouts, YWCA). I would argue that the evidence is already strong for this impact – see the chapter by John Falk and Lynn Dierking in this book. From the work of Adam Maltese and Robert Tai,[1,2] we know that most adults who have an interest in science developed that interest before eighth grade, and the greatest influencers are not from school, but what happens outside of school (e.g. parents, peers, playing outside, and museums).

The challenge is that while supporting data is essential, it is not sufficient to significantly impact decision and policy makers. What I learned during my many years working with the Washington State legislature is that while data is important, good "stories" have more influence, as it allows people to see – and emotionally connect to – the impact of programs. But even good stories are not enough. Relationships with and among decision/policy makers are

Key Elements to the Path for Successful Funding

FIGURE 34.3   Diagram of path to funding for out-of-school learning.

critical to affecting policy and funding decisions. We must find and develop relationships with decision-makers sympathetic to our cause and who can use their relationships with others to bring about change (Figure 34.3).

Thus, what is more important than making the case, is to develop an effective public affairs strategy (i.e., lobbying) to bring about the change we desire. We also need to recognize that just as it is easier to get funding for capital development projects vs. operating support, we need to think about creating highly visible programs that excite decision-makers. We need the equivalent of the *100K Science Teachers in 10 Years (110Kin10)* or *Read to Your Child Every Day* efforts to spark the interest of both decision-makers and the public.

Impact on formal education

I would be remiss if I did not include the impact that science museums have had on the formal education system. I think we can argue that the inquiry method and student-direct approach to science learning that occurs in museums changed science teaching in the formal classroom. Some the popular science curricula, especially at the elementary level came from efforts led by science museums (e.g., Amplify and Full Option Science System [FOSS] from the Lawrence Hall of Science, Engineering is Elementary [EIE] from the Boston Museum of Science). Lynn Rankin (The Exploratorium) echoed this impact.

The museums that have had the interest and capacity to provide in-depth professional learning experiences & activities and curriculum (i.e., Lawrence Hall of Science, Exploratorium, Boston Museum of Science, etc.) have made major contributions to science education reform over the past several decades.

A significant number of respondents echoed the role science museums play in supporting formal education. This will certainly be a continuing need, especially at the elementary school level.

Summary of reflections and recommendations for the future

The science museum field should be proud of many things: The field developed a participant-centered and interactive engagement strategy that is the industry standard and influenced non-science-based museums. Formative evaluation and front-end evaluation of exhibits and programs are now pervasive in the industry. Taking exhibit and programs "on the road" to distant locations and non-science-based gatherings is offered by many science museums. The participant-center and interactive approach to science learning by science museums changed the way science is taught in schools. The growth in the number of science museums is outstanding, making it easier for many smaller communities to have science museum experiences. The field is generous in sharing resources, program/exhibit ideas, and lessons learned.

But there are also many challenges that science museums face in the coming years: One of the highest priorities is to serve a more diverse audience, which requires the field to be more culturally responsive and inclusive. This will require more co-development of programs and exhibits where the audience is engaged throughout the development process and has an authentic influence on what is developed. Closely related to the first two challenges is the need for science museum staff and board to represent the diversity of the community that each science museum serves. Science museums need to broaden the audience served by offering more programs for adults. In addition, more programs/exhibits need to address important societal issues. A new financial model for how science museums operate is essential to identify, which will be especially important if we want to serve a more diverse audience and focus on important societal issues. Identifying a new financial model will require making the case for the positive impact science museums have on individuals and the community. But it will not be sufficient to have data to share with decision-makers. It will be important develop an effective public affairs effort (i.e., lobbying) to share "good stories" about programs that impact the community, and to build effective relationships with and among decision-makers. To accomplish all of this, it will be imperative for science museums to work with other organizations their community to develop a broad, community-wide STEM learning ecosystem.

I look forward to seeing how these perceptions, predications, and recommendations play out in the years to come.

Science Museum Professionals Sharing Perspectives:

Al DeSena – Formerly at Carnegie Science Center and Exploration Place
Andrea Bandelli – Science Gallery International
Antonio Gomes da Costa – Director of Scientific Mediation and Education, Universcience, Paris.
Barry Van Deman – El Paso Children's Museum and Science Center
Bonnie Sachatello-Sawyer – Hopa Mountain, Inc.
Bronwyn Bevan – The Wallace Foundation
Cary Sneider – Portland State University
Charlie Walter – Mayborn Museum Complex, Baylor University
Christian Greer – Michigan Science Center
Claire Pillsbury – Bay Area Discovery Center
Darryl Williams – The Franklin Institute
David Chesebrough – Formerly at COSI
David Ucko – Museums+*more*
Diane Miller – Detroit Zoo
Elaine Reynoso Haynes - Dirección General de Divulgación de la Ciencia, National Autonomous University of Mexico
Ganga Rautela – Formerly at National Council of Science Museums, India
Ganigar Chen – National Science Museum of Thailand
Gillian Thomas – Formerly at Patricia and Phillip Frost Museum of Science
Jamie Bell – CAISE (Center for Advancing Informal STEM Education)
Jen DeWitt – Research and Evaluation Consultant
Joanne Jones Rizzi – Science Museum of Minnesota
John Falk – Institute for Learning Innovation
Judith Koke – Institute for Learning Innovation
Judy Diamond – University of Nebraska State Museum and University Libraries
Kris Morrissey – Formerly at Museology Department, U of Washington
Kristin Leigh – Explora
Laura Huerta Migus – Institute for Museum and Library Services
Lesley Lewis – Formerly at Ontario Science Centre
Lynn Dierking – Institute for Learning Innovation
Lynn Rankin – The Exploratorium
Margaret Honey – New York Hall of Science
Mark St. John – Inverness Research, Inc.
Marsha Semmel – Marsha Semmel Consulting
Martin Storksdieck – STEM Research Center, Oregon State University
Meena Selvakumar – Museology Department, University of Washington
Mikko Myllykoski – Heureka, the Finish science centre

Paul Martin – Center for Innovation in Informal STEM Learning, Arizona State University
Rabiah Mayas – Museum of Science and Industry, Chicago
Randi Korn – Formerly at RK&A
Steve Pizzey – Science Projects
Sue Allen – Allen and Associates
Tit Meng Lim – Science Centre Singapore
Tom Rockwell – The Exploratorium

Notes

1 Adam Maltese, Christina Melki and Heidi, Wiebke (2014), The nature of experiences responsible for the generation and maintenance of interest in STEM. *Science Education*, *98*(6), 937–962.
2 Adam V. Maltese and Robert H. Tai (2010), Eyeballs in the Fridge: Sources of early interest in science. *International Journal of Science Education*, *32*(5), 669–685.

CONTRIBUTORS

Marianne Achiam, PhD, Science Education, University of Copenhagen, Denmark

Nirit Lavie Alon, PhD, Education in Science and Technology, the Technion, Haifa, Israel

Florencia K. Anggoro, PhD, Psychology, College of the Holy Cross, Worcester, Massachusetts

Heidi Ballard, PhD, University of California at Davis, Davis, California

Hila Shefet Barkae, MSc, Education in Science and Technology, the Technion, Haifa, Israel

Bradley Barker, PhD, 4-H Youth Development, University of Nebraska-Lincoln, Lincoln, Nebraska

David Begay, PhD, Cherokee/Navajo, Indigenous Education Institute at Friday Harbor, Washington

Jamie Bell, EdM, Center for Advancement of Informal Science Education (CAISE), Association of Science and Technology Centers (ASTC), Washington DC

Larry Bell, MA, Boston Museum of Science, Boston, Massachusetts

Abby Bergman, Partnerships in Education and Resilience (PEAR) and NOAM Institute, Cambridge, Massachusetts

Ron Blonder, PhD, Department of Science Teaching, Weizmann Institute of Science, Rehovot, Israel

Franz X. Bogner, PhD, Center for the Promotion and Mathematics and Science Education, University of Bayreuth, Bayreuth, Germany

Mike Bruton, PhD, Imagineering, Cape Town, South Africa

Hsin-Yi Chien, PhD, Exploratorium, San Francisco, California

Knowledge Chikundi, Africa Science Busker's Festival, Harare, Zimbabwe

Julie Cleverdon, Cape Town Science Centre, Cape Town, South Africa

Roberta Cooks, MD, Museum consultant

Anne Dhanaraj, PhD, Science Center Singapore, Singapore

Judy Diamond, PhD, University of Nebraska Libraries and State Museum, Lincoln, Nebraska

Lynn D. Dierking, PhD, Institute for Learning Innovation and Education, Oregon State University, Corvallis, Oregon

Megan Ennes, PhD, Florida Museum of Natural History, Tallahassee, Florida

Henry James Evans, PhD, Science Education, University of Copenhagen, Denmark

John H. Falk, PhD, Institute for Learning Innovation, Beaverton, Oregon

John Fraser, PhD, Knology, New York, New York

Aneta Gop, PhD, Centrum Nauki Kopernik/Copernicus Science Center, Warsaw, Poland

Jeffry Gottfried, PhD, formerly Oregon Museum of Science and Industry, Portland, Oregon

Joshua P. Gutwill, PhD, Visitor Research and Evaluation, Exploratorium, San Francisco, California

Eileen G. Harrington, PhD, Health and Life Sciences Library, University of Maryland College Park, Maryland; Formerly Naturalist Center, California Academy of Sciences, San Francisco, California

Mary Grace Harris, College of the Holy Cross, Worcester, Massachusetts

Devra Hock, PhD, Earth and Atmospheric Sciences, University of Nebraska-Lincoln, Lincoln, Nebraska

Michael Horn, PhD, Computer Science and Learning Sciences, Northwestern University, Evanston, Illinois

Ilona Iłowiecka-Tańska, PhD, Centrum Nauki Kopernik/Copernicus Science Center, Warsaw, Poland

Fatema Jasim, MPS, Georgetown University, Washington DC

Benjamin D. Jee, PhD, Psychology, Worcester State University, Worcester, Massachusetts

M. Gail Jones, PhD, Science Education, North Carolina State University, Raleigh, North Carolina

Ted M. Kahn, PhD, DesignWorlds for Learning, Inc., Cupertino, California

Kathayoon Khalil, PhD, New England Aquarium, Boston, Massachusetts

Shawn Lani, MA, Studio for Public Spaces, Exploratorium, San Francisco, California

Meghan Leadabrand, University of Nebraska-Lincoln, Lincoln, Nebraska

Tit Meng Lim, PhD, Science Center Singapore, Singapore

Nancy C. Maryboy, PhD, Cherokee/Navajo, Indigenous Education Institute at Friday Harbor, and the University of Washington Seattle, Washington

Camillia Matuk, PhD, Educational Communication & Technology, New York University, New York, New York

Dawn McDaniel, PhD, Partnerships in Education and Resilience (PEAR), McLean Hospital and Harvard Medical School, Cambridge, Massachusetts

Hooley McLaughlin, PhD, Information, University of Toronto, Toronto, Canada

Julia McQuillan, PhD, Sociology, University of Nebraska-Lincoln, Lincoln. Nebraska

Kenneth Monjero, PhD, Science Centre Kenya, Nairobi, Kenya

Taiji Nelson, Climate and Rural Systems Partnership, Carnegie Museum of Natural History, Pittsburgh, Pennsylvania

Gil Noam, EdD, Dr. Habil, Partnerships in Education and Resilience (PEAR), McLean Hospital and Harvard Medical School, Cambridge, Massachusetts

Kathryn Nock, Knology, New York, New York

Rebecca Joy Norlander, PhD, Knology, New York, New York

Yao Teck Ong, Science Center Singapore, Singapore

Tina Phillips, PhD, Cornell Lab of Ornithology, Ithaca, New York

Julia Plummer, PhD, Science Education, Pennsylvania State University, State College, Pennsylvania

Wendy Pollock, MA, formerly Association of Science and Technology Centers, Washington DC

Shelley Rap, PhD, Department of Science Teaching, Weizmann Institute of Science, Rehovot, Israel

Ganga S. Rautela, MSc, Advisor, Science City, Dehradun, and Former Director General, Indian Science Centre Network, National Council of Science Museums (NCSM), India

Mihir K. Ravel, formerly Portland State University, Oregon, National Instruments, and Tektronix

Sherman Rosenfeld, PhD, Department of Science Teaching, Weizmann Institution of Science, Rehovot, Israel

Dennis Schatz, PhD, formerly Lawrence Hall of Science and Pacific Science Center; Institute for Learning Innovation, Beaverton, Oregon

Cary I. Sneider, PhD, formerly Portland State University and the National Assessment Governing Board, Portland, Oregon

Sofoklis Sotiriou, PhD, R & D Department, Ellinogermaniki Agogi, Athens, Greece

Amy N. Spiegel, PhD, Methodology and Evaluation Research Core, University of Nebraska-Lincoln, Lincoln, Nebraska

Nancy L. Staus, PhD, STEM Research Center, Oregon State University, Corvalis, Oregon

Mary Ann Steiner, PhD, University of Pittsburgh Center for Learning in Out of School Environments (UPCLOSE), Pittsburgh, Pennsylvania

Martin Storksdieck, PhD, Center for Research on Lifelong STEM Learning, Oregon State University, Corvallis, Oregon

Tali Tal, PhD, Education in Science and Technology, the Technion, Haifa, Israel

Bailey Triggs, MS, Partnerships in Education and Resilience (PEAR), McLean Hospital and Harvard Medical School, Cambridge, Massachusetts

Izaiah Wallace, PhD, Computer Science and Learning Sciences, Northwestern University, Evanston, Illinois

Martin Weiss, PhD, New York Hall of Science, New York, New York

Trish Wonch Hill, PhD, Center for Science, Mathematics, and Computer Education, University of Nebraska-Lincoln, Lincoln. Nebraska

Malka Yayon, PhD, Department of Science Teaching, Weizmann Institute of Science, Rehovot, Israel

Katarzyna Potęga vel Żabik, MA, Centrum Nauki Kopernik/Copernicus Science Center, Warsaw, Poland

INDEX

Note: Page references in *italics* denote figures, in **bold** tables and with "n" endnotes.